Fracture Mechanics

Fracture Mechanics

2nd Edition

M. Janssen, J. Zuidema and R. J. H. Wanhill

LONDON AND NEW YORK

First published 1983 by Delftse Uitgevers Maatschappij
on behalf of Vereniging voor Studie- en Studentenbelangen te Delft
Second edition published 2002

Second edition republished 2004 by Spon Press
2 Park Square, Milton Park, Abingdon, Oxon OX14 4RN

Simultaneously published in the USA and Canada
by Spon Press
270 Madison Avenue, New York, NY 10016

Transferred to Digital Printing 2009

Spon Press is an imprint of the Taylor & Francis Group, an informa business

British Library Cataloguing in Publication Data
A catalogue record for this book is available from the British Library

Library of Congress Cataloging in Publication Data
Janssen, M. (Michael)
Fracture mechanics/M. Janssen, J. Zuidema, and R.J.H. Wanhill.– 2nd ed.
p. cm.
Includes bibliographical references and index.
ISBN 0-415-34622-3 (pb: alk. paper)
1. Fracture mechanics. I. Zuidema, J. (Jan) II. Wanhill, R. J. H. III. Title.
TA409.J36 2004
620.1'126–dc22

2004005903

ISBN 10: 0-415-34622-3
ISBN 13: 978-0-415-34622-1

Contents

Preface to the First Edition

While teaching a course on fracture mechanics at Delft University of Technology we discovered that although there are a few excellent textbooks, their subject matter covers developments only up to the early 1970s. Consequently there was no systematic treatment of the concepts of elastic-plastic fracture mechanics. Also the description of fracture mechanics characterisation of crack growth needed updating, especially for sustained load fracture and unstable dynamic crack growth.

In the present textbook we have attempted to cover the basic concepts of fracture mechanics for both the linear elastic and elastic-plastic regimes, and three chapters are devoted to the fracture mechanics characterisation of crack growth (fatigue crack growth, sustained load fracture and dynamic crack growth).

There are also two chapters concerning mechanisms of fracture and the ways in which actual material behaviour influences the fracture mechanics characterisation of crack growth. The reader will find that this last topic is treated to some way beyond that of a basic course. This is because to our knowledge there is no reference work that systematically covers it. A consequence for instructors is that they must be selective here. However, any inconvenience thereby entailed is, we feel, outweighed by the importance of the subject matter.

This textbook is intended primarily for engineering students. We hope it will be useful to practising engineers as well, since it provides the background to several new design methods, criteria for material selection and guidelines for acceptance of weld defects.

Many people helped us during preparation of the manuscript. We wish to thank particularly J. Zuidema, who made vital contributions to uniform treatment of the energy balance approach for both the linear elastic and elastic-plastic regimes; R.A.H. Edwards, who assisted with the chapter on sustained load fracture; A.C.F. Hagedorn, who drew the figures for the first seven chapters; and the team of the VSSD, our publisher, whose patience was sorely tried but who remained unbelievably co-operative.

Finally, we wish to thank the National Aerospace Laboratory NLR and the Boiler and Pressure Vessel Authority 'Dienst voor het Stoomwezen' for providing us the opportunity to finish this book, which was begun at the Delft University of Technology.

H.L. Ewalds
R.J.H. Wanhill

September 1983

Preface to the Second Edition

In 1991, the fifth reprint of the first edition of the textbook "Fracture Mechanics", by H.L. Ewalds and R.J.H. Wanhill, was published. Obviously the field of fracture mechanics has developed further since that time. A new edition was needed. The task fell mainly to the new authors, M. Janssen and J. Zuidema, both in the Department of Materials Science at Delft University of Technology, with assistance by R.J.H. Wanhill, of the National Aerospace Laboratory NLR. The original first author, H.L. Ewalds, indicated that he no longer wished to be involved with this textbook. We respect his decision, and thank him for his major contribution to the First Edition, which has been very successful.

This second edition is the result of numerous revisions, updates and additions. These were driven by the ongoing development of fracture mechanics, but also by teaching the course on fracture mechanics at Delft University of Technology. The fracture mechanics parameters K, G and J are now treated in a more basic manner. Test methods for J_{Ic} and for crack arrest toughness are updated. The development of failure assessment based on elastic-plastic fracture mechanics is reflected in a comprehensive treatment. On the subject of subcritical crack growth more attention is paid to the important topic of the initiation and growth of short fatigue cracks.

Throughout the book some paragraphs are typeset in a smaller font. This text is intended to provide additional background information on certain subjects, but is not considered essential for a basic understanding.

We would like to acknowledge the assistance of colleagues in preparing this second edition. With critical reading and profound discussions A.R. Wachters helped considerably in drawing up the part on the J integral. G. Pape did the preparatory work necessary for updating the chapter on dynamic fracture. A. Bakker contributed to the treatment of the R6 failure assessment procedure. Finally, A.H.M. Krom provided useful comments and suggestions on various subjects.

The authors wish to thank our publisher, J.E. Schievink of the VSSD, for his encouragement and co-operation in creating this new edition.

M. Janssen
J. Zuidema
R.J.H. Wanhill

March 2002

Part I
Introduction

1
An Overview

1.1 About this Course

This course is intended as a basic grounding in fracture mechanics for engineering use. In order to compile the course we have consulted several textbooks and numerous research articles. In particular, the following books have been most informative and are recommended for additional reading:

- D. Broek, "Elementary Engineering Fracture Mechanics", Martinus Nijhoff (1986) The Hague;
- J.F. Knott, "Fundamentals of Fracture Mechanics", Butterworths (1973) London;
- Richard W. Hertzberg, "Deformation and Fracture Mechanics of Engineering Materials", John Wiley and Sons (1988) New York;
- T.L. Anderson, "Fracture Mechanics, Fundamentals and Applications", CRC Press (1991) Boston.

Four international journals are also recommended:

- Fatigue and Fracture of Engineering Materials and Structures;
- International Journal of Fatigue;
- International Journal of Fracture;
- Engineering Fracture Mechanics.

As indicated in the table of contents the course has been divided into five parts. Part I, consisting of this chapter, is introductory. In Part II the well established subject of Linear Elastic Fracture Mechanics (LEFM) is treated, and this is followed in Part III by the more recent and still evolving topic of Elastic-Plastic Fracture Mechanics (EPFM). In Part IV the applicability of fracture mechanics concepts to crack growth behaviour is discussed: namely subcritical, stable crack growth under cyclic loading (fatigue) or sustained load, and dynamic crack growth beyond instability. Finally, in Part V the mechanisms of fracture in actual materials are described together with the influence of material behaviour on fracture mechanics-related properties.

1.2 Historical Review

Strength failures of load bearing structures can be either of the yielding-dominant or fracture-dominant types. Defects are important for both types of failure, but those of primary importance to fracture differ in an extreme way from those influencing yielding and the resistance to plastic flow. These differences are illustrated schematically in figure 1.1.

For yielding-dominant failures the significant defects are those which tend to warp and interrupt the crystal lattice planes, thus interfering with dislocation glide and providing a resistance to plastic deformation that is essential to the strength of high strength metals. Examples of such defects are interstitial and out-of-size substitutional atoms, grain boundaries, coherent precipitates and dislocation networks. Larger defects like inclusions, porosity, surface scratches and small cracks may influence the effective net section bearing the load, but otherwise have little effect on resistance to yielding.

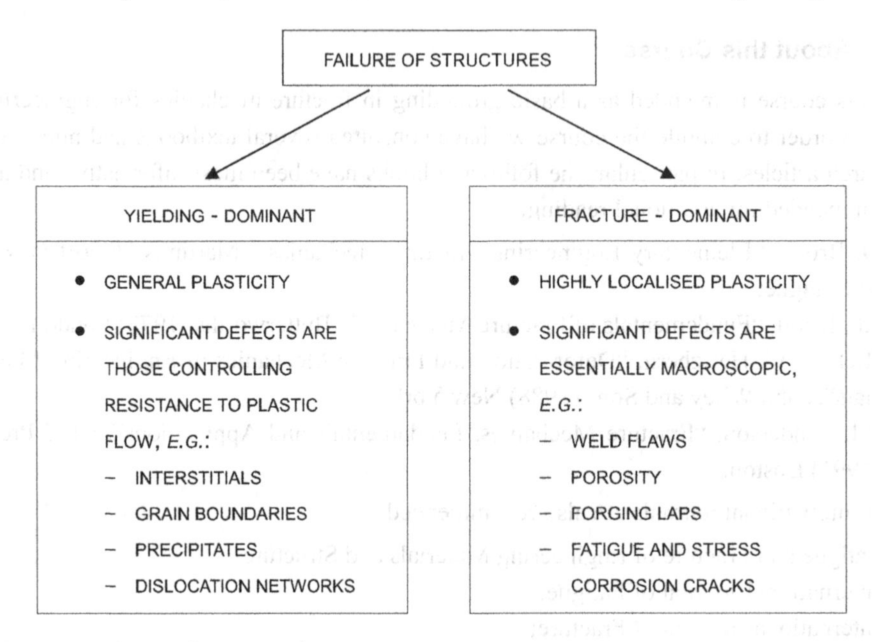

Figure 1.1. Types of structural failure.

For fracture-dominant failures, *i.e.* fracture before general yielding of the net section, the size scale of the defects which are of major significance is essentially macroscopic, since general plasticity is not involved but only the local stress-strain fields associated with the defects. The minute lattice-related defects which control resistance to plastic flow are not of direct concern. They are important insofar as the resistance to plastic flow is related to the material's susceptibility to fracture.

Fracture mechanics, which is the subject of this course, is concerned almost entirely with fracture-dominant failure. The commonly accepted first successful analysis of a fracture-dominant problem was that of Griffith in 1920, who considered the propagation of brittle cracks in glass. Griffith formulated the now well-known concept that an existing crack will propagate if thereby the total energy of the system is lowered, and he assumed that there is a simple energy balance, consisting of a decrease in elastic strain energy within the stressed body as the crack extends, counteracted by the energy needed to create the new crack surfaces. His theory allows the estimation of the theoretical strength of brittle solids and also gives the correct relationship between fracture strength and defect size.

The Griffith concept was first related to brittle fracture of metallic materials by Zener and Hollomon in 1944. Soon after, Irwin pointed out that the Griffith-type energy balance must be between (i) the stored strain energy and (ii) the surface energy plus the work done in plastic deformation. Irwin defined the 'energy release rate' or 'crack driving force', G, as the total energy that is released during cracking per unit increase in crack size. He also recognised that for relatively ductile materials the energy required to form new crack surfaces is generally insignificant compared to the work done in plastic deformation.

In the middle 1950s Irwin contributed another major advance by showing that the energy approach is equivalent to a stress intensity (K) approach, according to which fracture occurs when a critical stress distribution ahead of the crack tip is reached. The material property governing fracture may therefore be stated as a critical stress intensity, K_c, or in terms of energy as a critical value G_c.

Demonstration of the equivalence of G and K provided the basis for development of the discipline of Linear Elastic Fracture Mechanics (LEFM). This is because the form of the stress distribution around and close to a crack tip is always the same. Thus tests on suitably shaped and loaded specimens to determine K_c make it possible to determine what cracks or crack-like flaws are tolerable in an actual structure under given conditions. Furthermore, materials can be compared as to their utility in situations where fracture is possible. It has also been found that the sensitivity of structures to subcritical cracking such as fatigue crack growth and stress corrosion can, to some extent, be predicted on the basis of tests using the stress intensity approach.

The beginnings of Elastic-Plastic Fracture Mechanics (EPFM) can be traced to fairly early in the development of LEFM, notably Wells' work on Crack Opening Displacement (COD), which was published in 1961. In 1968 Rice introduced an elastic-plastic fracture parameter with a more theoretical basis: the J integral. Although both COD and J are now well established concepts, EPFM is still very much an evolving discipline. The reason is the greater complexity of elastic-plastic analyses. Important topics are:

- the description of stable ductile crack growth (tearing),
- the development of failure assessment methods that combine the effects of plasticity and fracture.

As opposed to using the above-mentioned global fracture mechanics parameters, fracture problems are also increasingly being tackled by means of local fracture criteria. Here the mechanical conditions that actually exist in the crack tip region are being determined and are being related to the material properties.

1.3 The Significance of Fracture Mechanics

In the nineteenth century the Industrial Revolution resulted in an enormous increase in the use of metals (mainly irons and steels) for structural applications. Unfortunately, there also occurred many accidents, with loss of life, owing to failure of these structures. In particular, there were numerous accidents involving steam boiler explosions and railway equipment.

Some of these accidents were due to poor design, but it was also gradually discovered that material deficiencies in the form of pre-existing flaws could initiate cracking and fracture. Prevention of such flaws by better production methods reduced the number of failures to more acceptable levels.

A new era of accident-prone structures was ushered in by the advent of all-welded designs, notably the Liberty ships and T-2 tankers of World War II. Out of 2500 Liberty ships built during the war, 145 broke in two and almost 700 experienced serious failures. Many bridges and other structures also failed. The failures often occurred under very low stresses, for example even when a ship was docked, and this anomaly led to extensive investigations which revealed that the fractures were brittle and that flaws and stress concentrations were responsible. It was also discovered that brittle fracture in the types of steel used was promoted by low temperatures. This is depicted in figure 1.2: above a certain transition temperature the steels behave in a ductile manner and the energy required for fracture increases greatly.

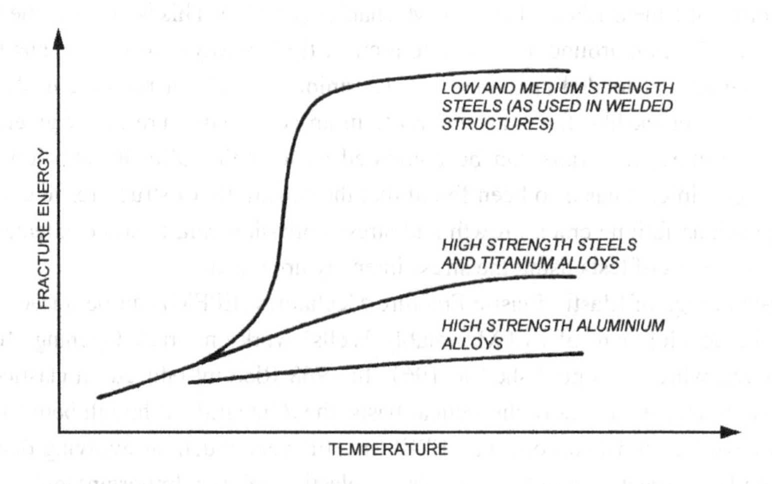

Figure 1.2. Schematic of the general effect of temperature on the fracture energy of structural metals.

Current manufacturing and design procedures can prevent the intrinsically brittle fracture of welded steel structures by ensuring that the material has a suitably low transition temperature and that the welding process does not raise it. Nevertheless, service-induced embrittlement, for example hydrogen embrittlement in the petrochemical industries, irradiation effects in nuclear pressure vessels and corrosion fatigue in offshore platforms, remains a cause for concern.

Looking at the present situation it may be seen from figure 1.3 that since World War II the use of high strength materials for structural applications has greatly increased.

These materials are often selected to obtain weight savings — aircraft structures are an obvious example. Additional weight savings have come from refinements in stress analysis, which have enabled design allowables to be raised. However, it was not recognised until towards the end of the 1950s that although these materials are not intrinsi-

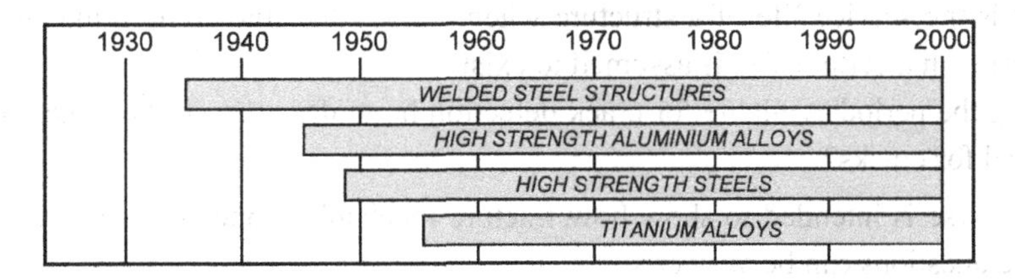

Figure 1.3. Introduction of high strength materials for structural applications.

cally brittle, the energy required for fracture is comparatively low, as figure 1.2 shows. The possibility, and indeed occurrence, of this low energy fracture in high strength materials stimulated the modern development of fracture mechanics.

The object of fracture mechanics is to provide quantitative answers to specific problems concerning cracks in structures. As an illustration, consider a structure containing pre-existing flaws and/or in which cracks initiate in service. The cracks may grow with time owing to various causes (for example fatigue, stress corrosion, creep) and will generally grow progressively faster, figure 1.4.a. The residual strength of the structure, which is the failure strength as a function of crack size, decreases with increasing crack size, as shown in figure 1.4.b. After a time the residual strength becomes so low that the structure may fail in service.

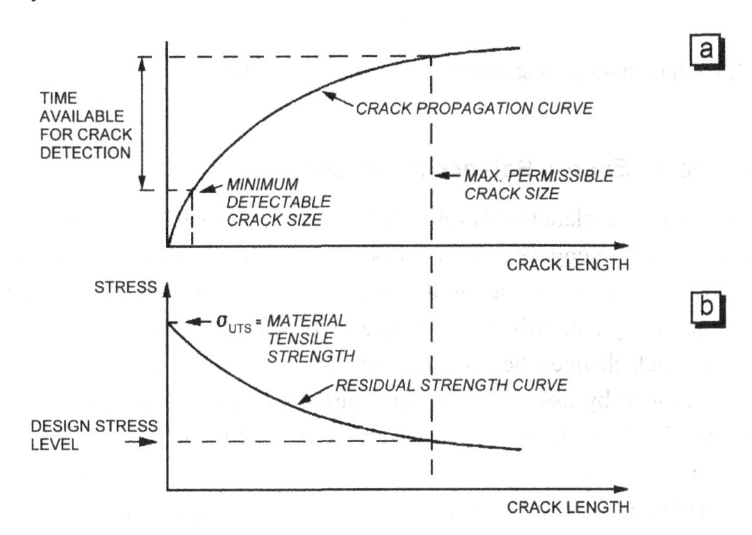

Figure 1.4. The engineering problem of a crack in a structure.

With respect to figure 1.4 fracture mechanics should attempt to provide quantitative answers to the following questions:

1. What is the residual strength as a function of crack size?
2. What crack size can be tolerated under service loading, *i.e.* what is the maximum permissible crack size?
3. How long does it take for a crack to grow from a certain initial size, for example the minimum detectable crack size, to the maximum permissible crack size?

4. What is the service life of a structure when a crack-like flaw (*e.g.* a manufacturing defect) with a certain size is assumed to exist?
5. During the period available for crack detection how often should the structure be inspected for cracks?

This course is intended to show how fracture mechanics concepts can be applied so that these questions can be answered.

In the remaining sections 1.4 – 1.11 of this introductory chapter an overview of the basic concepts and applications of LEFM and EPFM are given in preparation for more detailed treatment in subsequent chapters.

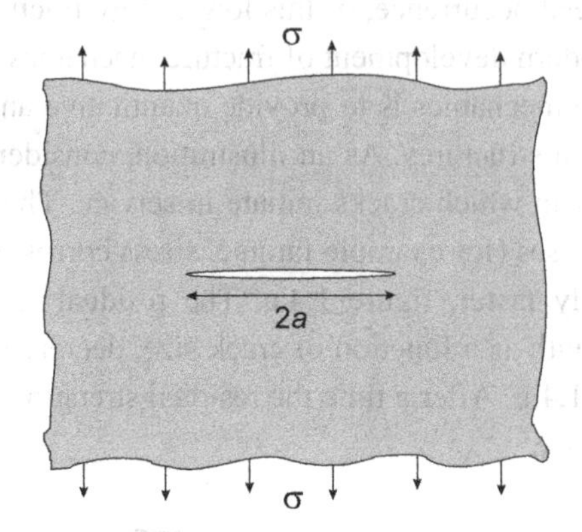

Figure 1.5. A through-thickness crack in a loaded infinite plate.

1.4 The Griffith Energy Balance Approach

Consider an *infinite* plate that is subjected to a uniform tensile stress, σ, applied at infinity (see figure 1.5). Suppose that we introduce a through-thickness crack of length $2a$. In the area directly above and below the crack the stress (in the loading direction) will decrease significantly and will even become zero along the crack flanks. Hence introduction of the crack changes the elastic strain energy stored in the plate. We can roughly estimate this change by assuming that in a circle-shaped area of radius a around the crack the stress has become zero, while the remainder of the plate experiences the same stress as before. In this case the elastic energy in the plate has decreased by an amount equal to the volume of the stress-free material times the original elastic energy per unit volume, *i.e.* ½×stress×strain. Assuming linear elastic material behaviour, *i.e.* a Young's modulus E, the elastic energy change would be[1]:

$$\pi a^2 \cdot \frac{\sigma^2}{2E} = \frac{1}{2}\frac{\pi\sigma^2 a^2}{E} . \tag{1.1}$$

Obviously, this is only an approximation because the stress field becomes non-

[1] In this section we consider two-dimensional geometries only and all energies and forces are defined per unit thickness.

homogeneous near the crack, as will be shown in chapter 2. Griffith used a stress analysis developed by Inglis to show that for an infinite plate the elastic energy change is actually given by

$$U_a = -\frac{\pi\sigma^2 a^2}{E}, \tag{1.2}$$

where U_a = change in the elastic strain energy of the plate caused by introducing a crack with length $2a$. The minus sign shows this change is a decrease in elastic energy.

The introduction of a crack will require a certain amount of energy. Griffith assumed that for ideally brittle materials this is in the form of surface energy. A crack with length $2a$ in a plate involves the creation of a crack surface area (defined per unit thickness) equal to $2\cdot(2a) = 4a$, leading to an increase in surface energy of

$$U_\gamma = 4a\cdot\gamma_e, \tag{1.3}$$

where U_γ = change in surface energy of the plate due to introduction of a crack with length $2a$,

γ_e = surface energy per unit area, *i.e.* the surface tension.

Griffith postulated that a crack will extend when the *potential* energy decreases. He considered the surface energy as a part of this potential energy. In practice the energy involved in creating crack surfaces will not be reversible due to several reasons (oxidation etc.) and strictly speaking is not part of the potential energy. However, as long as only growing cracks are considered, the irreversibility of the surface energy is not relevant. Here, the potential energy according to Griffith will be referred to as the *total* energy.

For a real plate, *i.e.* one with finite dimensions, the total energy U is that of the plate *and* its loading system. When a crack is present the total energy U is composed of

$$U = U_o + U_a + U_\gamma - F, \tag{1.4}$$

where U_o = total energy of the plate *and* its loading system before introducing a crack (a constant),

F = work performed by the loading system during the introduction of the crack = load × displacement.

The combination of plate and loading system is assumed to be isolated from its surroundings, *i.e.* no work is performed on the plate or on the loading system from outside. This explains why F must be subtracted in equation (1.4): if the loading system performs work it goes at the expense of the energy content of the loading system and therefore lowers the total energy U. A more extensive treatment will be given in section 4.2.

In this introductory chapter we will conveniently assume that no work is done by the loading system. This is the case if the specimen is loaded by a constant displacement, a so-called *fixed grip* condition. Then the term F in equation (1.4) will vanish. Introducing a crack now leads to a decrease in elastic strain energy of the plate, *i.e.* U_a is negative, because the plate loses stiffness and the load applied by the fixed grips will drop. A plate with finite dimensions resembles an infinite plate when $2a << W$, the plate width. Consequently, the total energy U of a finite plate loaded with fixed grips and containing a small crack is approximately

$$U = U_o + U_a + U_\gamma = U_o - \frac{\pi\sigma^2 a^2}{E} + 4a\gamma_e \,. \tag{1.5}$$

Following Griffith, crack extension will occur when U decreases. In order to formulate a criterion for crack extension, we consider an increase of the crack length by d($2a$). Since U_o is constant, it will not change and $\mathrm{d}U_o/\mathrm{d}(2a)$ is zero. Also, since no work is done by the loading system, the driving force for crack extension can be delivered only by the decrease in elastic energy $\mathrm{d}U_a$ due to the crack length increase d($2a$). The crack will extend when the *available* energy $\mathrm{d}U_a$ is larger than the energy *required* $\mathrm{d}U_\gamma$. Thus the criterion for crack extension is

$$\frac{\mathrm{d}U}{\mathrm{d}(2a)} = \frac{\mathrm{d}}{\mathrm{d}(2a)}(U_a + U_\gamma) < 0 \quad \text{or} \quad \frac{\mathrm{d}}{\mathrm{d}(2a)}\left(-\frac{\pi\sigma^2 a^2}{E} + 4a\gamma_e\right) < 0 \,. \tag{1.6}$$

This is illustrated in figure 1.6. Figure 1.6.a schematically represents the two energy terms in equation (1.6) and their sum as functions of the introduced crack length, $2a$. Figure 1.6.b represents the derivative, $\mathrm{d}U/\mathrm{d}(2a)$. When the elastic energy release due to a potential increment of crack growth, d($2a$), outweighs the demand for surface energy for the same crack growth, the introduction of a crack will lead to its unstable propagation.

From the criterion for crack extension, equation (1.6), one obtains

$$\frac{\pi\sigma^2 a}{E} > 2\gamma_e \,, \tag{1.7}$$

which can be rearranged to

$$\sigma\sqrt{a} > \sqrt{\frac{2E\gamma_e}{\pi}} \,. \tag{1.8}$$

Equation (1.8) indicates that crack extension in ideally brittle materials is governed by the product of the remotely applied stress and the square root of the crack length and by material properties. Because E and γ_e are material properties the right-hand side of equation (1.8) is equal to a constant value characteristic of a given ideally brittle material. Consequently, equation (1.8) indicates that crack extension in such materials occurs when the product $\sigma\sqrt{a}$ attains a certain critical value.

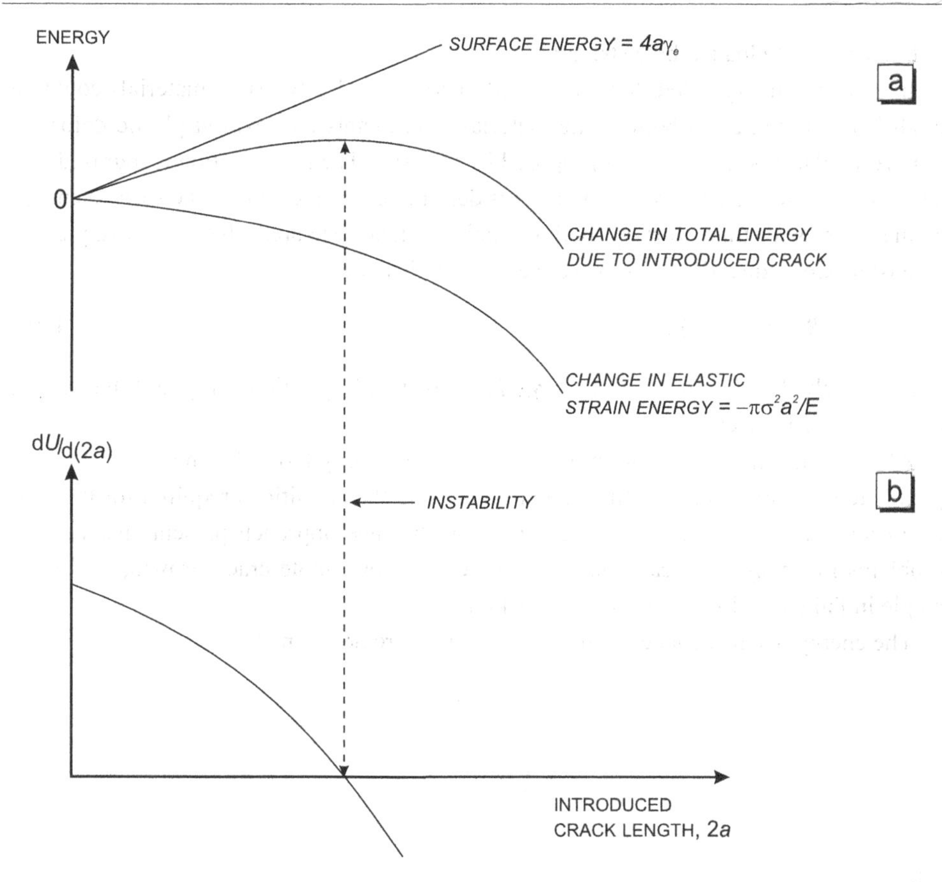

Figure 1.6. Energy balance for a small crack in a large plate loaded under fixed grip conditions.

1.5 Irwin's Modification to the Griffith Theory

Irwin designated the left-hand side of equation (1.7) as the *energy release rate*, G, representing the energy per unit new crack area that is available for infinitesimal crack extension.[2] The right-hand side of equation (1.7) represents the surface energy increase per unit new crack area that would occur owing to infinitesimal crack extension and is designated the *crack resistance*, R. It follows that G must be larger than R before crack growth occurs. If R is a constant, this means that G must exceed a critical value $G_c = R$ = constant. Thus fracture occurs when

$$G = \frac{\pi\sigma^2 a}{E} > G_c = R = 2\gamma_e \,. \tag{1.9}$$

The critical value G_c can be determined by measuring the critical stress σ_c required to fracture a plate with a crack of size $2a$ or by measuring the critical crack size $2a_c$ needed

[2] The crack area is defined as the projected area, normal to the crack plane, of the newly formed surfaces.

to fracture a plate loaded by a stress σ.

In 1948 Irwin suggested that the Griffith theory for ideally brittle materials could be modified and applied to both brittle materials and metals that exhibit plastic deformation. A similar modification was proposed by Orowan. The modification recognised that a material's resistance to crack extension is determined by the sum of the surface energy γ_e and the plastic strain work γ_p (both per unit crack surface area) that accompany crack extension. Consequently, in this case the crack resistance is

$$R = 2(\gamma_e + \gamma_p) \,. \tag{1.10}$$

For relatively ductile materials $\gamma_p >> \gamma_e$, *i.e.* R is mainly plastic energy and the surface energy can be neglected.

Although Irwin's modification includes a plastic energy term, the energy balance approach to crack extension is still limited to defining the conditions required for instability of an ideally sharp crack. Also, the energy balance approach presents insuperable problems for many practical situations, especially slow stable crack growth, as for example in fatigue and stress corrosion cracking.

The energy balance concept will be treated in more detail in chapter 4.

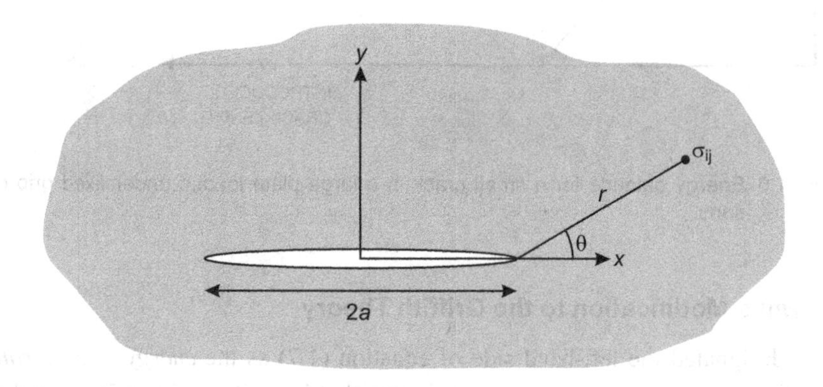

Figure 1.7. Stresses at a point ahead of a crack tip.

1.6 The Stress Intensity Approach

Owing to the practical difficulties of the energy approach a major advance was made by Irwin in the 1950s when he developed the stress intensity approach. First, from linear elastic theory Irwin showed that the stresses in the vicinity of a crack tip take the form

$$\sigma_{ij} = \frac{K}{\sqrt{2\pi r}} f_{ij}(\theta) + \ldots \,, \tag{1.11}$$

where r, θ are the cylindrical polar co-ordinates of a point with respect to the crack tip, figure 1.7.

K is a quantity which gives the magnitude of the elastic stress field. It is called the *stress intensity factor*.[3] Dimensional analysis shows that K must be linearly related to stress and directly related to the square root of a characteristic length. Equation (1.8) from Griffith's analysis indicates that this characteristic length is the crack length, and it turns out that the general form of the stress intensity factor is given by

$$K = \sigma\sqrt{\pi a} \cdot f(a/W)\,, \tag{1.12}$$

where $f(a/W)$ is a dimensionless parameter that depends on the geometries of the specimen and crack, and σ is the (remotely) applied stress. For an infinite plate with a central crack with length $2a$, $f(a/W) = 1$ and thus $K = \sigma\sqrt{\pi a}$. For this case we also have $G = \pi\sigma^2 a/E$, see equation (1.9). Combining the two formulae for K and G yields the relation:

$$G = \frac{K^2}{E}\,, \tag{1.13}$$

which Irwin showed to be valid for any geometry.

Since $K = \sigma\sqrt{\pi a}$ for a central crack in an infinite plate, it follows from the result of Griffith's energy balance approach, equation (1.8), that crack extension will occur when K reaches a certain critical value. This value, K_c, is equal to $\sqrt{2E\gamma_e}$ or, after applying Irwin's modification, $\sqrt{2E(\gamma_e + \gamma_p)}$. The criterion for crack extension in terms of K is

$$K = \sigma\sqrt{\pi a} > K_c\,. \tag{1.14}$$

The parameter governing fracture may therefore be stated as either a critical energy release rate, G_c, or a critical stress intensity, K_c. For tensile loading the relationships between G_c and K_c are

$$G_c = \frac{K_c^2}{E}\,. \tag{1.15}$$

The value of the critical stress intensity K_c can be determined experimentally by measuring the fracture stress for a large plate that contains a through-thickness crack of known length. This value can also be measured by using other specimen geometries, or else can be used to predict critical combinations of stress and crack length in these other geometries. This is what makes the stress intensity approach to fracture so powerful, since values of K for different specimen geometries can be determined from conventional elastic stress analyses: there are several handbooks giving relationships between the stress intensity factor and many types of cracked bodies with different crack sizes, orientations and shapes, and loading conditions. Furthermore, the stress intensity factor, K, is applicable to stable crack extension and does to some extent characterize processes of subcritical cracking like fatigue and stress corrosion, as will be mentioned in section 1.10 of this chapter and in greater detail in chapters 9 and 10.

[3] The stress intensity factor is essentially different from the well-known stress concentration factor. The latter is a dimensionless ratio that describes the increase in stress level relative to the nominal stress.

It is the use of the stress intensity factor as the characterizing parameter for crack extension that is the fundamental principle of Linear Elastic Fracture Mechanics (LEFM). The theory of Linear Elastic Fracture Mechanics is well developed and will be discussed in chapter 2.

1.7 Crack Tip Plasticity

The elastic stress distribution in the vicinity of a crack tip, equation (1.11), shows that as r tends to zero the stresses become infinite, *i.e.* there is a stress singularity at the crack tip. Since structural materials deform plastically above the yield stress, there will in reality be a plastic zone surrounding the crack tip. Thus the elastic solution is not unconditionally applicable.

Irwin considered a circular plastic zone to exist at the crack tip under tensile loading. As will be discussed in chapter 3, he showed that such a circular plastic zone has a diameter $2r_y$, figure 1.8, with

$$r_y = \frac{1}{2\pi}\left(\frac{K}{\sigma_{ys}}\right)^2 , \qquad (1.16)$$

where σ_{ys} is the yield stress.

Irwin argued that the occurrence of plasticity makes the crack behave as if it were longer than its physical size — the displacements are larger and the stiffness is lower than in the elastic case. He showed that the crack may be viewed as having a notional tip at a distance r_y ahead of the real tip, *i.e.* in the centre of the circular plastic zone (see figure 1.8). Beyond the plastic zone the elastic stress distribution is described by the K corresponding to the notional crack size. As shown in figure 1.8, this elastic stress distribution takes over from the yield stress at a distance $2r_y$ from the actual crack tip.

Since the same K always gives the same plastic zone size for materials with the same

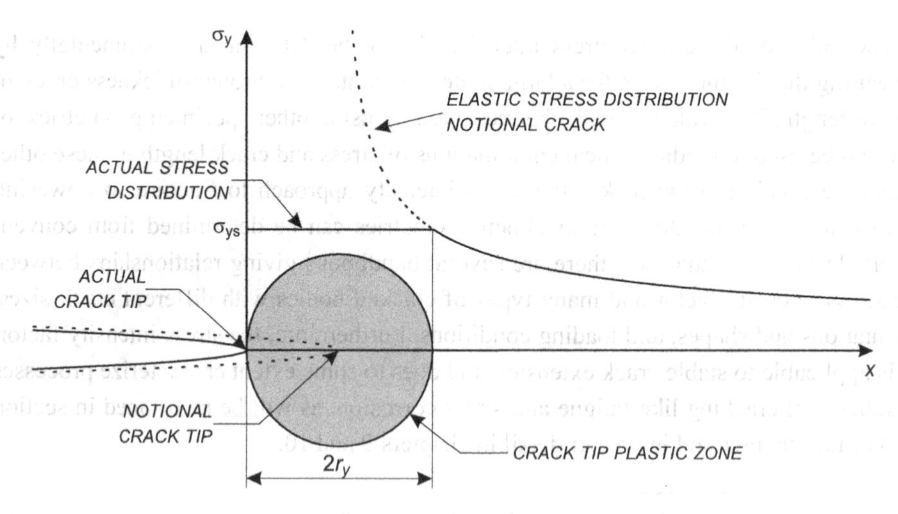

Figure 1.8. The crack tip plastic zone according to Irwin.

yield stress, equation (1.16), the stresses and strains both within and outside the plastic zone will be determined by K and the stress intensity approach can still be used. In short, the effect of crack tip plasticity corresponds to an apparent increase of the elastic crack length by an increment equal to r_y.

A plastic zone at the tip of a through-thickness crack will inevitably tend to contract in the thickness direction along the crack front. If the plate thickness is of the order of the plastic zone size or smaller, this contraction can occur freely and a *plane stress* state will prevail. On the other hand, if the plate thickness is much larger than the plastic zone size, contraction is constrained by the elastic material surrounding the plastic zone. The strain in the thickness direction will then be small, meaning that a *plane strain* state is present.[4]

The occurrence at the crack tip of either a plane stress or plane strain state has a large effect on the plastic behaviour of the material. In plane strain the plastic deformation occurs only when the stresses amply exceed the yield stress. Actually, equation (1.16) is valid for a plane stress state only. For plane strain

$$r_y = \frac{1}{2\pi}\left(\frac{K}{C\sigma_{ys}}\right)^2, \qquad (1.17)$$

where C is usually estimated to be about 1.7. Thus in plane strain the plastic zone size is considerably smaller.

1.8 Fracture Toughness

From sections 1.6 and 1.7 it follows that under conditions of limited crack tip plasticity the parameter governing tensile fracture can be stated as a critical stress intensity, K_c. The value of K_c at a particular temperature depends on the amount of thickness constraint and thus on specimen thickness. It is customary to write the limiting value of K_c for maximum constraint (plane strain) as K_{Ic}.[5]

K_{Ic} can be considered a material property characterizing the crack resistance, and is therefore called the plane strain fracture toughness. Thus the same value of K_{Ic} should be found by testing specimens of the same material with different geometries and with critical combinations of crack size and shape and fracture stress. Within certain limits this is indeed the case, and so a knowledge of K_{Ic} obtained under standard conditions can be used to predict failure for different combinations of stress and crack size and for different geometries.

K_c can also be determined under standard conditions, and the value thus found may also be used to predict failure, but only for situations with the same material thickness and constraint.

---

[4] In all formulae up to this point a plane stress state was implicitly assumed.

[5] The subscript I refers to the loading mode where the crack flanks are pulled straight apart (see section 2.1). In fracture mechanics it is customary to include this subscript in expressions that contain the stress intensity factor as a variable, i.e. K_I. However, in this introductory chapter this is not yet done.

As an introductory numerical example of the design application of LEFM, consider the equation for a through-thickness crack in a wide plate, *i.e.*

$$K = \sigma\sqrt{\pi a}\,. \tag{1.18}$$

Assume that the test results show that for a particular steel the K_c is 66 MPa$\sqrt{m}$ for the plate thickness and temperature in service. Using equation (1.18) a residual strength curve for this steel can be constructed relating K_c and nominal stress and crack size. This is shown in figure 1.9. Also assume that the design stress is 138 MPa. It follows from equation (1.18) and figure 1.9 that the tolerable crack size would be about 145 mm. For a design stress of 310 MPa the same material could tolerate a crack size of only about 28 mm. Note from figure 1.9 that if a steel with a higher fracture toughness is used, for example one with a K_c of 132 MPa$\sqrt{m}$, the permissible design stress for a given crack size is significantly increased. *Thus a material with a higher fracture toughness permits a longer crack at a given stress or a higher stress at a given crack length.*

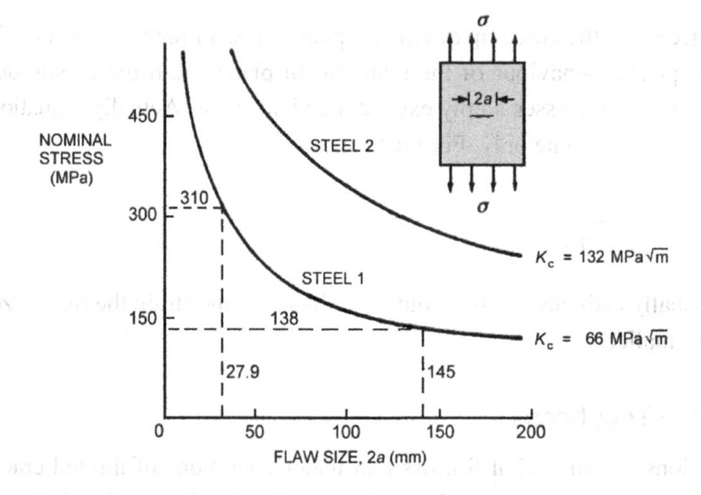

Figure 1.9. Residual strength curves for two steels.

1.9 Elastic-Plastic Fracture Mechanics

Linear Elastic Fracture Mechanics can deal with only limited crack tip plasticity, *i.e.* the plastic zone must remain small compared to the crack size and the cracked body as a whole must still behave in an approximately elastic manner. If this is not the case then the problem has to be treated elasto-plastically. Due to its complexity the concepts of Elastic-Plastic Fracture Mechanics (EPFM) are not so well developed as LEFM theory, a fact that is reflected in the approximate nature of the eventual solutions.

In 1961 Wells introduced the crack opening displacement (COD) approach. This approach focuses on the strains in the crack tip region instead of the stresses, unlike the stress intensity approach. In the presence of plasticity a crack tip will blunt when it is loaded in tension. Wells proposed to use the crack flank displacement at the tip of a blunting crack, the so-called crack tip opening displacement (CTOD) as a fracture parameter (see figure 1.10).

Even for tougher materials exhibiting considerable plasticity critical CTOD values could be defined corresponding to the onset of fracture. Such critical CTOD values

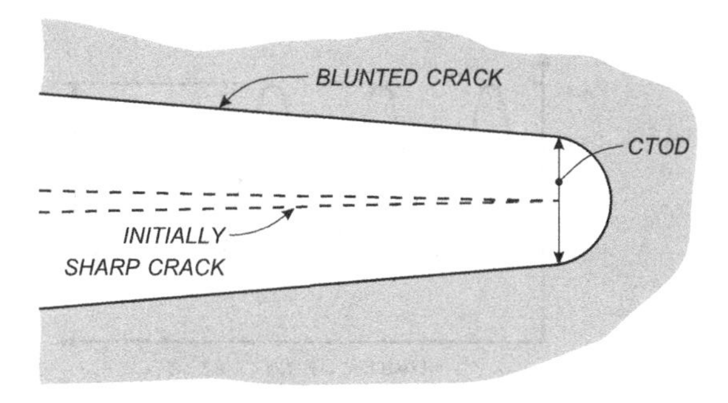

Figure 1.10. Crack tip opening displacement.

could then be used to qualify the materials concerned for a given application. However, initially it proved difficult to determine the required CTOD for a given load and geometry or alternatively to calculate critical crack lengths or loads for a given material.

In 1968 Rice considered the potential energy changes involved in crack growth in non-linear elastic material. Such non-linear elastic behaviour is a realistic approximation for plastic behaviour provided no unloading occurs in any part of the material. Rice derived a fracture parameter called J, a contour integral that can be evaluated along any arbitrary path enclosing the crack tip, as illustrated in figure 1.11. He showed J to be equal to the energy release rate for a crack in non-linear elastic material, analogous to G for linear elastic material.

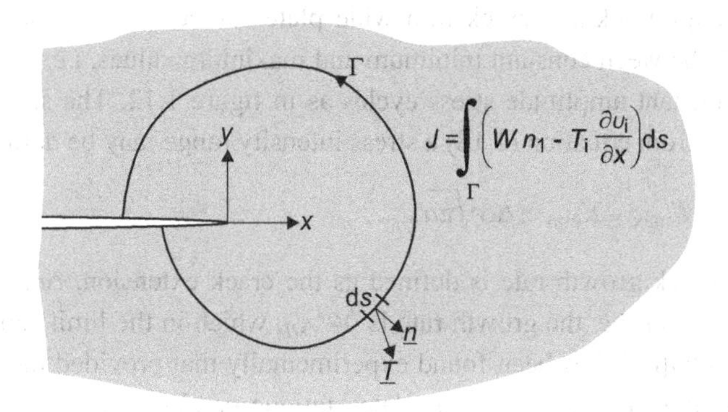

Figure 1.11. J contour integral along arbitrary path Γ enclosing a crack tip in non-linear elastic material. W is strain energy density along Γ, $\underline{n}$ is outward-directed unit vector normal to Γ, $\underline{T}$ is traction acting on Γ and $\underline{u}$ is the displacement along Γ.

For simple geometries and load cases the J integral can be evaluated analytically. However, in practice finite element calculations are often required. In spite of this J has found widespread application as a parameter to predict the onset of crack growth in elastic-plastic problems. Later it was found that J could also be used to describe a limited amount of stable crack growth.

In chapter 6 the background to the J and COD approaches are discussed, while chapter 7 deals with the procedures to measure critical values of these parameters in

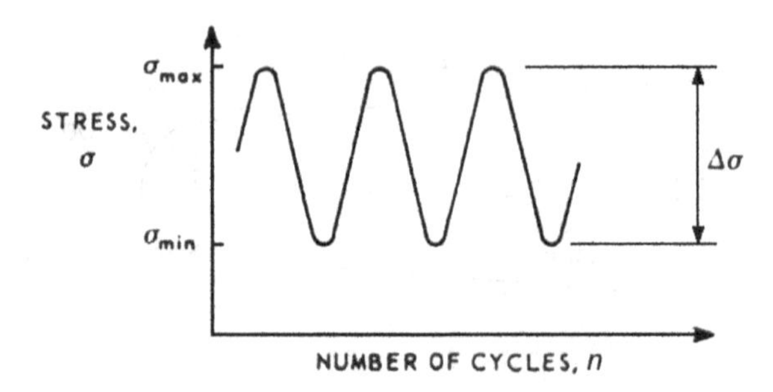

Figure 1.12. Stress-cycle parameters in constant amplitude fatigue.

actual materials. In chapter 8 some specific aspects of EPFM are discussed.

1.10 Subcritical Crack Growth

In section 1.7 it was mentioned that the stress intensity factor can still be used when crack tip plasticity is limited. This latter condition holds for some important kinds of subcritical crack growth, where most of the crack extension usually takes place at stress intensities well below K_c and K_{Ic}. In particular the stress intensity approach can provide correlations of data for fatigue crack growth and stress corrosion cracking.

Fatigue

Consider a through-thickness crack in a wide plate subjected to remote stressing that varies cyclically between constant minimum and maximum values, *i.e.* a fatigue loading consisting of constant amplitude stress cycles as in figure 1.12. The stress range $\Delta\sigma = \sigma_{max} - \sigma_{min}$, and from equation (1.18) a stress intensity range may be defined:

$$\Delta K = K_{max} - K_{min} = \Delta\sigma\sqrt{\pi a} . \qquad (1.19)$$

The fatigue crack growth rate is defined as the crack extension, Δa, during a small number of cycles, Δn, *i.e.* the growth rate is $\Delta a/\Delta n$, which in the limit can be written as the differential $\mathrm{d}a/\mathrm{d}n$. It has been found experimentally that provided the stress ratio, $R = \sigma_{min}/\sigma_{max}$, is the same then ΔK correlates fatigue crack growth rates in specimens with different stress ranges and crack lengths and also correlates crack growth rates in specimens of different geometry, *i.e.*

$$\frac{\mathrm{d}a}{\mathrm{d}n} = f(\Delta K, R) . \qquad (1.20)$$

This correlation is shown schematically in figure 1.13. Note that it is customary to plot $\mathrm{d}a/\mathrm{d}n - \Delta K$ data on a double logarithmic scale. The data obtained with a high stress range, $\Delta\sigma_h$, correspond to a lower critical crack length and commence at relatively high values of $\mathrm{d}a/\mathrm{d}n$ and ΔK. The data for a low stress range, $\Delta\sigma_l$, commence at lower values of $\mathrm{d}a/\mathrm{d}n$ and ΔK, but reach the same high values as in the high stress range case. The data frequently show a sigmoidal trend, and this will be discussed in chapter 9 together

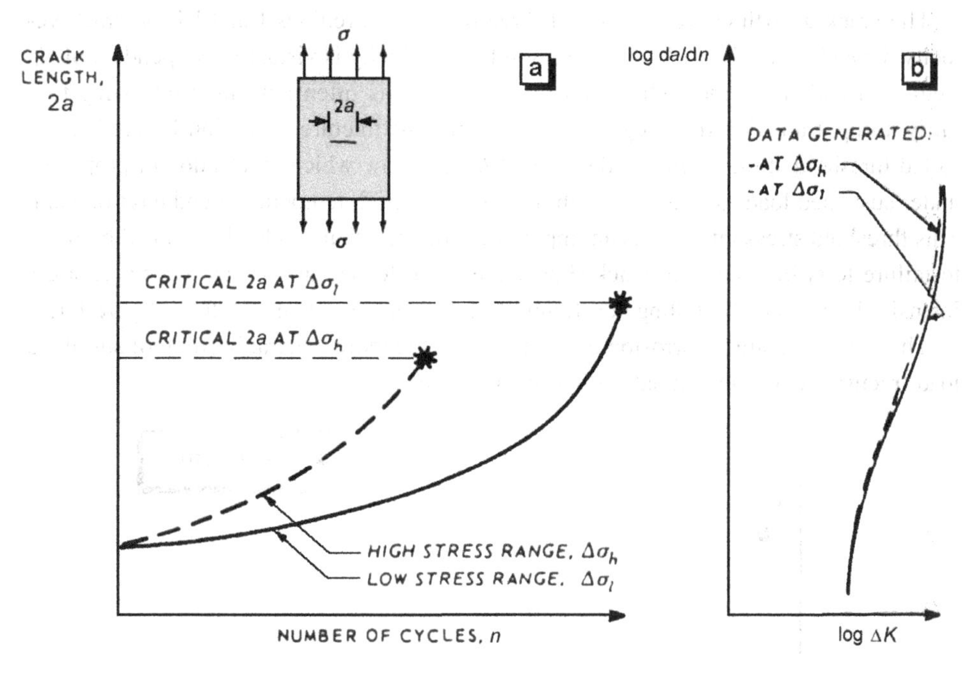

Figure 1.13. Correlation of fatigue crack propagation data by ΔK when the stress ratio, R, is the same.

with additional aspects of fatigue crack growth.

Stress Corrosion

It has also been found that stress corrosion cracking data may be correlated by the stress intensity approach. Figure 1.14 gives a generalised representation of the stress corrosion crack growth rate, da/dt, as a function of K, where t is time.

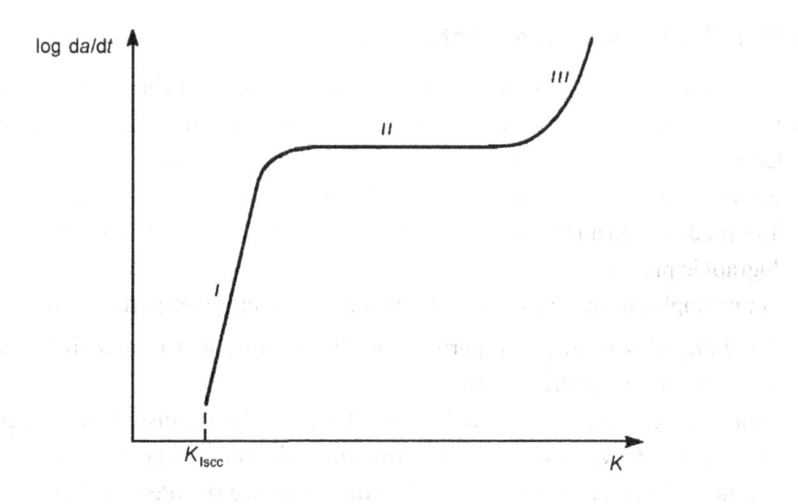

Figure 1.14. Stress corrosion crack growth rate as a function of K.

The crack growth curve consists of three regions. In regions I and III the crack velocity depends strongly on K_I, but in region II the velocity is virtually independent of K_I. Regions I and II are most characteristic. Region III is often not observed owing to a fairly abrupt transition from region II to unstable fast fracture. In region I there is a so-called threshold stress intensity, designated $K_{I_{scc}}$, below which cracks do not propagate under sustained load for a given combination of material, temperature and environment. This threshold stress intensity is an important parameter that can be determined by time-to-failure tests in which pre-cracked specimens are loaded at various (constant) stress intensity levels, thereby failing at different times as shown schematically in figure 1.15.

The subject of stress corrosion cracking, under the more general heading of sustained load fracture, will be examined further in chapter 10.

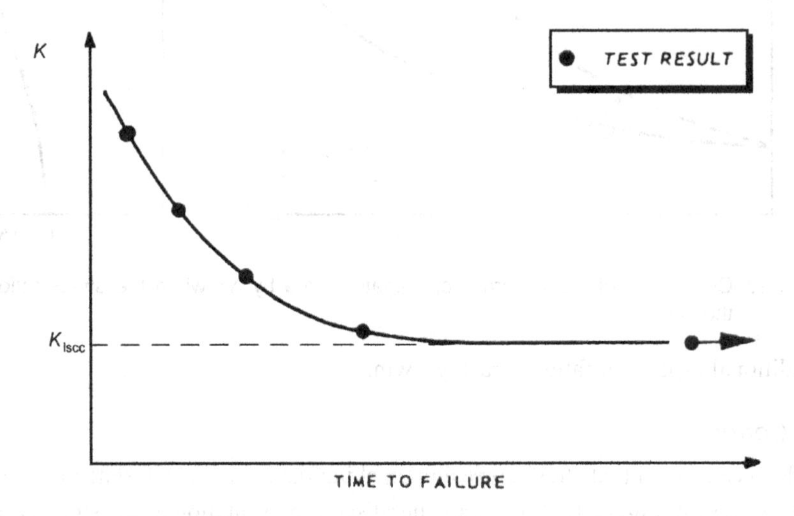

Figure 1.15. Schematic time-to-failure curve with $K_{I_{scc}}$.

1.11 Influence of Material Behaviour

So far, this overview of the use of fracture mechanics to characterize crack extension has not taken account of actual material behaviour, the influence of which may be considerable. For example, the fracture toughness of a material is much less when crack extension occurs by cleavage instead of ductile fracture. Cleavage is an intrinsically brittle mode of fracture involving separation of atomic bonds along well-defined crystallographic planes.

Other examples of material behaviour that affect fracture properties are:

1. Cracking of second phase particles in the metallic matrix and formation of microvoids at particle/matrix interfaces.
2. Anisotropic deformation and fracture. This may be intrinsic (crystallographic) as in the case of cleavage, or may result from material processing (texture).
3. Choice of fracture path, *i.e.* whether transgranular or intergranular, or a mixture of both.

4. Crack blunting and branching.

In fact, fracture often depends on combinations of such types of material behaviour. For this reason we consider that a basic course in fracture mechanics should include information concerning mechanisms of fracture and the influence of material behaviour on fracture mechanics-related properties. These topics are considered in chapters 12 and 13.

Part II
Linear Elastic Fracture Mechanics

2 The Elastic Stress Field Approach

2.1 Introduction

In the overview given in chapter 1 it was stated that the stress intensity factor K describes the magnitude of the elastic crack tip stress field. Also, K can be used to describe crack growth and fracture behaviour of materials provided that the crack tip stress field remains predominantly elastic. This correlating ability makes the stress intensity factor an extremely important fracture mechanics parameter. For this reason its derivation is treated in some detail in section 2.2.

All stress systems in the vicinity of a crack tip may be divided into three basic types, each associated with a local mode of crack surface displacements, figure 2.1. In what follows the derivation of elastic stress field equations will be limited to mode I, since this is the predominant mode in many practical cases. Once this derivation is understood it is possible to obtain a number of useful expressions for stresses and displacements in the crack tip region. However, use of the stress intensity factor approach for practical geometries does involve some difficulties. For example, actual cracks may be very irregular in shape as compared to the often highly idealised cracks considered in theoretical treatments. Moreover, assumptions such as the infinite width of a sheet or plate

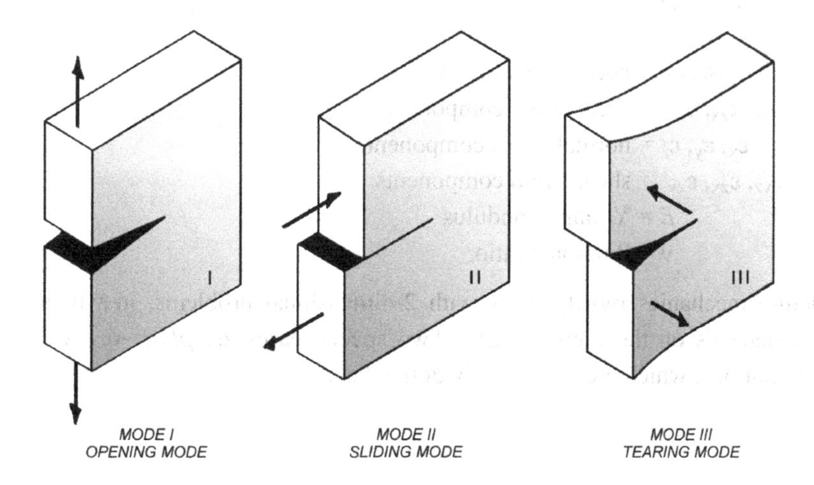

Figure 2.1. The three modes of crack surface displacements.

frequently cannot be maintained if an accurate result is required. The consequences of necessary deviations from the theoretical solutions will also be discussed in this chapter.

2.2 Derivation of the Mode I Elastic Stress Field Equations

This section gives an overview of the derivation of the mode I stress field equations. More rigorous treatments may be found in references 1 and 2 of the bibliography at the end of this chapter. The derivation covers the following topics:

- the concepts of plane stress and plane strain;
- the equilibrium equations of stress;
- the compatibility equation of strain;
- Airy stress functions;
- a general introduction to complex functions;
- Westergaard complex stress functions;
- the case of a biaxially loaded plate;
- the mode I stress intensity factor.

Finally, some additional remarks are made concerning the derivation and the stress intensity concept.

Plane Stress - Plane Strain

For an *isotropic* material the 3-dimensional form of Hooke's law can be written as:

$$\begin{aligned}
\varepsilon_x &= \frac{1}{E}\left\{\sigma_x - \nu(\sigma_y + \sigma_z)\right\}, & \varepsilon_{yz} &= \frac{1+\nu}{E}\tau_{yz}, \\
\varepsilon_y &= \frac{1}{E}\left\{\sigma_y - \nu(\sigma_z + \sigma_x)\right\}, & \varepsilon_{zx} &= \frac{1+\nu}{E}\tau_{zx}, \\
\varepsilon_z &= \frac{1}{E}\left\{\sigma_z - \nu(\sigma_x + \sigma_y)\right\}, & \varepsilon_{xy} &= \frac{1+\nu}{E}\tau_{xy}.
\end{aligned} \tag{2.1}$$

where $\sigma_x, \sigma_y, \sigma_z$ = normal stress components
$\tau_{yz}, \tau_{zx}, \tau_{xy}$ = shear stress components
$\varepsilon_x, \varepsilon_y, \varepsilon_z$ = normal strain components
$\varepsilon_{yz}, \varepsilon_{zx}, \varepsilon_{xy}$ = shear strain components [1]
E = Young's modulus
ν = Poisson's ratio.

Fracture mechanics mostly deals with 2-dimensional problems, in which case no quantity depends on the z co-ordinate. Two special cases are *plane stress* and *plane strain* conditions, which are respectively defined as:

[1] Sometimes shear strains are expressed as $\gamma_{yz} = \tau_{yz}/G$ etc., where G is the shear modulus. It then follows that $\gamma_{yz} = 2\varepsilon_{yz}$, etc. and $G = E/2(1+\nu)$

Plane stress **Plane strain**

$$\sigma_z = \tau_{yz} = \tau_{zx} = 0\,, \qquad \varepsilon_z = \varepsilon_{yz} = \varepsilon_{zx} = 0\,. \tag{2.2}$$

For these conditions the in-plane components of equations (2.1), *i.e.* those not involving the z coordinate, can be expanded as:

Plane stress **Plane strain**

$$\begin{aligned}
\varepsilon_x &= \frac{1}{E}(\sigma_x - \nu\sigma_y)\,, & \varepsilon_x &= \frac{1-\nu^2}{E}\left(\sigma_x - \frac{\nu}{1-\nu}\sigma_y\right),\\
\varepsilon_y &= \frac{1}{E}(\sigma_y - \nu\sigma_x)\,, & \varepsilon_y &= \frac{1-\nu^2}{E}\left(\sigma_y - \frac{\nu}{1-\nu}\sigma_x\right),\\
\varepsilon_{xy} &= \frac{1+\nu}{E}\tau_{xy}\,, & \varepsilon_{xy} &= \frac{1+\nu}{E}\tau_{xy} = \frac{1-\nu^2}{E}\left(1 + \frac{\nu}{1-\nu}\right)\tau_{xy}\,.
\end{aligned} \tag{2.3}$$

From this it can be seen that the strains in the plane strain case can be derived from the strains in the plane stress case as follows:

1) replace E by $\dfrac{E}{1-\nu^2}$,

2) replace ν by $\dfrac{\nu}{1-\nu}$.

The reverse, *i.e.* transforming from plane strain to plane stress, is also possible. To do this, we first have to separate a factor $E/(1-\nu^2)$ and replace it by E. Next, ν must be replaced by $\nu/(1+\nu)$. For example, the transition of ν for a plane stress $\rightarrow$ plane strain $\rightarrow$ plane stress cycle reads:

$$\nu \rightarrow \frac{\nu}{1-\nu} \rightarrow \frac{\nu/1+\nu}{1-\nu/1+\nu} \quad (=\nu)\,.$$

These transitions apply only to the expressions for the in-plane components. The thickness stress σ_z and thickness strain ε_z follow from:

Plane stress **Plane strain**

$$\begin{aligned}
\sigma_z &= 0\,, & \sigma_z &= \nu(\sigma_x+\sigma_y)\,,\\
\varepsilon_z &= -\frac{\nu}{E}(\sigma_x+\sigma_y)\,, & \varepsilon_z &= 0\,.
\end{aligned} \tag{2.4}$$

The plane stress $\leftrightarrow$ plane strain transitions enable the separate expressions for plane stress and plane strain conditions, equations (2.3), to be written in a concise manner:

$$\varepsilon_x = \frac{1}{E'}(\sigma_x - \nu'\sigma_y)\,,$$

$$\varepsilon_y = \frac{1}{E'}(\sigma_y - \nu'\sigma_x)\,, \tag{2.5}$$

$$\varepsilon_{xy} = \frac{1+\nu'}{E'}\tau_{xy}\,,$$

where $E' = E$ and $\nu' = \nu$ for plane stress,
and $E' = E/(1-\nu^2)$ and $\nu' = \nu/(1-\nu)$ for plane strain.

Equilibrium Equations of Stress

From figure 2.2 it can be seen that there is an equilibrium of forces in the x direction if:

$$\left(\sigma_x + \frac{\partial \sigma_x}{\partial x}dx\right)dydz - \sigma_x dydz + \left(\tau_{xy} + \frac{\partial \tau_{xy}}{\partial y}dy\right)dzdx - \tau_{xy}dzdx + \left(\tau_{xz} + \frac{\partial \tau_{xz}}{\partial z}dz\right)dxdy - \tau_{xz}dxdy = 0.$$

Analogous formulae follow from the equilibrium of forces in the y and z directions.

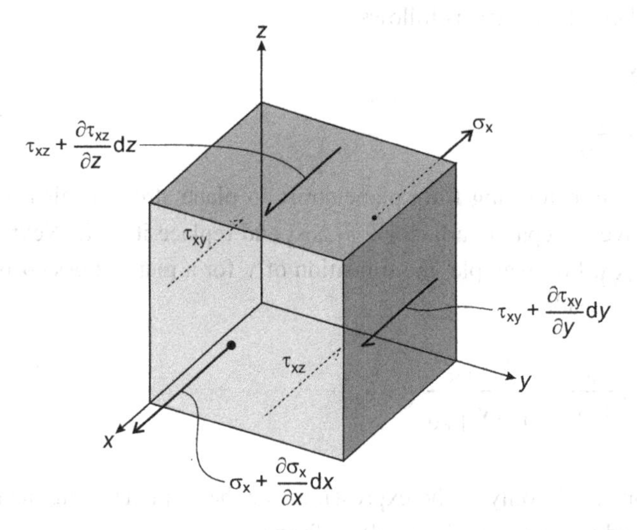

Figure 2.2. The stress components acting in the x direction on an infinitesimal material element.

This leads to the equilibrium equations of stress:

$$\frac{\partial \sigma_x}{\partial x} + \frac{\partial \tau_{xy}}{\partial y} + \frac{\partial \tau_{xz}}{\partial z} = 0\,,$$

$$\frac{\partial \sigma_y}{\partial y} + \frac{\partial \tau_{yz}}{\partial z} + \frac{\partial \tau_{yx}}{\partial x} = 0\,, \tag{2.6}$$

$$\frac{\partial \sigma_z}{\partial z} + \frac{\partial \tau_{zx}}{\partial x} + \frac{\partial \tau_{zy}}{\partial y} = 0\,.$$

If we confine ourselves to the two-dimensional cases of plane stress and plane strain, for which $\tau_{zy} = \tau_{zx} = 0$ and $\partial/\partial z = 0$, the equilibrium equations (2.6) reduce to:

$$\frac{\partial \sigma_x}{\partial x} + \frac{\partial \tau_{xy}}{\partial y} = 0 ,$$
$$\frac{\partial \sigma_y}{\partial y} + \frac{\partial \tau_{yx}}{\partial x} = 0 . \tag{2.7}$$

Compatibility Equation of Strain

Again we consider only the 2-dimensional case. Small strains in the x-y plane can be expressed in terms of the displacements in the x and y direction, u and v respectively, according to:

$$\varepsilon_x = \frac{\partial u}{\partial x} \quad , \quad \varepsilon_y = \frac{\partial v}{\partial y} \quad , \quad \varepsilon_{xy} = ½\left(\frac{\partial v}{\partial x} + \frac{\partial u}{\partial y}\right) . \tag{2.8}$$

Obviously, the three strain components are determined by only two displacement components u and v. Therefore, the three strain components cannot be independent. An extra relation must exist between them. This relation is the *compatibility equation of strain* and is obtained by eliminating u and v through differentiation of equations (2.8):

$$\frac{\partial^2 \varepsilon_x}{\partial y^2} + \frac{\partial^2 \varepsilon_y}{\partial x^2} = 2\frac{\partial^2 \varepsilon_{xy}}{\partial x \partial y} . \tag{2.9}$$

Using equations (2.5) the compatibility equation can also be expressed in terms of stress components:

$$\frac{\partial^2}{\partial y^2}(\sigma_x - \nu' \sigma_y) + \frac{\partial^2}{\partial x^2}(\sigma_y - \nu' \sigma_x) - 2(1+\nu')\frac{\partial^2 \tau_{xy}}{\partial x \partial y} = 0 , \tag{2.10}$$

where again $\nu' = \nu$ for plane stress
$\nu' = \nu/(1-\nu)$ for plane strain

Airy Stress Functions

Any stress field solution for an elastic problem must fulfil both the equilibrium and the compatibility equations. For this purpose, Airy introduced a way of describing two-dimensional stress fields using a function $\Phi(x,y)$:

$$\sigma_x = \frac{\partial^2 \Phi}{\partial y^2} \quad , \quad \sigma_y = \frac{\partial^2 \Phi}{\partial x^2} \quad , \quad \tau_{xy} = -\frac{\partial^2 \Phi}{\partial x \partial y} . \tag{2.11}$$

Straightforward substitution shows that this stress field

- always fulfils the equilibrium equations of stress (2.7);
- only fulfils the compatibility equation (2.10) if the stress function Φ is a solution of the so-called *biharmonic equation*:

$$\frac{\partial^4\Phi}{\partial x^4}+2\frac{\partial^4\Phi}{\partial x^2\partial y^2}+\frac{\partial^4\Phi}{\partial y^4}=0 \quad \text{or} \quad \nabla^2\nabla^2\Phi=\nabla^4\Phi=0\,. \tag{2.12}$$

A function $\Phi(x,y)$ that fulfils the biharmonic equation and defines a stress field according to equations (2.11) is called an *Airy stress function*. Both equilibrium of stress as well as compatibility of strain are guaranteed. Note that Airy stress functions are real and have the dimension of force.

A General Introduction to Complex Functions

A complex function $f(z)$, where z is a complex variable, is called *analytic* if its derivative $f'(z)$ exists. The consequence of this definition is that if a primitive of $f(z)$ exists, then this primitive is also analytic.

For an analytic function the *Cauchy-Riemann equations* can be derived. Defining $z = x + \mathrm{i}\cdot y$, we may write:

$$\frac{\partial f}{\partial x}=\frac{\mathrm{d}f}{\mathrm{d}z}\frac{\partial z}{\partial x}=\frac{\mathrm{d}f}{\mathrm{d}z}=f'(z)\,,$$

$$\frac{\partial f}{\partial y}=\frac{\mathrm{d}f}{\mathrm{d}z}\frac{\partial z}{\partial y}=\frac{\mathrm{d}f}{\mathrm{d}z}\cdot\mathrm{i}=\mathrm{i}\cdot f'(z)$$

and therefore

$$\frac{\partial f}{\partial x}=\frac{\partial \mathrm{Re} f(z)}{\partial x}+\mathrm{i}\cdot\frac{\partial \mathrm{Im} f(z)}{\partial x}=f'(z)\,,$$

$$\frac{\partial f}{\partial y}=\frac{\partial \mathrm{Re} f(z)}{\partial y}+\mathrm{i}\cdot\frac{\partial \mathrm{Im} f(z)}{\partial y}=\mathrm{i}\cdot f'(z)\,.$$

Eliminating $f'(z)$, leads to

$$\mathrm{i}\cdot\left(\frac{\partial \mathrm{Re} f(z)}{\partial x}+\mathrm{i}\cdot\frac{\partial \mathrm{Im} f(z)}{\partial x}\right)=\frac{\partial \mathrm{Re} f(z)}{\partial y}+\mathrm{i}\cdot\frac{\partial \mathrm{Im} f(z)}{\partial y}\,.$$

Now the Cauchy-Riemann equations follow as

$$\frac{\partial \mathrm{Re} f(z)}{\partial x}=\frac{\partial \mathrm{Im} f(z)}{\partial y}\,,\quad \frac{\partial \mathrm{Re} f(z)}{\partial y}=-\frac{\partial \mathrm{Im} f(z)}{\partial x}\,. \tag{2.13}$$

In the Cauchy-Riemann equations the real or the imaginary part of $f(z)$ can be eliminated. For example, equations (2.13) can be differentiated as follows:

$$\frac{\partial^2 \mathrm{Re} f(z)}{\partial x^2}=\frac{\partial^2 \mathrm{Im} f(z)}{\partial x\partial y}\,,\quad \frac{\partial^2 \mathrm{Re} f(z)}{\partial y^2}=-\frac{\partial^2 \mathrm{Im} f(z)}{\partial x\partial y}\,. \tag{2.14}$$

Consequently:

$$\frac{\partial^2 \text{Re} f(z)}{\partial x^2} + \frac{\partial^2 \text{Re} f(z)}{\partial y^2} = 0$$

or

$$\nabla^2 \text{Re} f(z) = 0. \tag{2.15}$$

An identical relation can be derived for $\text{Im} f(z)$ by eliminating $\text{Re} f(z)$. It is said that $\text{Re} f(z)$ and $\text{Im} f(z)$ are *conjugate harmonic functions*. They both satisfy Laplace's equation:

$$\nabla^2 \text{Re} f(z) = \nabla^2 \text{Im} f(z) = 0\,. \tag{2.16}$$

Westergaard Complex Stress Functions

Westergaard (reference 1) introduced a specific type of Airy stress function Φ using an analytic *complex stress function* $\phi(z)$ of which the first and second order integrals are assumed to exist:

$$\Phi = \text{Re}\,\bar{\bar{\phi}}(z) + y \cdot \text{Im}\,\bar{\phi}(z)\,, \tag{2.17}$$

where $\bar{\phi}(z)$, $\bar{\bar{\phi}}(z)$ = first and second order integrals (primitives) of $\phi(z)$ respectively and $z = x + \text{i} \cdot y$.

For Φ to qualify as an Airy stress function it must fulfil the biharmonic equation (2.12). Note that Φ has a real value. Because $\bar{\bar{\phi}}(z)$ is itself an analytic function, it is obvious that the first part of (2.17), Re $\bar{\bar{\phi}}(z)$, is a harmonic function fulfilling Laplace's equation (2.16) and thus also the biharmonic equation (2.12). For the second part of equation (2.17) we consider the function $y \cdot \varphi$, where φ is harmonic, *i.e.* $\nabla^2 \varphi = 0$:

$$\nabla^2 (y \cdot \varphi) = \frac{\partial^2 (y \cdot \varphi)}{\partial x^2} + \frac{\partial^2 (y \cdot \varphi)}{\partial y^2} = \frac{\partial}{\partial x}\left(y \frac{\partial \varphi}{\partial x}\right) + \frac{\partial}{\partial y}\left(\varphi + y \frac{\partial \varphi}{\partial y}\right) = y \frac{\partial^2 \varphi}{\partial x^2} + \frac{\partial \varphi}{\partial y} + \frac{\partial \varphi}{\partial y} + y \frac{\partial^2 \varphi}{\partial y^2} = 2 \frac{\partial \varphi}{\partial y}\,.$$

Thus:

$$\nabla^2 \nabla^2 (y \cdot \varphi) = \nabla^2 \left(2 \frac{\partial \varphi}{\partial y}\right) = 2 \frac{\partial}{\partial y} \nabla^2 \varphi = 0\,.$$

Consequently, the function Φ defined in terms of the complex stress function $\phi(z)$ introduced by Westergaard does qualify as an Airy stress function.

Now we can express the stress components σ_x, σ_y and τ_{xy} in terms of the complex stress function $\phi(z)$. For example σ_x becomes:

$$\begin{aligned}
\sigma_x &= \frac{\partial^2 \Phi}{\partial y^2} \\
&= \frac{\partial}{\partial y}\left\{\frac{\partial}{\partial y}\text{Re}\,\bar{\bar{\phi}}(z) + \frac{\partial}{\partial y} y \cdot \text{Im}\,\bar{\phi}(z)\right\} = \frac{\partial}{\partial y}\left\{-\frac{\partial}{\partial x}\text{Im}\,\bar{\bar{\phi}}(z) + \text{Im}\,\bar{\phi}(z) + y \frac{\partial}{\partial x}\text{Re}\,\bar{\phi}(z)\right\} \\
&= \frac{\partial}{\partial y}\left\{-\text{Im}\,\bar{\phi}(z) + \text{Im}\,\bar{\phi}(z) + y \cdot \text{Re}\,\phi(z)\right\} = \text{Re}\,\phi(z) + y \cdot \frac{\partial}{\partial y}\text{Re}\,\phi(z)
\end{aligned}$$

$$= \text{Re}\ \phi(z) - y \cdot \text{Im}\ \phi'(z)\ . \tag{2.18.a}$$

Use has been made of the Cauchy-Riemann equations (2.13). The components σ_y and τ_{xy} can be derived analogously, leading to:

$$\sigma_y = \text{Re}\ \phi(z) + y \cdot \text{Im}\ \phi'(z)\ , \tag{2.18.b}$$

$$\tau_{xy} = -\ y \cdot \text{Re}\ \phi'(z)\ . \tag{2.18.c}$$

These expressions have a general nature, *i.e.* they give the stress components for any (Westergaard) complex stress function $\phi(z)$. The stress field solution corresponding to a particular two-dimensional elastic problem is found by choosing the stress function $\phi(z)$ in such a way that all boundary conditions are fulfilled. Note that using a Westergaard complex stress function limits the boundary conditions for problems that can be solved. From equations (2.18) it is easily seen that for $y = 0$ it is required that $\sigma_x = \sigma_y$ and $\tau_{xy} = 0$.

Biaxially Loaded Plate

We will consider an infinite plate containing a crack. The plate is biaxially loaded in tension by a stress σ_∞ (see figure 2.3). The boundary conditions for this problem are:

1) $\sigma_y = 0$ for $-a < x < +a$ *and* $y = 0$,
2) $\sigma_x \rightarrow \sigma_\infty$ *and* $\sigma_y \rightarrow \sigma_\infty$ for $x \rightarrow \pm\infty$ *and/or* $y \rightarrow \pm\infty$,
3) $\sigma_y \rightarrow \infty$ for $x = \pm a$ *and* $y = 0$.

The first condition arises from the fact that the crack flanks are free surfaces. The

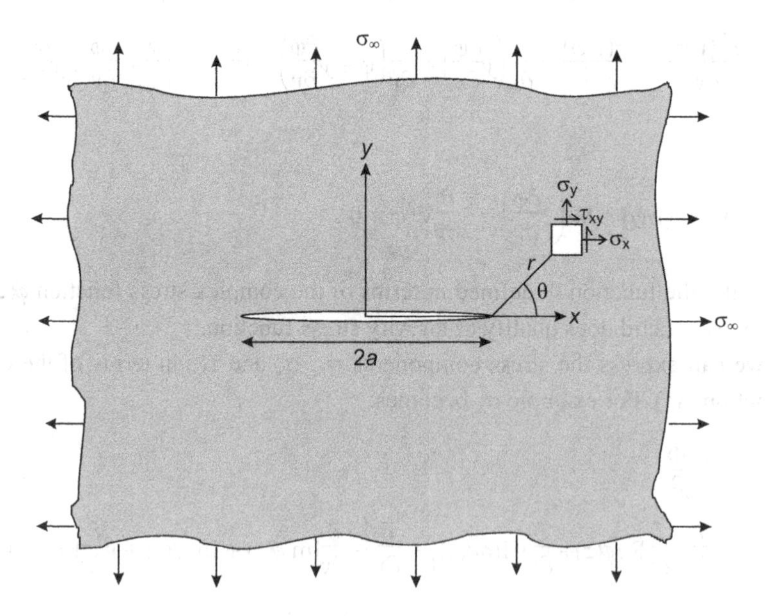

Figure 2.3. A biaxially loaded infinite plate containing a crack.

second condition states that at an infinite distance from the crack the stress will be equal to the applied stress σ_∞. The last condition accounts for the stress raising effect of a crack. At a crack tip with zero radius the stress σ_y becomes singular.

We have to find a function $\phi(z)$ such that σ_x and σ_y, defined according to equations (2.18), fulfil the above boundary conditions. A function that does this is:

$$\phi(z) = \frac{\sigma_\infty}{\sqrt{1 - a^2/z^2}} . \tag{2.19}$$

In the following the boundary conditions are checked:

1) If $y = 0$ it follows that $z = x$ and thus:

$$\phi(z) = \phi(x) = \frac{\sigma_\infty}{\sqrt{-1\left(a^2/x^2 - 1\right)}} = -\mathrm{i}\cdot\frac{\sigma_\infty}{\sqrt{a^2/x^2 - 1}} .$$

Therefore, as $|x| < a$ along the crack flanks, the function $\phi(z)$ is purely imaginary, *i.e.* $\mathrm{Re}\,\phi(z) = 0$. Consequently:

$$\sigma_y = \mathrm{Re}\,\phi(z) + y\cdot\mathrm{Im}\,\phi'(z) = 0 .$$

2) For $x \rightarrow \pm\infty$ and/or $y \rightarrow \pm\infty$, *i.e.* $|z| \rightarrow \infty$, equation (2.19) becomes $\phi(z) = \sigma_\infty$. Because now $\mathrm{Im}\,\phi(z) = 0$, it follows that:

$$\sigma_x = \sigma_y = \mathrm{Re}\,\phi(z) = \sigma_\infty .$$

3) For $z = \pm a$ we obtain $\phi(z) \rightarrow \infty$. So at the crack tips ($y = 0$, thus $z = x$) we get:

$$\sigma_y = \mathrm{Re}\,\phi(z) \rightarrow \infty .$$

The Mode I Stress Intensity Factor

It will prove convenient to translate the origin of the co-ordinate system to the crack tip at $z = +a$ by introducing the variable $\eta = z - a$. The complex stress function $\phi(z)$ now becomes:

$$\phi(\eta) = \frac{\sigma}{\sqrt{1 - a^2/(a+\eta)^2}} = \frac{\sigma(a+\eta)}{\sqrt{(a+\eta)^2 - a^2}} . \tag{2.20}$$

Note that the suffix ∞ is now omitted from σ_∞. Near the crack tip, *i.e.* $|\eta| << a$, the stress function may be approximated as:

$$\phi(\eta) \approx \frac{\sigma a}{\sqrt{2a\eta}} = \sigma\sqrt{\frac{a}{2}}\eta^{-1/2} . \tag{2.21}$$

In polar co-ordinates, *i.e.* $\eta = r\,\mathrm{e}^{\mathrm{i}\theta}$, we may write:

$$\phi(\eta) = \sigma\sqrt{\frac{a}{2}}\, r^{-1/2} e^{-1/2 i\theta} = \frac{\sigma\sqrt{\pi a}}{\sqrt{2\pi r}} e^{-1/2 i\theta} . \tag{2.22}$$

In order to calculate the stress components using equations (2.18), the first derivative $\phi'(z)$ is required too:

$$\phi'(\eta) = \sigma\sqrt{\frac{a}{2}} \cdot -1/2\eta^{-3/2} = \sigma\sqrt{\frac{a}{2}} \cdot -1/2 (r e^{i\theta})^{-3/2} = \frac{\sigma}{2r}\frac{\sqrt{\pi a}}{\sqrt{2\pi r}} \cdot -e^{-3/2 i\theta} . \tag{2.23}$$

Using $r\,e^{i\theta} = r(\cos\theta + i\cdot\sin\theta)$, the expressions that appear in equations (2.18) may be written as:

$$\mathrm{Re}\,\phi(\eta) = \frac{\sigma\sqrt{\pi a}}{\sqrt{2\pi r}}\cos\theta/2 ,$$

$$\mathrm{Re}\,\phi'(\eta) = -\frac{\sigma}{2r}\frac{\sqrt{\pi a}}{\sqrt{2\pi r}}\cos 3\theta/2 ,$$

$$y\cdot\mathrm{Im}\,\phi'(\eta) = r\sin\theta\cdot\frac{\sigma}{2r}\frac{\sqrt{\pi a}}{\sqrt{2\pi r}}\sin 3\theta/2 = \frac{\sigma\sqrt{\pi a}}{\sqrt{2\pi r}}\sin\theta/2\cos\theta/2\sin 3\theta/2 .$$

After substitution we obtain the three stress components near the tip of a crack in a biaxially loaded plate:

$$\sigma_x = \frac{\sigma\sqrt{\pi a}}{\sqrt{2\pi r}}\cos\theta/2\left(1 - \sin\theta/2\sin 3\theta/2\right) ,$$

$$\sigma_y = \frac{\sigma\sqrt{\pi a}}{\sqrt{2\pi r}}\cos\theta/2\left(1 + \sin\theta/2\sin 3\theta/2\right) , \tag{2.24}$$

$$\tau_{xy} = \frac{\sigma\sqrt{\pi a}}{\sqrt{2\pi r}}\sin\theta/2\cos\theta/2\cos 3\theta/2 .$$

These expressions show that the stress components tend to infinity at the crack tip (r = 0), a so-called $1/\sqrt{r}$ singularity. The intensity of this stress singularity is given by the factor $\sigma\sqrt{\pi a}$, while the remaining parts of equations (2.24) are functions of the geometrical position relative to the crack tip. The intensity of the stress singularity is called the mode I *stress intensity factor*, K_I. For the configuration considered here, a biaxially loaded infinite plate, this factor is equal to $\sigma\sqrt{\pi a}$, and thus depends only on the remote stress and the crack length.

Some Additional Remarks

1) The foregoing derivation is only one of the methods for obtaining the stress field solution. There are several other, more general methods.

2) The case of a plate *uniaxially* loaded in the y direction cannot be solved using the complex Westergaard stress function, since then it is required that $\sigma_x = \sigma_y$ for $y = 0$, *cf.* equations (2.18). However, Irwin argued that this problem can be solved if the remote stress σ is subtracted from the expression for σ_x, equation (2.18.a), see reference 3. Then by using the complex stress function ϕ for the biaxial case, equation 2.19, the boundary conditions for the uniaxial case are satisfied. Thus the stress field in a uniaxially loaded plate is identical to that in a biaxially loaded plate with the exception of σ_x, which is reduced by the remote stress σ. In the near tip stress field, equations (2.24), this correction is usually omitted because near the crack tip σ_x is much larger than σ.
3) Owing to the approximation made in equation (2.21) these equations only apply to the stress field near the crack tip ($r << a$). For example, equations (2.24) suggest that if $r \to \infty$, the stress components σ_x and σ_y approach zero instead of σ (the applied load).
4) The stress in the proximity of the crack tip induced by a mode I load may be written as the product of a stress intensity factor and a geometrical function. Using the index notation (see section 6.3) we can write

$$\sigma_{ij} = K_I \cdot f_{ij}(r,\theta) . \tag{2.25}$$

Since in *linear elastic* material stresses are additive, multiple mode I loads will cause a total stress equal to:

$$(\sigma_{ij})_{total} = (\sigma_{ij})_1 + (\sigma_{ij})_2 + ... = (K_I)_1 \cdot f_{ij}(r,\theta) + (K_I)_2 \cdot f_{ij}(r,\theta) + ... \tag{2.26}$$

Consequently, as the geometrical functions are identical for all mode I loads, we may write:

$$(\sigma_{ij})_{total} = \left\{(K_I)_1 + (K_I)_2 + ...\right\} \cdot f_{ij}(r,\theta)$$

or

$$(\sigma_{ij})_{total} = (K_I)_{total} \cdot f_{ij}(r,\theta) , \tag{2.27}$$

where $(K_I)_{total} = (K_I)_1 + (K_I)_2 + ...$.

This is the principle of *superposition*: the total mode I stress intensity can be obtained by simply adding all mode I stress intensities caused by individual loads.
5) The stress field described by equations (2.24) only applies to cracks with infinitely sharp tips (see 3rd boundary condition). For a crack with a finite tip radius ρ, *i.e.* a *blunted* tip (figure 2.4), Creager and Paris obtained the following expressions by shifting the origin of the co-ordinate system over a distance of $\rho/2$ behind tip, see reference 4:

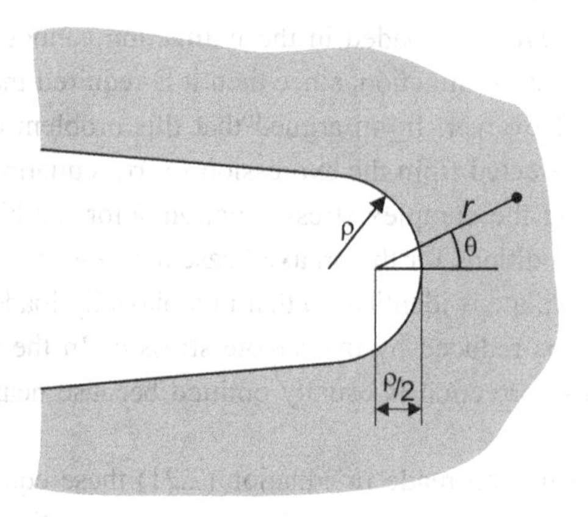

Figure 2.4. Crack with a finite tip radius, ρ.

$$\sigma_x = \frac{K_I}{\sqrt{2\pi r}} \cos \theta/2 \left(1 - \sin \theta/2 \sin 3\theta/2\right) - \frac{K_I}{\sqrt{2\pi r}} \left(\frac{\rho}{2r}\right) \cos 3\theta/2 ,$$

$$\sigma_y = \frac{K_I}{\sqrt{2\pi r}} \cos \theta/2 \left(1 + \sin \theta/2 \sin 3\theta/2\right) + \frac{K_I}{\sqrt{2\pi r}} \left(\frac{\rho}{2r}\right) \cos 3\theta/2 , \qquad (2.28)$$

$$\tau_{xy} = \frac{K_I}{\sqrt{2\pi r}} \sin \theta/2 \cos \theta/2 \cos 3\theta/2 - \frac{K_I}{\sqrt{2\pi r}} \left(\frac{\rho}{2r}\right) \sin 3\theta/2 .$$

At the blunted crack tip itself ($\theta = 0$, $r = \rho/2$), it follows that:

- $\sigma_x = 0$, which is obvious in view of the fact that the tip is a vertical free surface,
- $\sigma_y = 2K_I/\sqrt{2\pi\rho/2}$, which is a non-singular (*i.e.* finite) value.

These values clearly differ from the singular values at a sharp crack tip. On the other hand, for distances much greater than the tip radius ($r >> \rho/2$) the stress values near the blunted crack approach those for a sharp crack. This means that expressions (2.24) for an infinitely sharp crack are also valid in the neighbourhood of a blunted crack.

Note that when we differentiate the first of the equations (2.28) with respect to r for $\theta = 0$, *i.e.* along the positive x axis, we find a maximum value for σ_x at a distance ρ from the tip.

6) Using equations (2.28) we can examine the relation between the stress intensity factor K and the *stress concentration factor*. The latter describes the increase in stress level, relative to the nominal stress, owing to a stress concentration. Using the stress σ_y at the blunted tip, it follows that:

$$\sigma_y = \frac{2K_I}{\sqrt{\pi\rho}} = \frac{2\sigma\sqrt{\pi a}}{\sqrt{\pi\rho}} \Rightarrow \text{the stress concentration factor} = \frac{\sigma_y}{\sigma} = 2\sqrt{\frac{a}{\rho}} . \qquad (2.29)$$

Inglis (reference 5) analysed the case of a small elliptical hole in a large plate loaded perpendicular to the major axis by a remote tensile stress σ. He derived the stress at the tip of the major axis, σ_{tip}, as:

$$\sigma_{tip} = \sigma\left(1 + \frac{2a}{b}\right) = \sigma\left(1 + 2\sqrt{\frac{a}{\rho}}\right), \tag{2.30}$$

where $2a$, $2b$ = major and minor axes of the ellipse respectively,
ρ = radius of curvature of ellipse at the tip of its major axis = b^2/a.

An elliptical hole with a minor axis small compared to the major axis resembles a blunted crack. The radius ρ is small compared to the major axis. The stress concentration factor may now be approximated by:

$$\frac{\sigma_{tip}}{\sigma} \approx 2\sqrt{\frac{a}{\rho}} \quad \text{for } \rho \ll a\,. \tag{2.31}$$

This is consistent with the expression derived using the stress intensity factor, equation (2.29).

2.3 Useful Expressions

In this section some useful expressions for stresses and strains in the crack tip region will be given, namely

- The mode I stress field in terms of principal stresses:
  The use of principal stresses is convenient when considering yield criteria in order to estimate plastic zone sizes, as will be discussed in chapter 3.
- The elastic displacement field:
  The elastic displacement field enables calculation of the stored elastic energy (used in energy balance approaches, chapter 4) and also serves as a basis for displacement controlled fracture criteria, *e.g.* COD (chapter 7).
- The stress field for modes II and III:
  These expressions are required for studying crack problems in which mode II or mode III loading or combined mode loading apply.

Principal Stresses

Using Mohr's circle construction (figure 2.5), we can express the in-plane *principal* stresses σ_1 and σ_2 in terms of the stress components σ_x, σ_y and τ_{xy}:

$$\sigma_{1,2} = \frac{\sigma_x + \sigma_y}{2} \pm \sqrt{\left(\frac{\sigma_y - \sigma_x}{2}\right)^2 + \tau_{xy}^2} \tag{2.32}$$

Substitution of equations (2.24) gives:

$$\sigma_1 = \frac{K_I}{\sqrt{2\pi r}}\cos\theta/2\left(1 + \sin\theta/2\right) \tag{2.33.a}$$

$$\sigma_2 = \frac{K_I}{\sqrt{2\pi r}} \cos\theta/2 \left(1 - \sin\theta/2\right) \tag{2.33.b}$$

The remaining principal stress normal to the plane, σ_3, is either 0 for plane stress or $\nu(\sigma_1 + \sigma_2)$ for plane strain, *i.e.*

$$\sigma_3 = \frac{2\nu K_I}{\sqrt{2\pi r}} \cos\theta/2 \quad \text{for plane strain.} \tag{2.33.c}$$

The *principal directions* follow from:

$$\tan 2\varphi = \frac{\tau_{xy}}{(\sigma_y - \sigma_x)/2} = \frac{\cos 3\theta/2}{\sin 3\theta/2} \qquad \text{for } \theta \neq 0\,, \tag{2.34}$$

where φ = angle between σ_1 and y axis.

Straight ahead of the crack ($\theta = 0$) the shear stress $\tau_{xy} = 0$, the angle $\varphi = 0°$ and the principal stresses σ_1 and σ_2 are equal and also equal to σ_y and σ_x respectively.

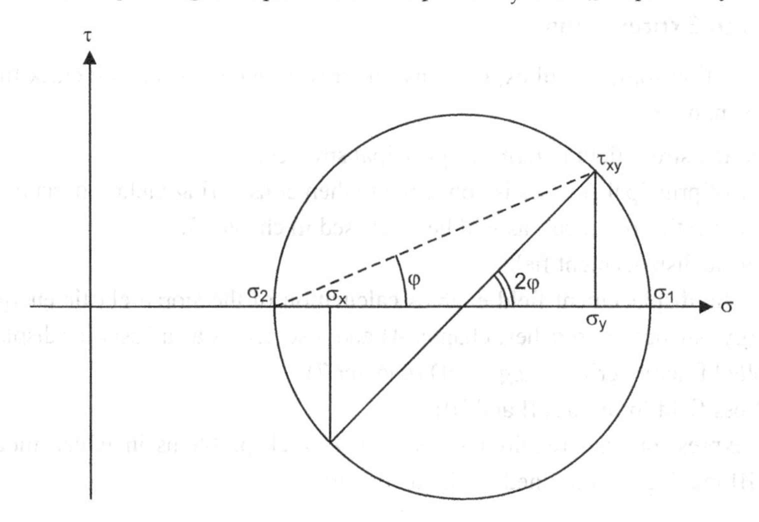

Figure 2.5. Mohr's circle construction.

Elastic Displacements

The elastic displacement field around a crack can be derived from the elastic stress field using the two-dimensional form of Hooke's law, equations (2.5), valid for either a plane stress or a plane strain condition. For example, the strain component ε_x can be written as:

$$\varepsilon_x = \frac{1}{E'}(\sigma_x - \nu'\sigma_y)\,. \tag{2.35}$$

The displacement in the x direction, u, is obtained by integrating the relation between strain and displacement, equations (2.8):

$$u = \int \varepsilon_x \mathrm{d}x = \frac{1}{E'} \int (\sigma_x - \nu' \sigma_y) \mathrm{d}x \, . \tag{2.36}$$

Substituting from equations (2.18), *i.e.* σ_x and σ_y in terms of the Westergaard complex stress function $\phi(z)$, leads to:

$$u = \frac{1}{E'} \left\{ (1-\nu') \, \mathrm{Re} \, \bar{\phi}(z) - y(1+\nu') \, \mathrm{Im} \, \phi(z) \right\} . \tag{2.37.a}$$

Similarly:

$$v = \frac{1}{E'} \left\{ 2 \, \mathrm{Im} \, \bar{\phi}(z) - y(1+\nu') \, \mathrm{Re} \, \phi(z) \right\} . \tag{2.37.b}$$

For the displacements *near* the crack tip, the approximate complex stress function $\phi(\eta)$ defined in equation (2.21) suffices. We have to integrate this function first in order to obtain the real and imaginary part of $\bar{\phi}(\eta)$:

$$\bar{\phi}(\eta) = \int \phi(\eta) \mathrm{d}\eta = \int \sigma \sqrt{\frac{a}{2}} \eta^{-1/2} \mathrm{d}\eta = \sigma \sqrt{\frac{a}{2}} \frac{1}{1/2} \eta^{1/2} = \sigma \sqrt{\pi a} \sqrt{\frac{2r}{\pi}} \mathrm{e}^{1/2 i\theta}$$

$$\Rightarrow \quad \mathrm{Re} \, \bar{\phi}(\eta) = K_I \sqrt{\frac{2r}{\pi}} \cos\theta/2 \quad ; \quad \mathrm{Im} \, \bar{\phi}(\eta) = K_I \sqrt{\frac{2r}{\pi}} \sin\theta/2 \, .$$

The real and imaginary parts of $\phi(\eta)$ follow in a straightforward manner from its polar representation in (2.22):

$$\phi(\eta) = \frac{K_I}{\sqrt{2\pi r}} \mathrm{e}^{-1/2 i\theta} \quad \Rightarrow \quad \mathrm{Re} \, \phi(\eta) = \frac{K_I}{\sqrt{2\pi r}} \cos\theta/2 \; ; \; \mathrm{Im} \, \phi(\eta) = -\frac{K_I}{\sqrt{2\pi r}} \sin\theta/2 \, .$$

After substituting these expressions into (2.37), the elastic displacements u and v are found. For *plane stress*

$$\begin{aligned} u &= 2 \frac{K_I}{E} \sqrt{\frac{r}{2\pi}} \cos\theta/2 \left(1 + \sin^2 \theta/2 - \nu \cdot \cos^2 \theta/2 \right) , \\ v &= 2 \frac{K_I}{E} \sqrt{\frac{r}{2\pi}} \sin\theta/2 \left(1 + \sin^2 \theta/2 - \nu \cdot \cos^2 \theta/2 \right) , \end{aligned} \tag{2.38}$$

while for *plane strain*, using $E' = E/(1-\nu^2)$ and $\nu' = \nu/(1-\nu)$,

$$u = 2(1+\nu)\frac{K_I}{E}\sqrt{\frac{r}{2\pi}}\cos\theta/_2\left(2 - 2\nu - \cos^2\theta/_2\right),$$

$$v = 2(1+\nu)\frac{K_I}{E}\sqrt{\frac{r}{2\pi}}\sin\theta/_2\left(2 - 2\nu - \cos^2\theta/_2\right). \tag{2.39}$$

The Crack Flank Displacement

Another useful expression is the displacement v at any position along the crack flank. Figure 2.6 shows a schematic definition of this *crack flank displacement*. Expressions (2.38) or (2.39) cannot be used for this case. The reason is that the approximate stress function $\phi(\eta)$ is valid only close to the crack tip. Therefore, we revert to the expression for the vertical displacement v in terms of the complex stress function $\phi(z)$:

$$v = \frac{1}{E'}\left\{2\,\mathrm{Im}\,\bar{\phi}(z) - y(1+\nu')\,\mathrm{Re}\,\phi(z)\right\}, \tag{2.37.b}$$

where $\phi(z)$ = complex stress function for biaxially loaded plate (eq. (2.19)) $= \sigma/\sqrt{1-a^2/z^2}$.

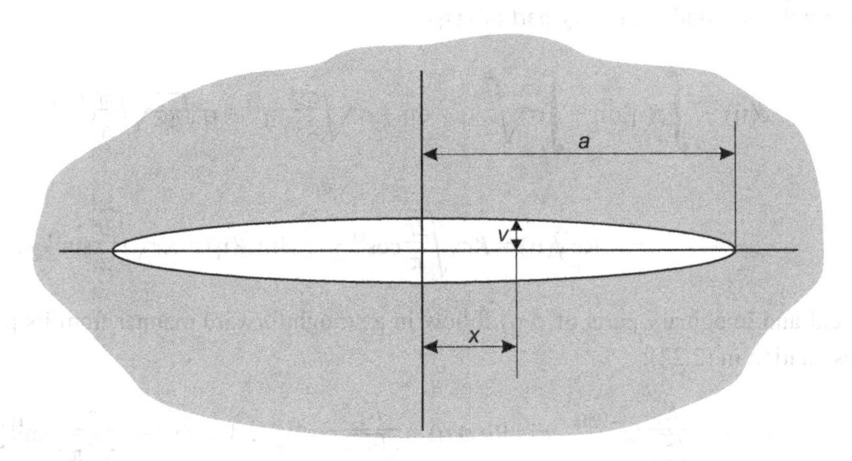

Figure 2.6. Crack flank displacement.

It was found earlier that Re $\phi(z) = 0$ along the crack flank. To obtain Im $\bar{\phi}(z)$ we have to integrate the stress function $\phi(z)$:

$$\bar{\phi}(z) = \int\phi(z)\mathrm{d}z = \int\frac{\sigma}{\sqrt{1-a^2/z^2}}\mathrm{d}z = \int\frac{\sigma z}{\sqrt{z^2-a^2}}\mathrm{d}z = \sigma\int\frac{z\mathrm{d}z}{\sqrt{z^2-a^2}} = \sigma\sqrt{z^2-a^2}.$$

At the crack flank $y = 0$ and $z = x$. Because $|x| < a$ also, it follows that:

$$\bar{\phi}(z) = \sigma\sqrt{x^2-a^2} = \mathrm{i}\cdot\sigma\sqrt{a^2-x^2} \Rightarrow \mathrm{Im}\,\bar{\phi}(z) = \sigma\sqrt{a^2-x^2}.$$

Substitution into equation (2.37.b), using the appropriate expressions for E' and ν' given in equation (2.5), leads to the *plane stress* crack flank displacement:

$$v = \frac{2\sigma}{E}\sqrt{a^2 - x^2} \tag{2.40}$$

and the *plane strain* crack flank displacement:

$$v = \frac{2\sigma(1-\nu^2)}{E}\sqrt{a^2 - x^2}\,. \tag{2.41}$$

The Stress Field Equations for Modes II and III

The stress field equations for mode II and mode III loading may be obtained in a similar manner to that outlined in section 2.2 for mode I. The results are (reference 6):

$$\begin{array}{c}\text{Mode II}\\ K_{II} = \tau\sqrt{\pi a}\end{array}\quad\left\{\begin{aligned}\sigma_x &= \frac{-K_{II}}{\sqrt{2\pi r}}\sin{}^{\theta}/_{2}\left(2 + \cos{}^{\theta}/_{2}\cos{}^{3\theta}/_{2}\right)\\ \sigma_y &= \frac{+K_{II}}{\sqrt{2\pi r}}\sin{}^{\theta}/_{2}\cos{}^{\theta}/_{2}\cos{}^{3\theta}/_{2}\\ \tau_{xy} &= \frac{+K_{II}}{\sqrt{2\pi r}}\cos{}^{\theta}/_{2}\left(1 - \sin{}^{\theta}/_{2}\sin{}^{3\theta}/_{2}\right).\end{aligned}\right. \tag{2.42}$$

$$\begin{array}{c}\text{Mode III}\\ K_{III} = \tau\sqrt{\pi a}\end{array}\quad\left\{\begin{aligned}\tau_{xz} &= \frac{-K_{III}}{\sqrt{2\pi r}}\sin{}^{\theta}/_{2}\\ \tau_{yz} &= \frac{+K_{III}}{\sqrt{2\pi r}}\cos{}^{\theta}/_{2}\,.\end{aligned}\right. \tag{2.43}$$

2.4 Finite Specimen Width

The solution for the stress intensity factor in section 2.2 is strictly valid only for an infinite plate. The geometry of finite size specimens has an effect on the crack tip stress field, and so expressions for stress intensity factors have to be modified by the addition of correction factors to enable their use in practical problems. A general form for such a modified expression is

$$K_I = C\sigma\sqrt{\pi a}\cdot f({}^{a}/_{W})\,, \tag{2.44}$$

where C and $f({}^{a}/_{W})$ have to be determined by stress analysis. There are very few closed form solutions to equation (2.44). Most expressions are obtained by numerical approximation methods.

As illustrations of the effect of finite geometry the derivation of modified expressions based on the solution for an infinite, remotely loaded plate will be discussed in some detail. These expressions concern the centre cracked specimen and the single and double edge notched specimens. (A compendium of these and additional expressions for a number of well known specimen geometries is given in section 2.8.)

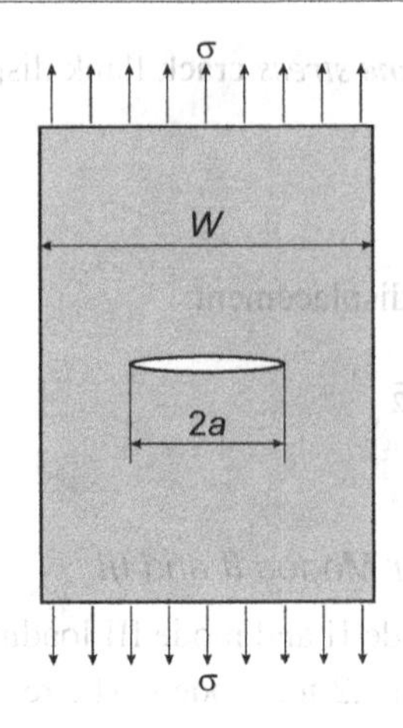

Figure 2.7. The finite width centre cracked specimen.

The Centre Cracked Specimen

The specimen geometry is depicted in figure 2.7. For this specimen there are several expressions for the stress intensity factor, *e.g.*

$$K_{\mathrm{I}} = \sigma\sqrt{\pi a}\sqrt{\frac{W}{\pi a}\tan\left(\frac{\pi a}{W}\right)}. \tag{2.45}$$

Equation (2.45) is the stress intensity factor analytically obtained by Irwin for one of a row of collinear cracks with interspacing W in an infinite plate (see reference 3). However, as long as a/W is sufficiently small, this can be considered a good approximation to a finite width centre cracked specimen: the accuracy is better than 5% for $a/W \leq 0.25$.

A virtually exact numerical solution was obtained by Isida. The geometric correction factor $f(a/W)$ was derived as a 36 term power series! However, Brown found a 4 term approximation to this power series with 0.5% accuracy for $a/W \leq 0.35$. This approximation is

$$f\left(\frac{a}{W}\right) = 1 + 0.256\left(\frac{a}{W}\right) - 1.152\left(\frac{a}{W}\right)^2 + 12.200\left(\frac{a}{W}\right)^3. \tag{2.46}$$

Another, purely empirical, correction factor is due to Feddersen. As an approximation to Isida's results he suggested that

$$K_{\mathrm{I}} = \sigma\sqrt{\pi a}\sqrt{\sec\left(\frac{\pi a}{W}\right)}. \tag{2.47}$$

This remarkably simple expression is accurate to within 0.3% for $a/W \leq 0.35$.

Figure 2.8 compares all the correction factors mentioned above in a graphical representation.

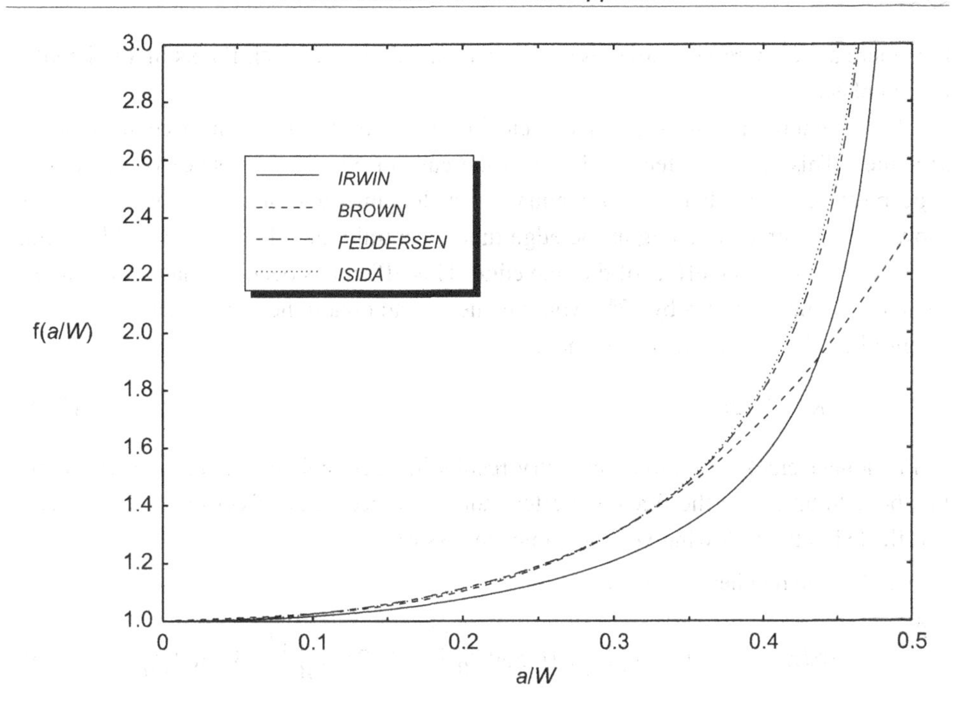

Figure 2.8. Comparison of correction factors for the centre cracked specimen.

Edge Notched Specimens

The geometries of single and double edge notched specimens are given in figure 2.9.

In remark 2 appended to section 2.2 it was stated that the elastic stress field for a uniaxially loaded plate can be obtained using the complex stress function $\phi(z)$ for a biaxially loaded plate after subtracting the remote stress σ from the expression for σ_x, equation (2.18.a). Along the crack flanks, where $-a < x < a$ and $y = 0$, the function $\phi(z)$ is purely imaginary. Equation (2.18.a) now yields $\sigma_x = 0$ for a biaxially loaded plate and

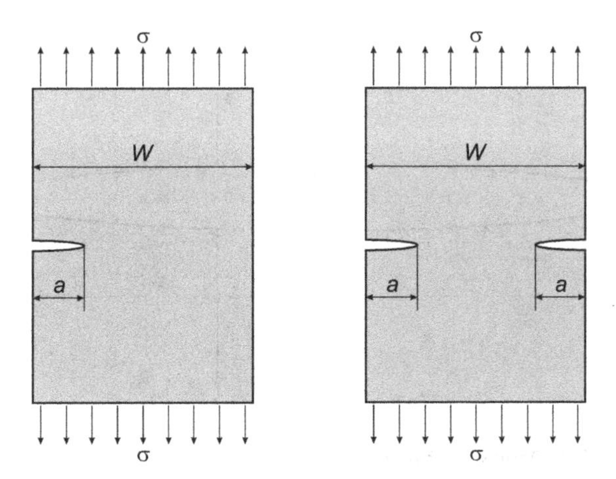

Figure 2.9. The single and double edge notched specimens.

thus there are compressive stresses $\sigma_x = -\sigma$ acting along the crack flanks in a uniaxially loaded plate.

These compressive stresses have a closing effect on the crack in a centre cracked specimen. This closing effect is absent in the case of edge cracks, since σ_x at the free edge must be zero. Therefore, for equal crack length a and stress σ, the edge crack shows a larger crack opening at the edge than a central crack does in the middle. Thus there is a stress raising effect of the free edge. This effect has been estimated to be about 12%, *i.e.* an increase of σ by 12% would be needed to obtain the same crack opening for a central crack. Thus for an edge crack:

$$K_I = 1.12\sigma\sqrt{\pi a}\,. \tag{2.48}$$

For longer cracks the finite geometry results in stress enhancement. Correction factors have to take both the free edge effect and finite geometry effect into account. This is reflected in the following very accurate expressions.

- Single edge notched specimen:

$$K_I = \sigma\sqrt{\pi a}\left\{1.122-0.231\left(\frac{a}{W}\right)+10.550\left(\frac{a}{W}\right)^2-21.710\left(\frac{a}{W}\right)^3+30.382\left(\frac{a}{W}\right)^4\right\} \tag{2.49}$$

  which has an accuracy of 0.5% for $a/W < 0.6$.
- Double edge notched specimen:

$$K_I = \sigma\sqrt{\pi a}\left\{\frac{1.122-1.122\left(\frac{a}{W}\right)-0.820\left(\frac{a}{W}\right)^2+3.768\left(\frac{a}{W}\right)^3-3.040\left(\frac{a}{W}\right)^4}{\sqrt{1-\frac{2a}{W}}}\right\} \tag{2.50}$$

  which is accurate to 0.5% for any a/W.

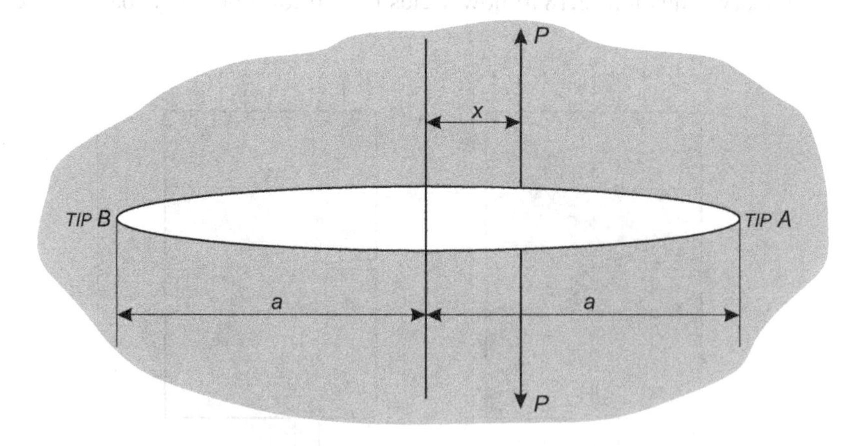

Figure 2.10. Crack-line loading by a point force *P*.

2.5 Two Additional Important Solutions for Practical Use

Besides the centre cracked and single and double edge notched geometries there are two other geometries which are important owing to their common occurrence in practice. These are:

- crack-line loading,
- elliptical cracks, either embedded or intersecting free surfaces. Elliptical cracks at free surfaces may be semi- or quarter-elliptical in shape.

Crack-line Loading

Consider a crack loaded by a point force as in figure 2.10. This is known as *crack-line loading*. Expressions for the stress intensity factors at crack tips A and B can be obtained by an analysis similar to that in section 2.2. The result in terms of the *force per unit thickness*, *P*, is:

$$K_{\mathrm{I_A}} = \frac{P}{\sqrt{\pi a}}\sqrt{\frac{a+x}{a-x}}; \quad K_{\mathrm{I_B}} = \frac{P}{\sqrt{\pi a}}\sqrt{\frac{a-x}{a+x}}. \qquad (2.51.a)$$

For a centrally located force ($x = 0$)

$$K_{\mathrm{I}} = \frac{P}{\sqrt{\pi a}}. \qquad (2.51.b)$$

Therefore for constant P an increase in crack length results in a decrease in stress intensity.

From equations (2.51.a) a stress intensity factor solution for a crack under internal pressure can be derived. Redefining P as the internal pressure, the point force per unit thickness in equation (2.51.a) is equal to $P\mathrm{d}x$. Integration yields:

$$K_{\mathrm{I}} = \frac{P}{\sqrt{\pi a}}\int_{-a}^{a}\sqrt{\frac{a+x}{a-x}}\,\mathrm{d}x = \frac{P}{\sqrt{\pi a}}\int_{0}^{a}\left(\sqrt{\frac{a+x}{a-x}}+\sqrt{\frac{a-x}{a+x}}\right)\mathrm{d}x$$

$$= \frac{P}{\sqrt{\pi a}}\int_{0}^{a}\frac{2a\,\mathrm{d}x}{\sqrt{a^2-x^2}} = \frac{2Pa}{\sqrt{\pi a}}\left[\arcsin\frac{x}{a}\right]_0^a = P\sqrt{\pi a}. \qquad (2.52)$$

Note that since the pressure P is a force per unit area the result in equation (2.52) is the same as that obtained by end loading, $K_{\mathrm{I}} = \sigma\sqrt{\pi a}$.

The usefulness of solutions for crack-line loading is twofold:

1) Expressions for point forces can be applied to cracks at loaded holes, *e.g.* riveted and bolted plates, provided that the holes are not too large with respect to the crack.

2) Integration of point force solutions to obtain expressions for cracks under internal pressure is particularly useful for analysing internal part-through wall thickness cracks in *e.g.* pressure vessels and piping. The solutions are also important for through wall thickness cracks when there is a break in a pressure vessel or a pipe filled with a gas. At the moment of breakage the full gas pressure must be assumed to work on the crack flanks.

Elliptical Cracks

Actual cracks often initiate at surface discontinuities or corners in structural components. If the components are fairly thick the cracks generally assume semi- or quarter-elliptical shapes as they grow in the thickness direction, for example as in figure 2.11. A knowledge of stress intensity factors for these geometries is thus of prime importance for practical application of LEFM.

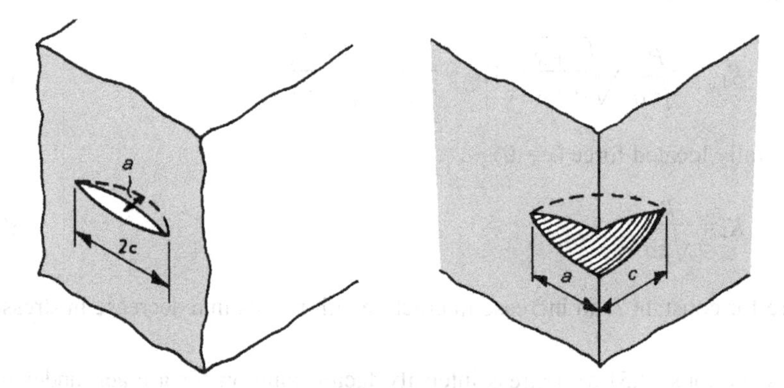

Figure 2.11. Semi-elliptical and quarter-elliptical cracks.

As a first step, Irwin derived an expression for the mode I stress intensity factor of an embedded slit-like elliptical crack. The loading situation is shown schematically in figure 2.12.

For this situation

$$K_\mathrm{I} = \frac{\sigma\sqrt{\pi a}}{\Phi}\left(\sin^2\varphi + \frac{a^2}{c^2}\cos^2\varphi\right)^{\frac{1}{4}}, \tag{2.53}$$

where the location along the crack front is represented by the parametric angle φ defined in figure 2.12 and Φ is an elliptic integral of the second kind, *i.e.*

$$\Phi = \int_0^{\pi/2} \sqrt{1 - \sin^2\alpha \sin^2\zeta}\, d\zeta ,$$

$\sin^2\alpha = (c^2 - a^2)/c^2$ and ζ is a vanishing integration variable. The solution for the integral is given in the table below for a number of a/c ratios.

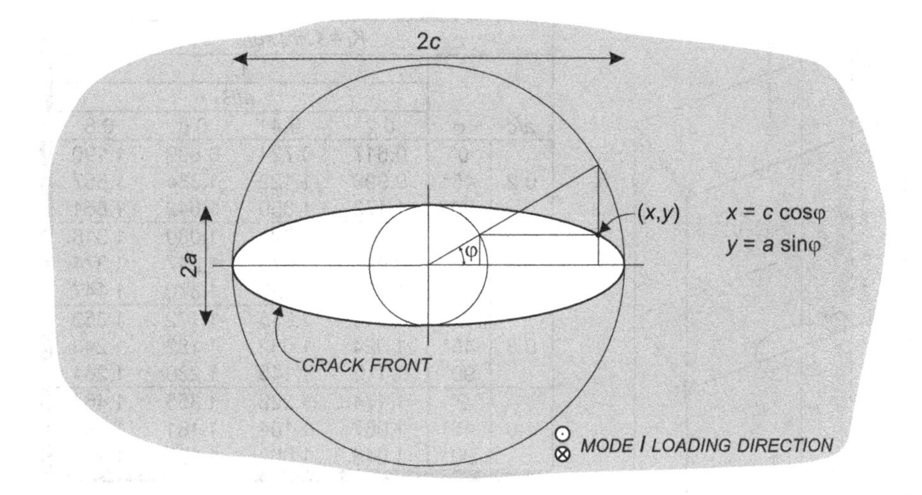

Figure 2.12. An embedded elliptical crack under mode I loading. The location along the crack front is represented by the parametric angle φ.

a/c	0	0.1	0.2	0.3	0.4	0.5	0.6	0.7	0.8	0.9	1.0
Φ	1.000	1.016	1.051	1.097	1.151	1.211	1.277	1.345	1.418	1.493	1.571

The integral can also be developed as a series expansion:

$$\Phi = \frac{\pi}{2}\left[1 - \frac{1}{4}\frac{c^2 - a^2}{c^2} - \frac{3}{64}\left(\frac{c^2 - a^2}{c^2}\right)^2 - \ldots\right].$$

Neglecting all terms beyond the second gives an accuracy of better than 5%. And so the stress intensity factor can be approximated by

$$K_I = \frac{\sigma\sqrt{\pi a}}{\frac{3\pi}{8} + \frac{\pi}{8}\left(\frac{a}{c}\right)^2}\left\{\sin^2\varphi + \left(\frac{a}{c}\right)^2\cos^2\varphi\right\}^{\frac{1}{4}}. \tag{2.54}$$

K_I varies along the elliptical crack front and has a maximum value $\sigma\sqrt{\pi a}/\Phi$ at the ends of the minor axis, a, and a minimum value $\sigma\sqrt{\pi a^2/c}/\Phi$ at the ends of the major axis, c. The implication is that during crack growth an embedded elliptical crack will tend to become circular. For an embedded circular crack

$$K_I = \frac{2}{\pi}\sigma\sqrt{\pi a}. \tag{2.55}$$

In practice elliptical cracks will generally occur as semi-elliptical surface cracks or quarter-elliptical corner cracks. The presence of free surfaces means that correction factors must be added to the expressions given for embedded cracks.

For a semi-elliptical surface crack the free front surface is generally accounted for by adding a correction factor of 1.12. For a quarter-elliptical corner crack, which intersects two surfaces, a correction factor of 1.2 is used. Besides these factors, corrections for

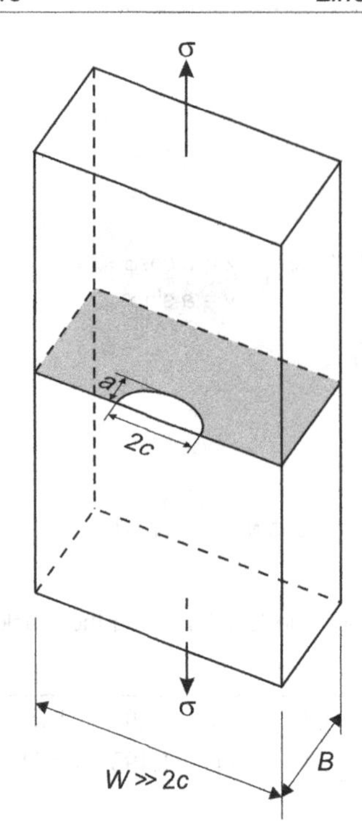

$K_I = C\sigma\sqrt{\pi a}/\phi$					
		C			
		a/B			
a/c	φ	0.2	0.4	0.6	0.8
0.2	0°	0.617	0.724	0.899	1.190
	45°	0.990	1.122	1.384	1.657
	90°	1.173	1.359	1.642	1.851
0.4	0°	0.767	0.896	1.080	1.318
	45°	0.998	1.075	1.247	1.374
	90°	1.138	1.225	1.370	1.447
0.6	0°	0.916	1.015	1.172	1.353
	45°	1.024	1.062	1.182	1.243
	90°	1.110	1.145	1.230	1.264
1.0	0°	1.174	1.229	1.355	1.464
	45°	1.067	1.104	1.181	1.193
	90°	1.049	1.062	1.107	1.112

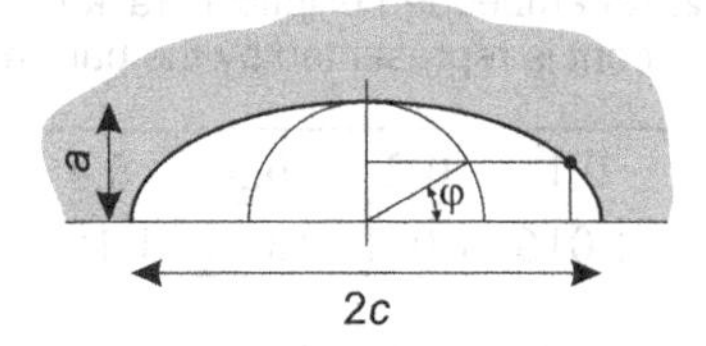

Figure 2.13. Stress intensity factor solutions for semi-elliptical surface cracks in a plate of finite dimensions, according to Raju and Newman (reference 8).

cracks approaching back surfaces (analogous to corrections for finite specimen width, section 2.4) must be considered. Back surface correction factors have been calculated by various authors: an overview can be found in reference 7 of the bibliography. The results are often combined with front free surface and finite width correction factors.

The best solutions available for semi-elliptical surface cracks are those based on the finite element calculations of Raju and Newman, reference 8 of the bibliography. For the crack of figure 2.13 Raju and Newman produced stress intensity factor solutions of the form $K_I = C\sigma\sqrt{\pi a}/\Phi$ where, for $W >> c$ the value of C depends only on a/c, a/B and φ. Values of C are also given in figure 2.13.

Raju and Newman also developed an empirical stress intensity factor equation based on their finite element results, reference 9 of the bibliography. For tension loading the solution is reproduced in section 2.8. For combined tension and bending load the reader is referred to reference 9.

2.6 Superposition of Stress Intensity Factors

In section 2.2 it was shown that in the vicinity of the crack tip the total stress field due to two or more different mode I loading systems can be obtained by an algebraic summation of the respective stress intensity factors. This is called the superposition principle.

It should be noted that the superposition principle is valid only for combinations of

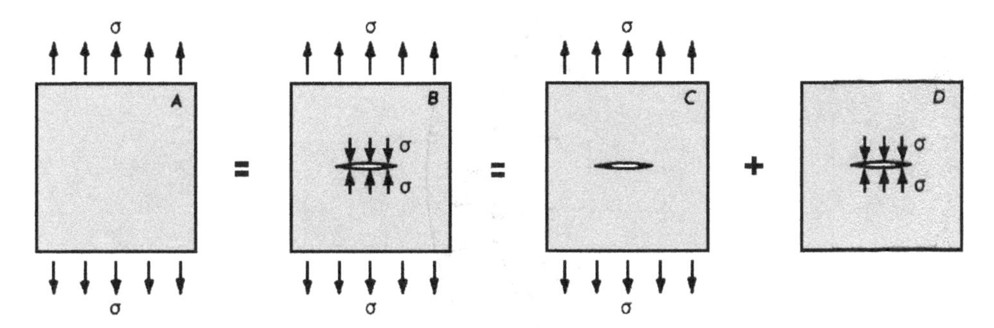

Figure 2.14. The superposition principle for a through crack under internal pressure.

the same mode of loading, *i.e.* all mode I, all mode II or all mode III. The different modes give different types of stress intensity factor solutions which cannot be superimposed.

By using the superposition principle the stress intensity factor for a number of seemingly complicated problems can be readily obtained. Two examples are given here. They are:

1) A through crack under internal pressure (already derived analytically in section 2.5).
2) A semi-elliptical surface crack in a cylindrical pressure vessel.

Through Crack under Internal Pressure

The solution for this problem was already given in equation (2.52). Here the solution is obtained by the superposition shown in figure 2.14. It is seen from this superposition that

$$K_I{}^A = K_I{}^B = K_I{}^C + K_I{}^D = 0 .$$

Therefore

$$K_I{}^D = -K_I{}^C = -\sigma\sqrt{\pi a} .$$

Substituting $P = -\sigma$ in case D gives the required result, $P\sqrt{\pi a}$.

Note that this superposition also holds when a correction for finite specimen size is required. Such a correction, of the form $f(a/W)$, can thus be applied to a finite width centre cracked specimen irrespective of whether uniaxial tension is applied to the plate or the crack is loaded by an internal pressure.

Semi-Elliptical Surface Crack in a Cylindrical Pressure Vessel

A section from a cylindrical pressure vessel with an internal semi-elliptical surface crack is given in figure 2.15. As long as the wall thickness, B, is small compared to the vessel radius, R, the hoop stress $\sigma_H = PR/B$. The maximum stress intensity factor will generally occur at the end of the minor axis of the semi-elliptical surface crack (φ = 90°). The contribution to this maximum stress intensity factor by the hoop stress is

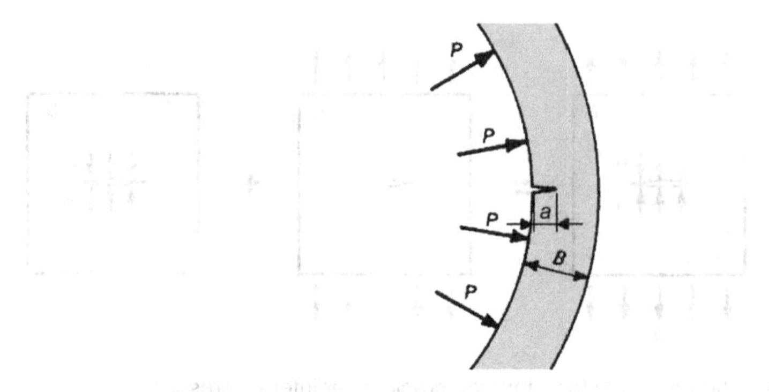

Figure 2.15. Cylindrical pressure vessel with internal surface crack.

$$K_{\mathrm{I}}^{\sigma_{\mathrm{H}}} = \frac{C\sigma_{\mathrm{H}}\sqrt{\pi a}}{\Phi} = \frac{CPR\sqrt{\pi a}}{B\Phi}, \tag{2.56}$$

where C can be obtained from figure 2.13 when a and c are known.

Since the crack is in a pressurised vessel the internal pressure will also act on the crack surfaces, with a contribution to the maximum stress intensity factor of

$$K_{\mathrm{I}}^{\mathrm{P}} = \frac{CP\sqrt{\pi a}}{\Phi}. \tag{2.57}$$

The maximum stress intensity factor is therefore

$$K_{\mathrm{I}_{\max}} = K_{\mathrm{I}}^{\sigma_{\mathrm{H}}} + K_{\mathrm{I}}^{\mathrm{P}} = \frac{CP\left(1 + \frac{R}{B}\right)\sqrt{\pi a}}{\Phi}. \tag{2.58}$$

2.7 Some Remarks Concerning Stress Intensity Factor Determinations

In attempting to use fracture mechanics it will often be found that there is no standard stress intensity factor solution for the particular crack shape and structural component geometry under consideration. Recourse must be made to one of a number of methods for obtaining the stress intensity factor. A detailed treatise of these methods is beyond the scope of this course, but some remarks on the subject will be made here.

Stress intensity factors are now available for many geometrical configurations and so the first step should always be a literature search. For example, references 6 and 10 of the bibliography and the indexes of well known journals such as the "International Journal of Fracture" and "Engineering Fracture Mechanics" can serve as starting points.

If no applicable solution is directly available, the next step is to assess the permissible effort to solve the problem. This effort depends on the seriousness of the problem, the desired accuracy, computational costs and how many times the solution will be useful. Limits to the amount of effort that can be justified will frequently rule out the use of sophisticated and expensive methods such as

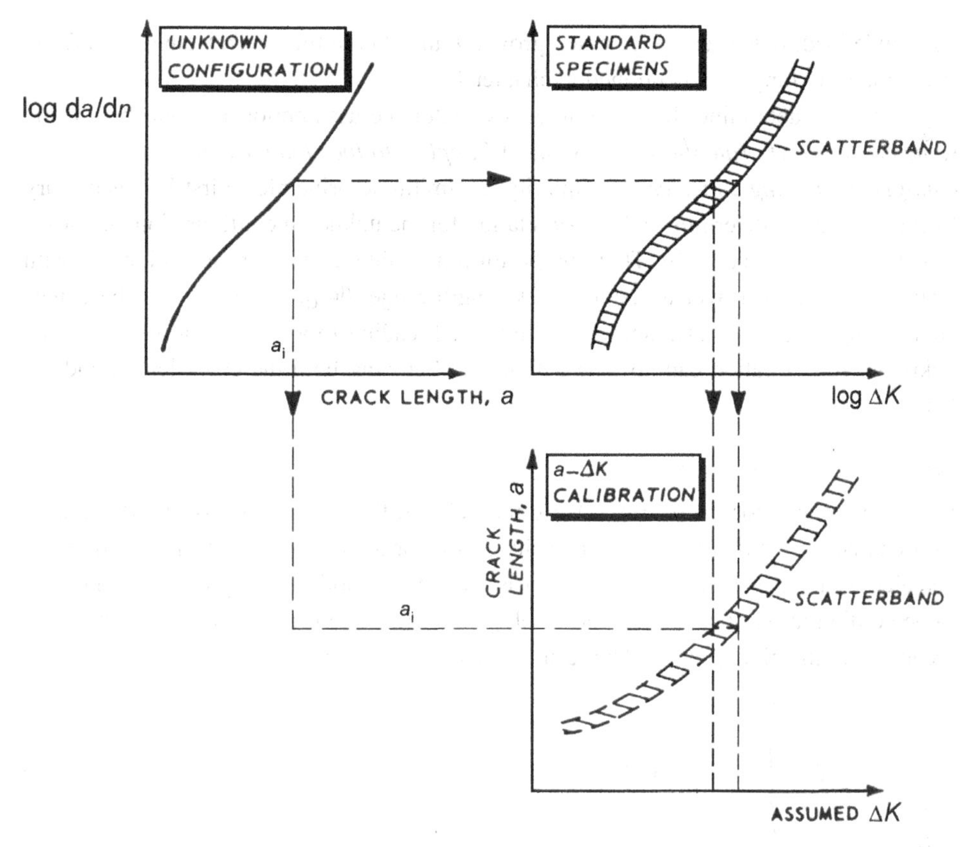

Figure 2.16. Schematic of the fatigue crack growth method for obtaining stress intensity factors.

- finite element calculations,
- boundary integral equations,
- conformal mapping.

The interested reader will find detailed information about these methods in reference 11 of the bibliography. A strong mathematical background is essential for using these methods.

However, in many cases one of several more straightforward methods can be applied. These include

- the superposition principle (already discussed in section 2.6) or the very similar compounding method (reference 12 of the bibliography),
- the experimental fatigue crack growth method,
- the weight function method.

The Experimental Fatigue Crack Growth Method

This method generally gives reliable results. It involves comparing fatigue crack growth rates in the configuration to be investigated with crack growth rates in standard specimens of the same material fatigued under exactly the same conditions. Constant amplitude loading is used, since crack growth rate data are then correlatable by ΔK, the stress intensity range, as mentioned in chapter 1.

In order to determine the stress intensity factor the assumption is made that *at the same crack growth rate* $^{da}/_{dn}$, *the same* ΔK *applies to the standard specimens and the configuration being considered* (similarity or similitude principle). First it is necessary to experimentally determine a $^{da}/_{dn}$-a relation for the unknown configuration, as shown schematically in figure 2.16. Then, by performing a fatigue crack propagation test with a standard specimen over a certain crack length range, $^{da}/_{dn}$-ΔK data must be generated. Using this, $^{da}/_{dn}$ can simply be eliminated leading to a ΔK-a relation. K_I for the unknown configuration can now be determined knowing both the crack length and the load.

The Weight Function Method

From expressions for the energy release rate, G, and the relation $G = {}^{K^2}/_{E}$ (see chapter 4) it can be shown that the stress intensity factor for a particular stress distribution can be calculated if the stress intensity factor is known for another stress distribution on the same configuration of specimen and crack geometries. The proof of this is given in reference 13 of the bibliography. The result for mode I loading is

$$K_I = \int_0^{2a} H(x,a)\cdot\sigma(x)\,dx\,, \tag{2.59}$$

$$\text{where } H(x,a) = \frac{E'}{2K_I^*}\frac{\partial v^*(x,a)}{\partial a}$$

is the weight function; E' is E for plane stress and $^{E}/_{(1-\nu^2)}$ for plane strain; K_I^* is the known stress intensity factor; v^* is the displacement at the loading point, x, of the known solution but in the direction of loading for the stress distribution to be analysed; and $\sigma(x)$ is a function describing this stress distribution with reference to the stress distribution for K_I^*.

A serious difficulty in using equation (2.59) is the part $\int{}^{\partial v^*}/_{\partial a}\,dx$ which often gives singularities in the stress field distribution around a crack tip. A better technique is to use special weight function solutions which give singularities only at the crack tip. Such special weight functions have been derived by Bueckner and, in a different way, by Paris. Discussion of these functions, including examples of their use, can be found in references 2 and 13 of the bibliography to this chapter.

2.8 A Compendium of Well-Known Stress Intensity Factor Solutions

A number of well known and widely used stress intensity factor solutions are presented here. Some of these solutions have already been discussed in sections 2.4 – 2.6. Others will be found in subsequent chapters of this course or else are mentioned because of their practical utility.

Elementary Solutions

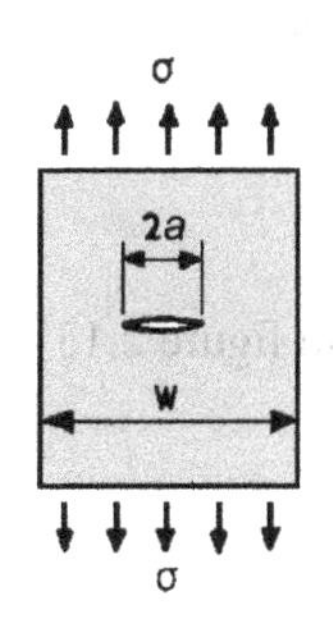

Centre cracked plate: $K_I = C\sigma\sqrt{\pi a}$

- Brown (accurate to 0.5% for $a/W \leq 0.35$):

$$C = 1 + 0.256\left(\frac{a}{W}\right) - 1.152\left(\frac{a}{W}\right)^2 + 12.200\left(\frac{a}{W}\right)^3$$

- Feddersen (accurate to 0.3% for $a/W \leq 0.35$):

$$C = \sqrt{\sec\frac{\pi a}{W}}$$

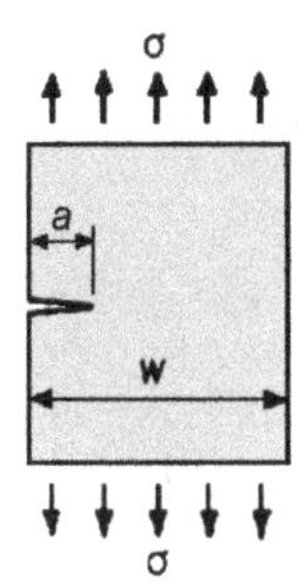

Single edge notched plate: $K_I = C\sigma\sqrt{\pi a}$

- Small cracks: $C = 1.12$
- Brown (accurate to 0.5% for $a/W \leq 0.6$):

$$C = 1.122 - 0.231\left(\frac{a}{W}\right) + 10.550\left(\frac{a}{W}\right)^2 - 21.710\left(\frac{a}{W}\right)^3 + 30.382\left(\frac{a}{W}\right)^4$$

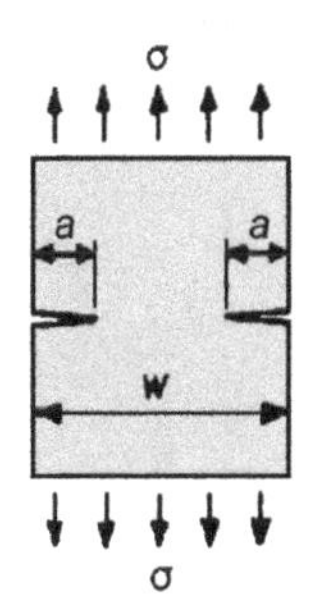

Double edge notched plate: $K_I = C\sigma\sqrt{\pi a}$

- Small cracks: $C = 1.12$
- Tada (accurate to 0.5% for any a/W)

$$C = \frac{1.122 - 1.122\left(\frac{a}{W}\right) - 0.820\left(\frac{a}{W}\right)^2 + 3.768\left(\frac{a}{W}\right)^3 - 3.040\left(\frac{a}{W}\right)^4}{\sqrt{1 - \frac{2a}{W}}}$$

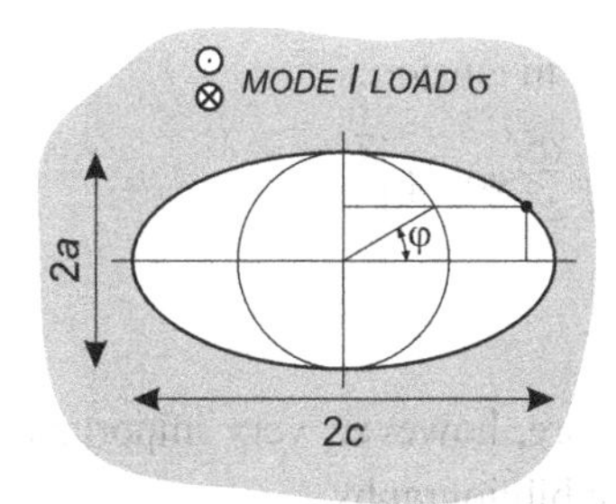

Embedded elliptical or circular slit crack:

Elliptical crack: $$K_I = \frac{\sigma\sqrt{\pi a}}{\frac{3\pi}{8} + \frac{\pi}{8}\frac{a^2}{c^2}}\left(\sin^2\varphi + \frac{a^2}{c^2}\cos^2\varphi\right)^{\frac{1}{4}}$$

Circular crack: $$K_I = \frac{2}{\pi}\sigma\sqrt{\pi a}$$

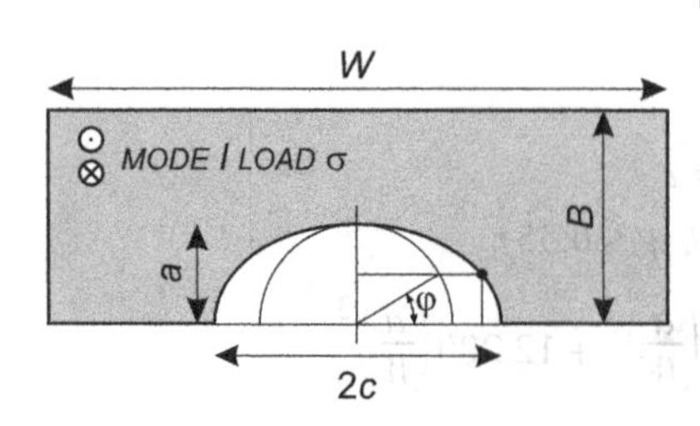

Semi-elliptical surface crack in tension:

$$K_I = C\frac{\sigma\sqrt{\pi a}}{\Phi}\left(\sin^2\varphi + \frac{a^2}{c^2}\cos^2\varphi\right)^{\frac{1}{4}}$$

$$\Phi = \frac{3\pi}{8} + \frac{\pi}{8}\frac{a^2}{c^2}$$

$$C = \begin{cases} 1.12 \text{ for shallow cracks} \\ \text{the correction for } K_I \text{ from figure 2.13} \end{cases}$$

or (reference 9):

$$K_I = \sigma\sqrt{\pi\frac{a}{Q}}\,F\left(\frac{a}{B}, \frac{a}{c}, \frac{c}{W}, \varphi\right)$$

$$Q = 1 + 1.464\left(\frac{a}{c}\right)^{1.65}$$

$$F = \left\{C_1 + C_2\left(\frac{a}{B}\right)^2 + C_3\left(\frac{a}{B}\right)^4\right\} f_\varphi\, C_4 f_W$$

$$C_1 = 1.13 - 0.09\frac{a}{c}$$

$$C_2 = -0.54 + \frac{0.89}{0.2 + a/c}$$

$$C_3 = 0.5 - \frac{1.0}{0.65 + a/c} + 14\left(1.0 - \frac{a}{c}\right)^{24}$$

$$C_4 = 1 + \left\{0.1 + 0.35\left(\frac{a}{B}\right)^2\right\}(1 - \sin\varphi)^2$$

$$f_\varphi = \left\{\sin^2\varphi + \left(\frac{a}{c}\right)^2\cos^2\varphi\right\}^{\frac{1}{4}}$$

$$f_W = \left\{\sec\left(\frac{\pi c}{W}\sqrt{\frac{a}{B}}\right)\right\}^{\frac{1}{2}}$$

for: $\frac{a}{B} < 1 \qquad \frac{a}{c} \leq 1$

$\frac{c}{W} < 0.25 \qquad 0 \leq \varphi \leq \pi$

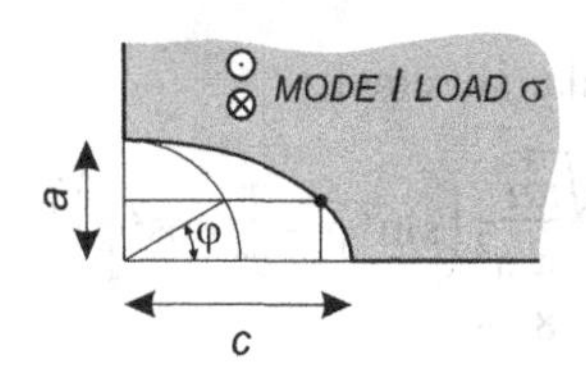

Quarter-elliptical corner crack in tension:

$$K_I = C\frac{\sigma\sqrt{\pi a}}{\frac{3\pi}{8} + \frac{\pi}{8}\frac{a^2}{c^2}}\left(\sin^2\varphi + \frac{a^2}{c^2}\cos^2\varphi\right)^{\frac{1}{4}}$$

$$C = 1.2$$

The solutions for corner cracks are not elementary. They are, however, very important and the reader is referred to references 14, 15 and 16 of the bibliography.

Solutions for Standard Test Specimens

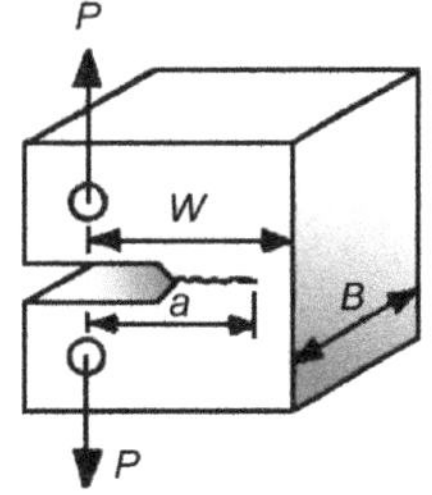

Compact tension specimen (CT):

$$K_I = \frac{P}{BW^{\frac{1}{2}}} \cdot f\left(\frac{a}{W}\right)$$

$$f\left(\frac{a}{W}\right) = \frac{\left(2 + \frac{a}{W}\right)\left\{0.886 + 4.64\left(\frac{a}{W}\right) - 13.32\left(\frac{a}{W}\right)^2 + 14.72\left(\frac{a}{W}\right)^3 - 5.6\left(\frac{a}{W}\right)^4\right\}}{\left(1 - \frac{a}{W}\right)^{\frac{3}{2}}}$$

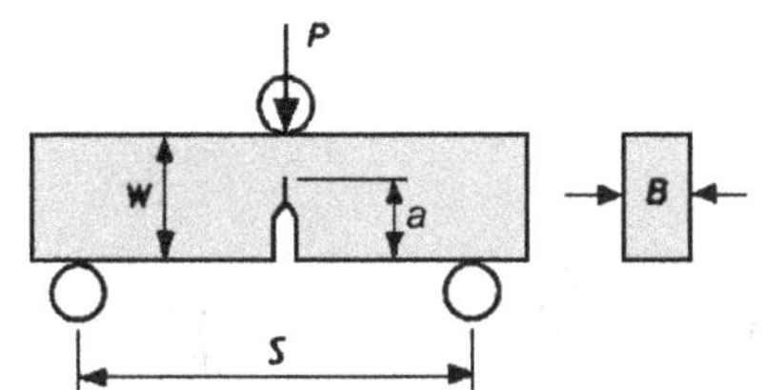

Single edge notched bend specimen (SENB):

$$K_I = \frac{P \cdot S}{BW^{\frac{3}{2}}} \cdot f\left(\frac{a}{W}\right)$$

$$f\left(\frac{a}{W}\right) = \frac{3\left(\frac{a}{W}\right)^{\frac{1}{2}}\left[1.99 - \frac{a}{W}\left(1 - \frac{a}{W}\right)\left\{2.15 - 3.93\left(\frac{a}{W}\right) + 2.7\left(\frac{a}{W}\right)^2\right\}\right]}{2\left(1 + 2\frac{a}{W}\right)\left(1 - \frac{a}{W}\right)^{\frac{3}{2}}}$$

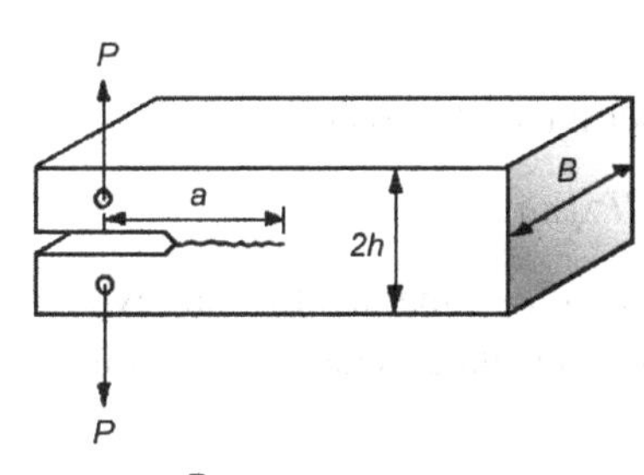

Double cantilever beam specimen (DCB):

$$K_I = 2\sqrt{3}\,\frac{Pa}{Bh^{\frac{3}{2}}} \qquad \text{(plane stress)}$$

$$K_I = \frac{2\sqrt{3}}{\sqrt{1 - \nu^2}}\,\frac{Pa}{Bh^{\frac{3}{2}}} \qquad \text{(plane strain)}$$

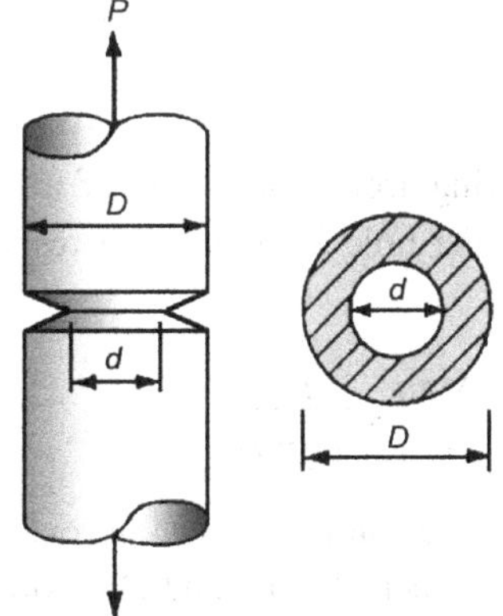

Circumferentially notched bar:

$$K_I = \frac{0.526\,P\sqrt{D}}{d^2}$$

over the range $1.2 \leq D/d \leq 2.1$

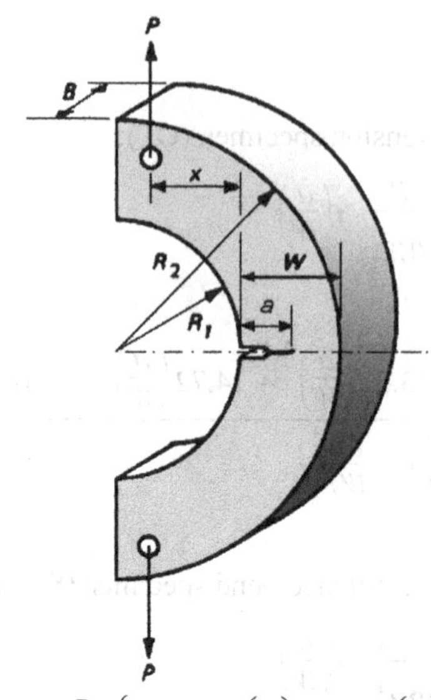

C-shaped specimen

$$K_I = \frac{P}{BW^{\frac{1}{2}}}\left\{1 + 1.54\left(\frac{x}{W}\right) + 0.50\left(\frac{a}{W}\right)\right\}\left\{1 + 0.221\left(1 - \sqrt{\frac{a}{W}}\right)\left(1 - \frac{R_1}{R_2}\right)\right\} \cdot f\left(\frac{a}{W}\right)$$

$$f\left(\frac{a}{W}\right) = 18.23\left(\frac{a}{W}\right)^{\frac{1}{2}} - 106.2\left(\frac{a}{W}\right)^{\frac{3}{2}} + 389.7\left(\frac{a}{W}\right)^{\frac{5}{2}} - 582.0\left(\frac{a}{W}\right)^{\frac{7}{2}} + 369.1\left(\frac{a}{W}\right)^{\frac{9}{2}}$$

Useful Solutions for Practical Applications

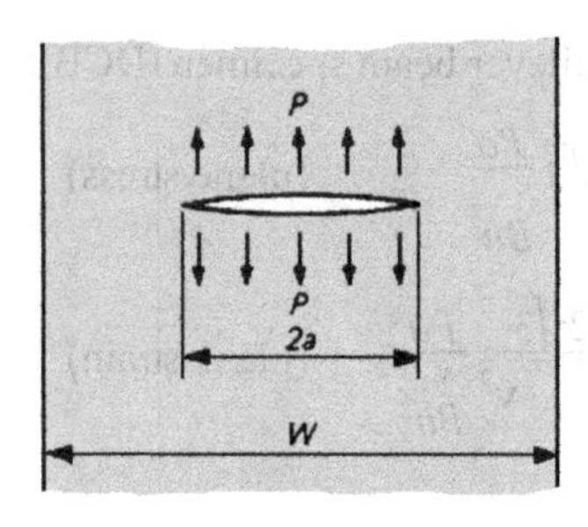

Crack under internal pressure:

$$K_I = CP\sqrt{\pi a}$$

where C is the same as for an externally loaded centre cracked plate and P is force per unit area.

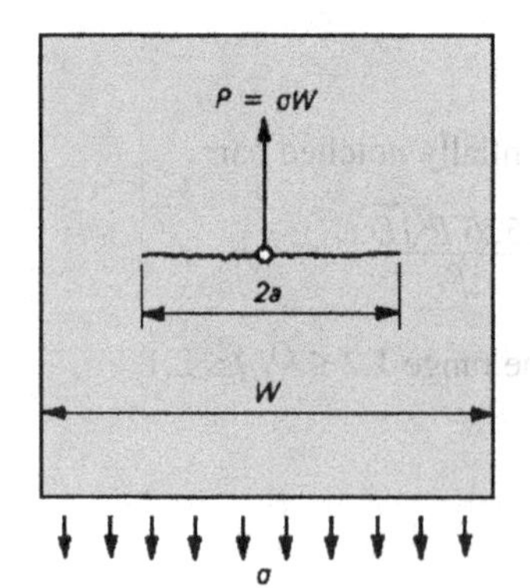

Cracks growing from both sides of a loaded hole, where the hole is small with respect to the crack:

$$K_I = C\left(\frac{\sigma\sqrt{\pi a}}{2} + \frac{P}{2\sqrt{\pi a}}\right)$$

where C is the same as for an externally loaded centre cracked plate and P is force per unit thickness.

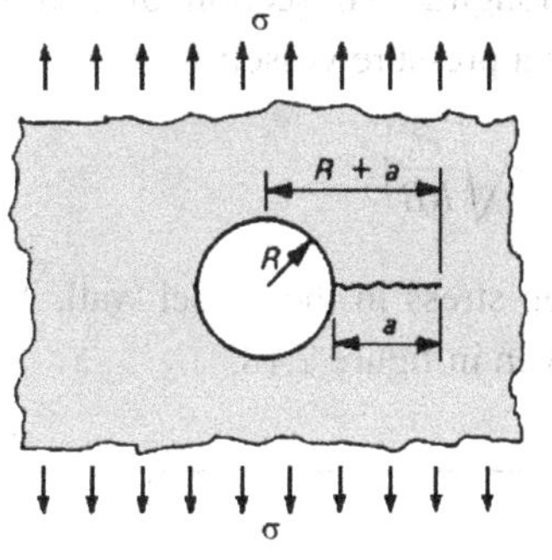

Through-thickness cracks from one or both sides of a remotely loaded hole:

$$K_I = \sigma\sqrt{\pi a}\, F_1\left(\frac{a}{R+a}\right) \quad \text{(single crack)}$$

$$K_I = \sigma\sqrt{\pi a}\, F_2\left(\frac{a}{R+a}\right) \quad \text{(double crack)}$$

The solutions for F_1 and F_2 are given in figure 2.17.

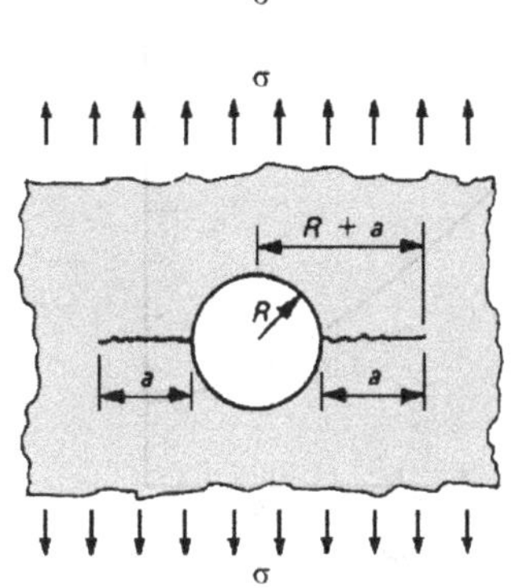

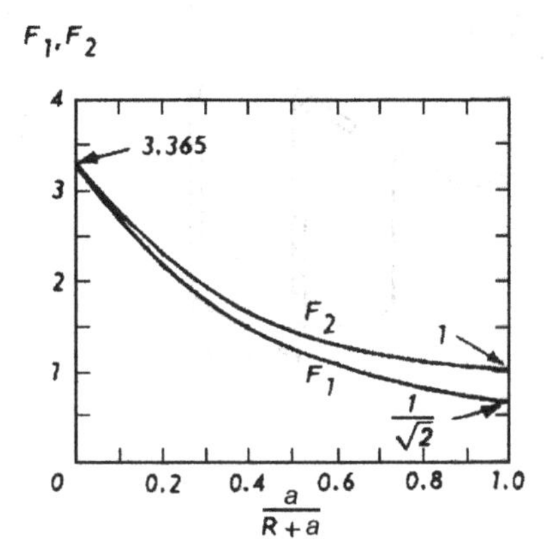

Figure 2.17. Correction factors for cracks growing from remotely loaded holes.

For long double cracks with $\frac{a}{R+a} > 0.3$ a good approximation is

$$K_I = C\sigma\sqrt{\pi(R+a)},$$

where C is the same as for a centre cracked plate.

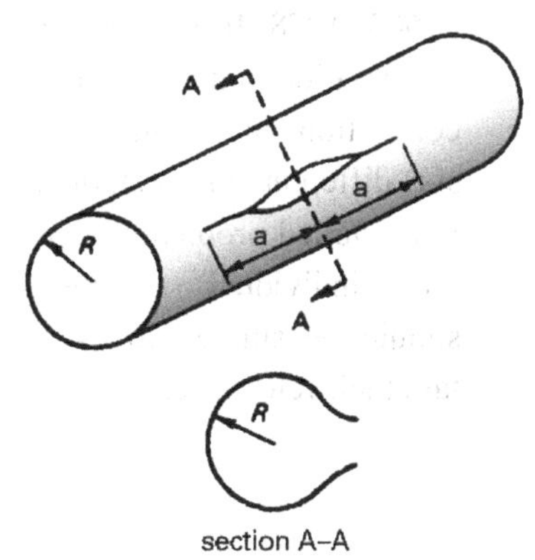

Axial through-thickness crack length $2a$ in a thin walled pipe with mean radius R, internal pressure P and wall thickness t:

$$K_I = \sigma_H M_f \sqrt{\pi a},$$

where $\sigma_H = PR/t$ and M_f is the Folias correction factor for bulging of the crack flanks:

$$M_f = \sqrt{1 + 1.225\frac{a^2}{Rt} - 0.0135\frac{a^4}{R^2t^2}}.$$

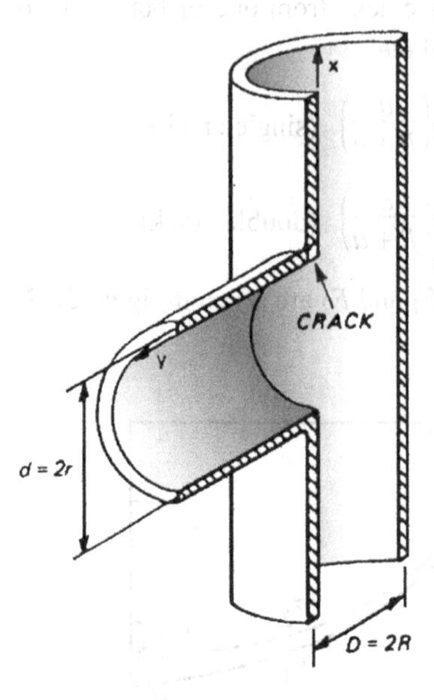

Corner crack in a longitudinal section of a pipe-vessel intersection in a pressure vessel:

$$\frac{K_I}{\sigma_H\sqrt{\pi a}} = F_m\left(1 + \sqrt{\frac{rt}{RB}}\right),$$

where σ_H is the hoop stress in the vessel wall. The solution for F_m is given in figure 2.18.

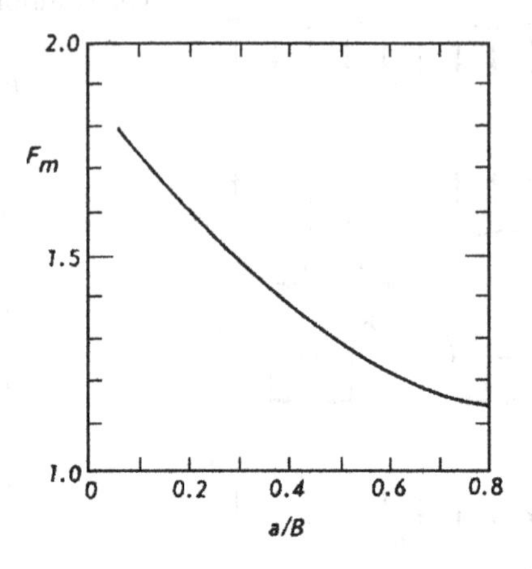

Figure 2.18. Correction factor for a corner crack in a longitudinal section of a pipe-vessel intersection on a pressure vessel.

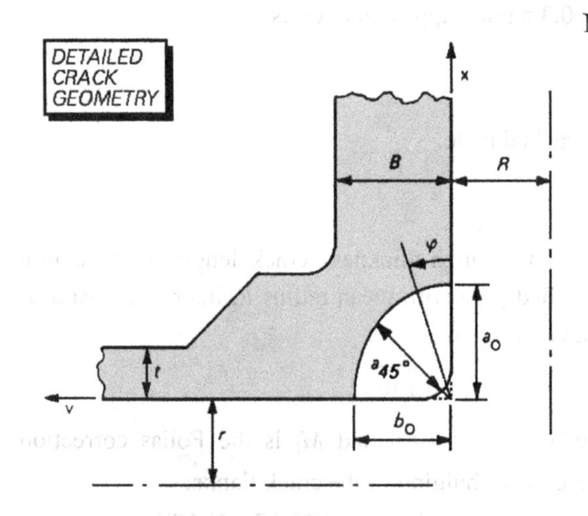

Note: This is a serious problem in pressure vessel technology. The solution given here is by M.A. Mohamed and J. Schroeder, International Journal of Fracture, Vol. 14, p. 605, 1978. It is a first approximation only, based on correlation of results from six different investigations. For detailed problem solving the individual solutions should be studied to ascertain their relevance.

2.9 Bibliography

1. Westergaard, H.M., *Bearing Pressures and Cracks*, Journal of Applied Mechanics, Vol. 61, pp. A49-A53 (1939).
2. Paris, P.C. and Sih, G.C., *Stress Analysis of Cracks,* Fracture Toughness Testing and its Applications, ASTM STP 381, American Society for Testing and Materials, pp. 63–77 (1965): Philadelphia.
3. Irwin, G.R., *Analysis of Stresses and Strains Near the End of a Crack Traversing a Plate*, Journal of Applied Mechanics, Vol. 24, pp. 361–364 (1957)
4. Creager, M. and Paris, P.C., *Elastic Field Equations for Blunt Cracks with Reference to Stress Corrosion Cracking*, International Journal of Fracture Mechanics, Vol. 3, pp. 247-252 (1967).
5. Inglis, C.E., *Stresses in a Plate due to the Presence of Cracks and Sharp Corners*, Transactions of the Institution of Naval Architects, Vol. 55, pp. 219-230 (1913).
6. Tada, H., Paris, P.C. and Irwin, G.R., The Stress Analysis of Cracks Handbook, Paris Productions Incorporated (1985): St. Louis, Missouri.
7. Maddox, S.J., *An Analysis of Fatigue Cracks in Fillet Welded Joints,* International Journal of Fracture, Vol. 11, pp. 221–243 (1975).
8. Raju, I.S. and Newman, J.C. Jr., *Stress Intensity Factors for a Wide Range of Semi-Elliptical Surface Cracks in Finite Thickness Plates*, Engineering Fracture Mechanics, Vol. 11, pp. 817–829 (1979).
9. Newman, J.C. Jr. and Raju, I.S., *An Empirical Stress Intensity Factor Equation for the Surface Crack,* Engineering Fracture Mechanics, Vol. 15, pp. 185–192 (1981).
10. Rooke, D.P. and Cartwright, D.J., Compendium of Stress Intensity Factors, Her Majesty's Stationery Office (1976): London.
11. Sih, G.C. (Editor), Methods of Analysis and Solutions of Crack Problems, Noordhoff International Publishing (1973): Leiden.
12. Cartwright, D.J. and Rooke, D.P., *Approximate Stress Intensity Factors Compounded from Known Solutions,* Engineering Fracture Mechanics, Vol. 6, pp. 563–571(1974).
13. Paris, P.C., McMeeking, R.M. and Tada, H., *The Weight Function Method for Determining Stress Intensity Factors,* Cracks and Fracture, ASTM STP 601, American Society for Testing and Materials, pp. 471–489 (1976): Philadelphia.
14. Pickard, A.C., *Stress Intensity Factors for Cracks with Circular and Elliptic Crack Fronts, Determined by 3D Finite Element Methods*, Numerical Methods in Fracture Mechanics, Pineridge Press, pp. 599–614 (1980): Swansea.
15. Raju, I.S. and Newman, J.C. Jr., *Stress Intensity Factors for Two Symmetric Comer Cracks,* Fracture Mechanics, ASTM STP 677, American Society for Testing and Materials, pp. 411–430 (1979): Philadelphia.
16. Murakami, Y., Stress Intensity Factors Handbook, Vols. 1, 2 and 3, Pergamon Press (1987 vols. 1 and 2, 1992 vol. 3): Oxford.

3 Crack Tip Plasticity

3.1 Introduction

In chapter 2 the elastic stress field equations for a sharp crack, equations (2.24), were obtained. These equations result in infinite stresses at the crack tip, *i.e.* there is a stress singularity. This solution is for a crack with zero crack tip radius. However, real materials have an atomic structure, and the minimum finite tip radius is about the interatomic distance. This limits the stresses at the crack tip. More importantly, structural materials deform plastically above the yield stress and so in reality there will be a plastic zone surrounding the crack tip.

Along the x-axis $\theta = 0$ and the expression for σ_y in equations (2.24) gives

$$\sigma_y = \frac{\sigma\sqrt{\pi a}}{\sqrt{2\pi r}} = \frac{K_I}{\sqrt{2\pi r}}. \tag{3.1}$$

By substituting the yield strength, σ_{ys}, for σ_y in equation (3.1) an estimate can be obtained of the distance r_y over which the material is plastically deformed ahead of the crack:

$$r_y = \frac{1}{2\pi}\left(\frac{K_I}{\sigma_{ys}}\right)^2. \tag{3.2}$$

Assuming as a first approximation that the plastic zone *size* along the x-axis, r_y,

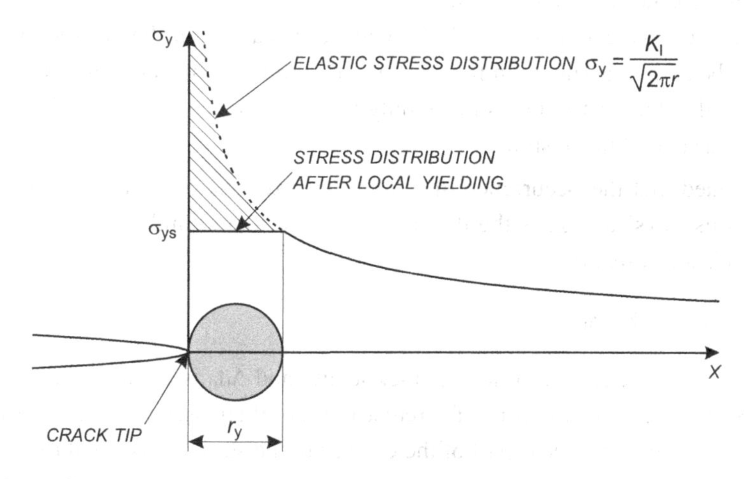

Figure 3.1. A first approximation to the crack tip plastic zone.

corresponds to the *diameter* of a circular plastic zone, the distribution of σ_y ahead of the crack tip will be as shown in figure 3.1. From this figure it is clear that the assumption is inaccurate, since part of the stress distribution (shown hatched in the figure) is simply cut off above σ_{ys}. Also, there is no a priori reason why the plastic zone should be circular.

In fact it turns out to be extremely difficult to give a proper description of plastic zone size and shape. For this reason the models most widely known from the literature have followed one of two approaches. Either they give a better approximation of the size but use a selected shape, *e.g.* the Irwin and Dugdale approaches discussed in sections 3.2 and 3.3, or they give an impression of the shape but retain the first size approximation, as in the derivations from classical yield criteria in section 3.4.

Besides these limitations there is the problem that the state of stress, *i.e.* plane stress or plane strain, will affect the plastic zone size and shape. It is well known that under plane strain conditions yielding need not occur until the applied stress is much higher than σ_{ys}, *i.e.* the plastic zone may be smaller. This and other effects of differing state of stress will be dealt with in sections 3.5 and 3.6.

Finally, in section 3.7 some remarks will be addressed to advanced methods of determining plastic zone size and shape.

3.2 The Plastic Zone Size According to Irwin

Irwin's analysis of plastic zone size attempts to account for the fact that the stress distribution cannot simply be cut off above σ_{ys} as in figure 3.1. For the analysis to be straightforward there are several restrictions:

1) The plastic zone shape is considered to be circular: however, circularity is not important, see restriction 2.
2) Only the situation along the *x*-axis ($\theta = 0$ in equations (2.24)) is analysed.
3) The material behaviour is considered to be elastic – perfectly plastic, *i.e.* it is assumed there is no strain hardening.
4) A plane stress state is considered. As will be discussed further in section 3.5, the material behaviour assumed in restriction 3 now implies that stresses will not exceed σ_{ys}. Note that this restriction is made only for convenience. The analysis can also be made for a state of plane strain.

Irwin argued that the occurrence of plasticity makes the crack behave as if it were longer than its physical size – the displacements are longer and the stiffness is lower than in the elastic case, *i.e.*

$$a_{\mathrm{eff}} = a + \Delta a_{\mathrm{n}} ,$$

where a_{eff} is the effective, or notional, crack length and Δa_{n} is the notional crack increment. This increment must account for redistribution of stresses that were above σ_{ys} in the elastic case. Δa_{n} behaves as part of the crack, but the stress σ_y is equal to σ_{ys}.

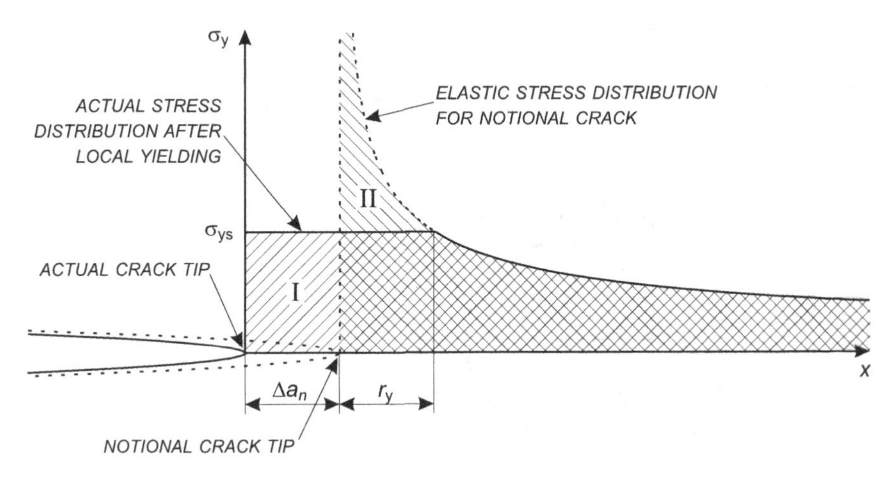

Figure 3.2. Schematic of Irwin's analysis.

Now consider figure 3.2, which shows the σ_y distribution that follows from the elastic solution for a crack of length $a + \Delta a_n$ as well as the actual σ_y distribution after local yielding. The stresses transmitted by these two distributions should be equal. This will be the case if area I is equal to area II, *i.e.*

$$\sigma_{ys} \cdot \Delta a_n = \int_0^{r_y} \frac{\sigma\sqrt{\pi(a+\Delta a_n)}}{\sqrt{2\pi r}}\, dr - \sigma_{ys} \cdot r_y \,. \tag{3.3}$$

or

$$\sigma_{ys}(\Delta a_n + r_y) = \int_0^{r_y} \frac{\sigma\sqrt{\pi(a+\Delta a_n)}}{\sqrt{2\pi}} \frac{dr}{\sqrt{r}} = \frac{2\sigma\sqrt{a+\Delta a_n}}{\sqrt{2}} \sqrt{r_y} \,. \tag{3.4}$$

For a crack of length $a + \Delta a_n$

$$r_y = \frac{1}{2\pi}\left(\frac{K_I}{\sigma_{ys}}\right)^2 = \frac{\sigma^2}{2\sigma_{ys}^2}(a + \Delta a_n) \,, \tag{3.5}$$

which we can use to substitute for $\sigma\sqrt{a+\Delta a_n}$ in equation (3.4). This gives

$$\sigma_{ys}(\Delta a_n + r_y) = \frac{2\sigma_{ys}\sqrt{2r_y}\sqrt{r_y}}{\sqrt{2}}$$

and therefore

$$\Delta a_n + r_y = 2r_y \,. \tag{3.6}$$

Thus the notional crack increment Δa_n is equal to the first approximation for the plastic

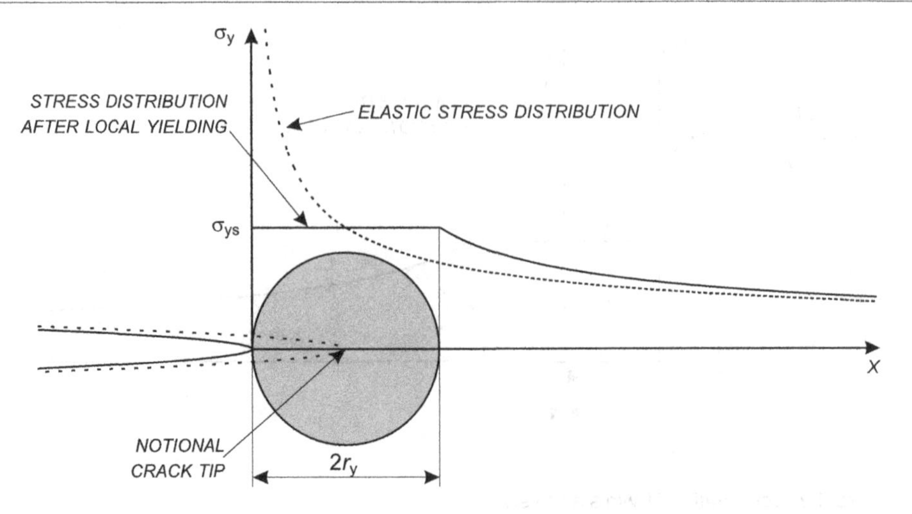

Figure 3.3. The Irwin plastic zone size.

zone size r_y.

It follows that Irwin's analysis results in a plastic zone diameter ($\Delta a_n + r_y$) *twice that obtained as a first approximation* (r_y). Furthermore, this result means that the notional crack length ($a+\Delta a_n$) = ($a + r_y$) extends to the centre of the circular plastic zone, figure 3.3, with a concomitant shift of the stress distribution over a distance r_y with respect to the elastic case. Note that the elastic stress distribution $\sigma_y = K_I/\sqrt{2\pi r}$ takes over from σ_{ys} at a distance $2r_y$ ahead of the actual crack tip. Thus K_I determines both the plastic zone size, equation (3.2), and the stresses and strains outside the plastic zone ($\sigma_y = K_I/\sqrt{2\pi r}$). It seems reasonable, therefore, that the stress intensity approach is still applicable for correlating crack growth and fracture behaviour.

Note that the expression used for K_I at the notional crack length, *i.e.*

$$K_I = \sigma\sqrt{\pi(a + r_y)}\,,$$

is only a first approximation. The reason is that

$$r_y = \frac{1}{2\pi}\left(\frac{K_I}{\sigma_{ys}}\right)^2,$$

which implies that K_I and r_y are mutually dependent. For this simple case, where $f(a/W) = 1$, the problem can be analytically solved by simple substitution. The result is

$$K_I = \frac{\sigma\sqrt{\pi a}}{\sqrt{1 - \frac{1}{2}\left(\frac{\sigma}{\sigma_{ys}}\right)^2}}\,, \tag{3.7}$$

an expression that approaches the usual $K_I = \sigma\sqrt{\pi a}$ for $\sigma << \sigma_{ys}$. For more complicated $f(a/W)$, however, a numerical computation is generally necessary. If K_I is not too large, only a few iterations are needed to find a converging result for K_I. At higher stresses the numerical procedure may lead to a diverging K_I value and LEFM is no longer applicable.

A Useful Expression: Crack Tip Opening Displacement (CTOD)

In section 2.3 the crack flank displacement, v, was defined. The expression obtained *for*

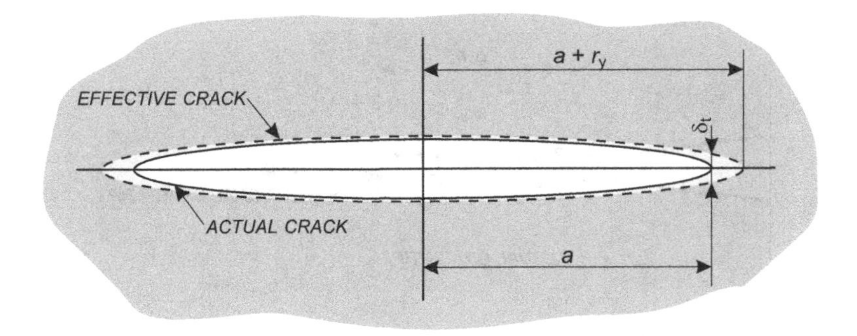

Figure 3.4. Crack Tip Opening Displacement according to Irwin

plane stress was $v = 2\sigma\sqrt{a^2 - x^2}/E$. The total Crack Opening Displacement (COD or δ) is equal to $2v$, *i.e.* $\delta = 4\sigma\sqrt{a^2 - x^2}/E$. For this purely elastic case the Crack Tip Opening Displacement (CTOD or δ_t) would be zero, since x is $\pm a$ at the tip. As shown in figure 3.4, crack tip plasticity can be accounted for by using Irwin's proposal for an effective crack length $2(a + r_y)$, and considering the crack opening displacement at the actual crack tip, *i.e.*

$$\delta_t = \frac{4\sigma}{E}\sqrt{a^2 + 2ar_y + r_y^2 - a^2} \approx \frac{4\sigma}{E}\sqrt{2ar_y}\,.$$

Substitution for r_y from equation (3.2) gives

$$\delta_t = \frac{4}{\pi}\frac{K_I^2}{E\sigma_{ys}}, \tag{3.8}$$

which is an approximation for the Crack Tip Opening Displacement (CTOD). This approximation will be compared in section 3.3 with the more usual expression for CTOD, which is derived from the Dugdale approach.

Equation (3.8) cannot be used for a situation with varying K_I, as for example in fatigue. Suppose we want to know δ_t for a value of $K_I < K_{max}$, the maximum value of K during the fatigue cycle. A derivation analogous to that leading to equation (3.8) now gives:

$$\delta_t = \frac{4}{\pi}\frac{K_I K_{max}}{E\sigma_{ys}}\,.$$

3.3 The Plastic Zone Size According to Dugdale: The Strip Yield Model

Dugdale's analysis assumes that all plastic deformation concentrates in a strip in front of the crack, the so-called *strip yield model*. This type of behaviour does indeed occur for a number of materials, but certainly not for all. Just as in Irwin's analysis, Dugdale argued that the effective crack length is longer than the physical length. The notional crack increment Δa_n is considered to carry the yield stress as shown in figure 3.5 (here the assumption of elastic – perfectly plastic material behaviour and a state of

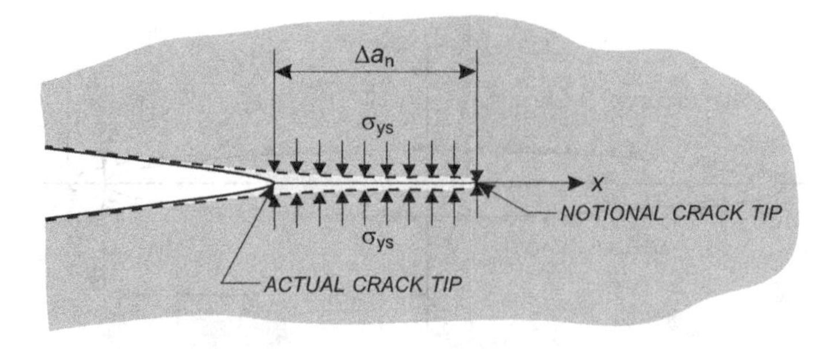

Figure 3.5. Schematic of Dugdale's analysis.

plane stress is also made). Note that Δa_n is not defined the same way as in section 3.2. Here Δa_n is the size of the *total* plastic zone.

The derivation of the Dugdale formula is made in two ways:

- straightforwardly using superposition,
- formally using (complex) stress functions.

Derivation using superposition

The superposition procedure is given in figure 3.6. In a plate with a crack of physical length $2a$ and plastic zone size $2\Delta a_n$ (plate A) the same stresses and displacements are present as in a plate with a physical crack length $2(a + \Delta a_n)$ where the crack is closed over the area $2\Delta a_n$ owing to a pressure σ_{ys} on the crack flanks (plate B). The two mode I stress systems acting on plate B can be split into the separate mode I stress systems shown in plates C and D. The approach is now that a finite value of σ_y, *i.e.* σ_{ys}, is required at the notional crack tip (at $a = a + \Delta a_n$). In other words $\sigma_y = K_I^{sum}/\sqrt{2\pi r}$ is finite, where $K_I^{sum} = K_I^B = K_I^C + K_I^D$. Because $r = 0$ at the notional tip, σ_y would become singular at that point unless K_I^{sum} is equal to zero. Using this knowledge, the value of

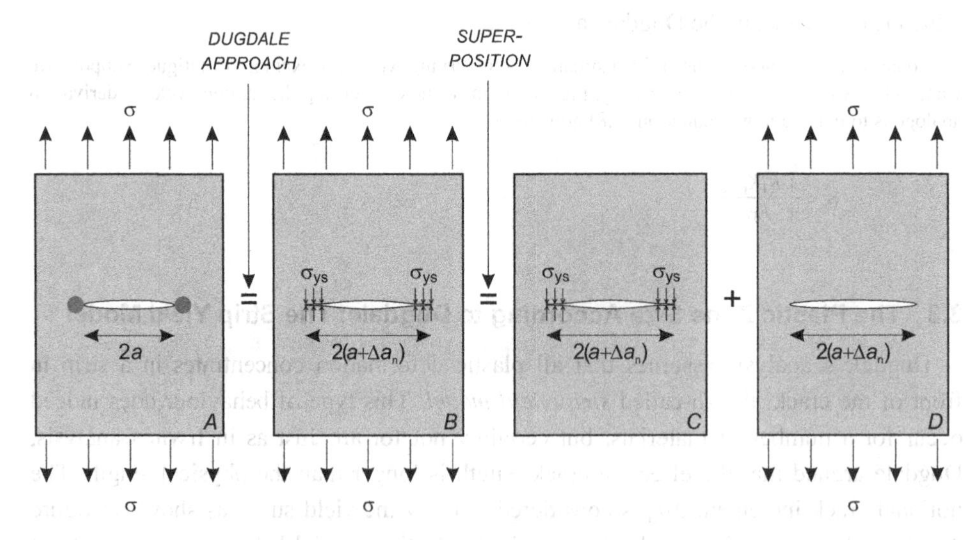

Figure 3.6. Obtaining the Dugdale formula using superposition.

Δa_n may be calculated as follows.

K_I^C can be solved with the formulae for crack-line loading given in section (2.5). Analogous to equation (2.52), we can write:

$$K_I^C = \frac{P}{\sqrt{\pi(a+\Delta a_n)}} \int_a^{a+\Delta a_n} \frac{2(a+\Delta a_n)\,dx}{\sqrt{(a+\Delta a_n)^2 - x^2}} = \frac{2P(a+\Delta a_n)}{\sqrt{\pi(a+\Delta a_n)}} \left[\arcsin \frac{x}{a+\Delta a_n} \right]_a^{a+\Delta a_n}$$

$$\Rightarrow \qquad K_I^C = -2\sigma_{ys} \sqrt{\frac{a+\Delta a_n}{\pi}} \arccos \frac{a}{a+\Delta a_n}, \tag{3.9}$$

where $P = -\sigma_{ys}$ was used. K_I^D follows from:

$$K_I^D = \sigma \sqrt{\pi(a+\Delta a_n)}\,. \tag{3.10}$$

Taking $K_I^{sum} = K_I^B = K_I^C + K_I^D = 0$ we find:

$$\cos \frac{\pi\sigma}{2\sigma_{ys}} = \frac{a}{a+\Delta a_n} \quad \text{or} \quad \sec \frac{\pi\sigma}{2\sigma_{ys}} = \frac{1}{\cos \frac{\pi\sigma}{2\sigma_{ys}}} = 1 + \frac{\Delta a_n}{a}\,. \tag{3.11}$$

Using the series expansion

$$\sec x = 1 + \frac{x^2}{2} + \frac{5x^4}{24} + \ldots \quad \text{for } |x| < \frac{\pi}{2}$$

and assuming $\sigma << \sigma_{ys}$, *i.e.* $x = \pi\sigma/2\sigma_{ys} << \pi/2$, only the first two terms of the expansion need to be considered and we find:

$$\Delta a_n = \frac{\pi^2\sigma^2 a}{8\sigma_{ys}^2} = \frac{\pi}{8}\left(\frac{K_I}{\sigma_{ys}}\right)^2. \tag{3.12}$$

Note that we cannot find the stress σ_y (= σ_{ys}) at the crack tip from the above derivation because $\sigma_y = K_I^{sum}/\sqrt{2\pi r}$ becomes indefinite when r goes to zero. We know only that σ_y is finite. From the derivation below, using stress functions, it is possible to obtain more information.

Derivation using (complex) stress functions

The approach proceeds as follows:

1) Obtain a Westergaard-type stress function for a crack of length $2(a + \Delta a_n)$ with the origin at the centre of the crack. From section 2.2 a suitable function is known to be $\phi_1(z) = \sigma/\sqrt{1-(a+\Delta a_n)^2/z^2}$.

2) Since Δa_n carries the yield stress, unlike a real crack, the elastic stress function $\phi_1(z)$ will overestimate the stress intensity at the notional crack tip. To obtain the correct estimate use is also made here of the superposition principle. Thus a stress function that describes the loading condition over the distance Δa_n must be found and this stress function must be subtracted from $\phi_1(z)$.
3) From reference 1 of the bibliography to this chapter the stress function $\phi_2(z)$ corresponding to two point forces P (per unit thickness) acting on both surfaces of a crack with length $2(a + \Delta a_n)$ at distances $+b$ and $-b$ from the centre is given by:

$$\phi_2(z) = \frac{2Pz\sqrt{(a+\Delta a_n)^2 - b^2}}{\pi\sqrt{z^2 - (a+\Delta a_n)^2}\,(z^2 - b^2)} \,. \tag{3.13}$$

Choosing a force P that corresponds to the yield stress, *i.e.* $P = \sigma_{ys}db$, and by integrating $\phi_2(z)$, the stress function $\phi_3(z)$ can be obtained that describes the loading condition over Δa_n:

$$\phi_3(z) = \int_a^{a+\Delta a_n} \frac{2\sigma_{ys} z}{\pi\sqrt{z^2 - (a+\Delta a_n)^2}} \frac{\sqrt{(a+\Delta a_n)^2 - b^2}}{z^2 - b^2} db$$

$$= \frac{2\sigma_{ys}}{\pi}\left[\frac{z}{\sqrt{z^2 - (a+\Delta a_n)^2}} \arccos\frac{a}{a+\Delta a_n} - \operatorname{arccot}\left\{\frac{a}{z}\sqrt{\frac{z^2 - (a+\Delta a_n)^2}{(a+\Delta a_n)^2 - a^2}}\right\}\right] . \tag{3.14}$$

4) The correct stress function $\phi_4(z) = \phi_1(z) - \phi_3(z)$, *i.e.*

$$\phi_4(z) = \frac{\sigma z}{\sqrt{z^2 - (a+\Delta a_n)^2}} - \frac{2\sigma_{ys}}{\pi}\frac{z}{\sqrt{z^2 - (a+\Delta a_n)^2}} \arccos\frac{a}{a+\Delta a_n} + \frac{2\sigma_{ys}}{\pi} \operatorname{arccot}\left\{\frac{a}{z}\sqrt{\frac{z^2 - (a+\Delta a_n)^2}{(a+\Delta a_n)^2 - a^2}}\right\} . \tag{3.15}$$

5) Dugdale continued with the argument that a stress singularity cannot exist at the notional crack tip, since at that point the elastic stress goes no higher than the yield stress for an elastic – perfectly plastic material. This means that the singular terms in equation (3.15) must cancel each other, *i.e.*

$$\frac{\sigma z}{\sqrt{z^2 - (a+\Delta a_n)^2}} - \frac{2\sigma_{ys}}{\pi}\frac{z}{\sqrt{z^2 - (a+\Delta a_n)^2}} \arccos\frac{a}{a+\Delta a_n} = 0 \,,$$

so

$$\sigma - \frac{2\sigma_{ys}}{\pi} \arccos\frac{a}{a+\Delta a_n} = 0 \,,$$

and

$$\frac{\pi\sigma}{2\sigma_{ys}} = \arccos\frac{a}{a+\Delta a_n} \quad \text{or} \quad \cos\frac{\pi\sigma}{2\sigma_{ys}} = \frac{a}{a+\Delta a_n}. \tag{3.16}$$

6) Equation (3.16) is identical to equation (3.11), which led to equation (3.12):

$$\Delta a_n = \frac{\pi^2\sigma^2 a}{8\sigma_{ys}^2} = \frac{\pi}{8}\left(\frac{K_I}{\sigma_{ys}}\right)^2. \tag{3.12}$$

Remarks

1) The Dugdale plastic zone size, equation (3.12), is

$$\Delta a_n = 0.393\left(\frac{K_I}{\sigma_{ys}}\right)^2.$$

This is somewhat larger than the diameter of the plastic zone according to Irwin. Irwin's analysis gives a plastic zone diameter $2r_y$, which from equation (3.2) is

$$2r_y = \frac{1}{\pi}\left(\frac{K_I}{\sigma_{ys}}\right)^2 = 0.318\left(\frac{K_I}{\sigma_{ys}}\right)^2.$$

2) The Dugdale approach has been given here in its most general form by using stress functions. Although this might appear unnecessarily complicated as compared to the more common treatment in terms of stress intensity factors as given above, there is an important advantage. Stress intensity factors can be used only as approximations consisting of singular terms. Thus in step 5) of the analysis, when all singular terms are required to cancel each other, the misleading impression is given that the stresses at the notional crack tip (the end of the plastic zone) should be zero. However, equation (3.15) shows that there is a non-singular term

$$\phi_5(z) = \frac{2\sigma_{ys}}{\pi}\operatorname{arccot}\left\{\frac{a}{z}\sqrt{\frac{z^2-(a+\Delta a_n)^2}{(a+\Delta a_n)^2-a^2}}\right\} \tag{3.17}$$

and this gives the elastic stress distribution *in* ($\sigma = \sigma_{ys}$) *and beyond* the plastic zone.

Equation (3.17) indeed predicts $\sigma_y = \sigma_{ys}$ in the whole of the Dugdale plastic zone, *i.e.* $a < x < a + \Delta a_n$ and $y = 0$. This can be seen as follows. The Taylor series for the arccot function is: $\operatorname{arccot} x = \pi/2 - (x - x^3/3 + x^5/5 - \ldots\ldots)$ for $|x| < 1$. Therefore, the argument of the arccot in equation (3.17) is purely imaginary within the plastic zone and has a value between 0 (for $z = a + \Delta a_n$) and i (for $z = a$). For this case equation (3.17) may be rewritten as $\phi_5(z) = 2\sigma_{ys}/\pi \operatorname{arccot}(p{\cdot}\mathrm{i})$ with $0 < p < 1$. From the Taylor series it is clear that the real part of $\operatorname{arccot}(p{\cdot}\mathrm{i}) = \pi/2$, because the other terms in the series have odd powers and will stay purely imaginary. Thus $\sigma_y = \mathrm{Re}\,\phi_5(z) = \sigma_{ys}$ for $a < z < a + \Delta a_n$, as would be expected.

The Dugdale Approach and COD

An important aspect of the Dugdale approach in terms of stress functions is that it enables a basic expression for the COD to be calculated. The crack flank displacement, v, in the region between a and $a + \Delta a_n$ can be obtained by substituting the non-singular term in equation (3.15), *i.e.* $\phi_5(z)$ from equation (3.17), into the plane stress version of equation (2.37.b), *i.e.*

$$v = \frac{1}{E}\left\{2\,\mathrm{Im}\,\bar{\phi}(z) - y(1+\nu)\mathrm{Re}\,\phi(z)\right\}.$$

Now $y = 0$, since the Dugdale plastic zone is a strip yield model along the x-axis. Thus

$$v = \frac{2\,\mathrm{Im}\,\bar{\phi}_5(z)}{E}. \tag{3.18}$$

The solution of equation (3.18) is fairly difficult and beyond the scope of this course. A full treatment is given in reference 2 of the bibliography. For our purpose it is sufficient to note that an expression for the physical CTOD, δ_t, is obtained by solving equation (3.18) for v and allowing z to tend to the limit a. Then

$$2v_t = \delta_t = \frac{8\sigma_{ys}a}{\pi E}\ln\sec\frac{\pi\sigma}{2\sigma_{ys}}. \tag{3.19}$$

Equation (3.19) is the starting point for many CTOD considerations in the literature. Further detailed attention is given to the COD concept in chapters 6, 7 and 8 of this course, but it is here informative to compare the results from Irwin's and Dugdale's analysis. To do this, it is convenient to first rewrite the ln sec expression in equation (3.19). As before, the sec function may be expanded as $\sec x = 1 + x^2/2 + 5x^4/24 + \ldots$ for $|x| < \pi/2$. If $\sigma/\sigma_{ys} << 1$, as is the case for LEFM conditions, the argument of the sec function in equation (3.19) may expected to be much smaller than unity. Therefore we can write

$$\ln(\sec x) \approx \ln\left(1 + \frac{x^2}{2}\right) \approx \frac{x^2}{2}$$

and thus

$$\delta_t = \frac{8\sigma_{ys}a}{\pi E}\frac{1}{2}\left(\frac{\pi\sigma}{2\sigma_{ys}}\right)^2 = \frac{\pi\sigma^2 a}{E\sigma_{ys}} = \frac{K_I^2}{E\sigma_{ys}}. \tag{3.20}$$

This value of CTOD is slightly less than that obtained via Irwin's analysis, equation (3.8):

$$\delta_t = \frac{4}{\pi}\frac{K_I^2}{E\sigma_{ys}} = 1.27\frac{K_I^2}{E\sigma_{ys}}.$$

Note: Plane Stress and Plane Strain

So far, all expressions in Irwin's and Dugdale's analyses have been derived for the state of plane stress (*i.e.* $\sigma_y = \sigma_{ys}$ in the plastic zone). Differences that arise owing to plane strain conditions will be discussed in section 3.5.

3.4 First Order Approximations of Plastic Zone Shapes

In the introduction to this chapter, section 3.1, it was mentioned that well known models describing the crack tip plastic zone fall into two categories. Either they estimate the size of a zone with an assumed shape, or else the shape is determined from a first order approximation to the size. Having dealt with the first category in sections 3.2 and 3.3, we shall now turn to methods for assessing the plastic zone shape.

The reason that a first order approximation to the plastic zone size is used in well known models for determining the shape is that the calculations employ classical yield criteria, *e.g.* those of Von Mises or Tresca, to give only the boundaries where the material starts to yield. No account is taken of the fact that the original elastic stress distribution above σ_{ys} must be redistributed and retransmitted. The procedure is similar to that in section 3.1, but instead of calculating r_y only for $\theta = 0$, the value of r_y over the range $(-\pi \leq \theta \leq +\pi)$ is determined.

Derivation of the plastic zone shape is thus simply a matter of substituting the appropriate stress equations into the yield criterion under consideration. Only the Von Mises criterion will be employed here, since the Tresca criterion gives similar results. However, both plane stress and plane strain plastic zone shapes will be analysed. The plane strain plastic zone shape is included in this section because the procedure for determining it is the same as for the plane stress case, and the result serves as a good basis for discussing the problem of differing states of stress in section 3.5.

Plastic Zone Shapes from the Von Mises Yield Criterion

The Von Mises yield criterion states that yielding will occur when

$$(\sigma_1 - \sigma_2)^2 + (\sigma_2 - \sigma_3)^2 + (\sigma_3 - \sigma_1)^2 = 2\sigma_{ys}^2\,, \tag{3.21}$$

where σ_1, σ_2 and σ_3 are the principal stresses[1].

In section 2.3 the mode I stress field equations were derived for a *two-dimensional case* in terms of the principal stresses, namely

$$\sigma_1 = \frac{K_I}{\sqrt{2\pi r}} \cos{}^{\theta}\!/_2 \,(1 + \sin{}^{\theta}\!/_2) \tag{2.33.a}$$

$$\sigma_2 = \frac{K_I}{\sqrt{2\pi r}} \cos{}^{\theta}\!/_2 \,(1 - \sin{}^{\theta}\!/_2) \tag{2.33.b}$$

and σ_3 is either 0 (plane stress) or $\nu(\sigma_1 + \sigma_2)$ for plane strain. Substitution into equation (3.21) gives for plane stress

$$\frac{K_I^2}{2\pi r}\left(1 + \frac{3}{2}\sin^2\theta + \cos\theta\right) = 2\sigma_{ys}^2$$

or

[1] Note that equation (3.21) describes a circular cylinder in the σ_1, σ_2, σ_3 space with a radius $\sqrt{2/3}\cdot\sigma_{ys}$ around the line $\sigma_1 = \sigma_2 = \sigma_3$.

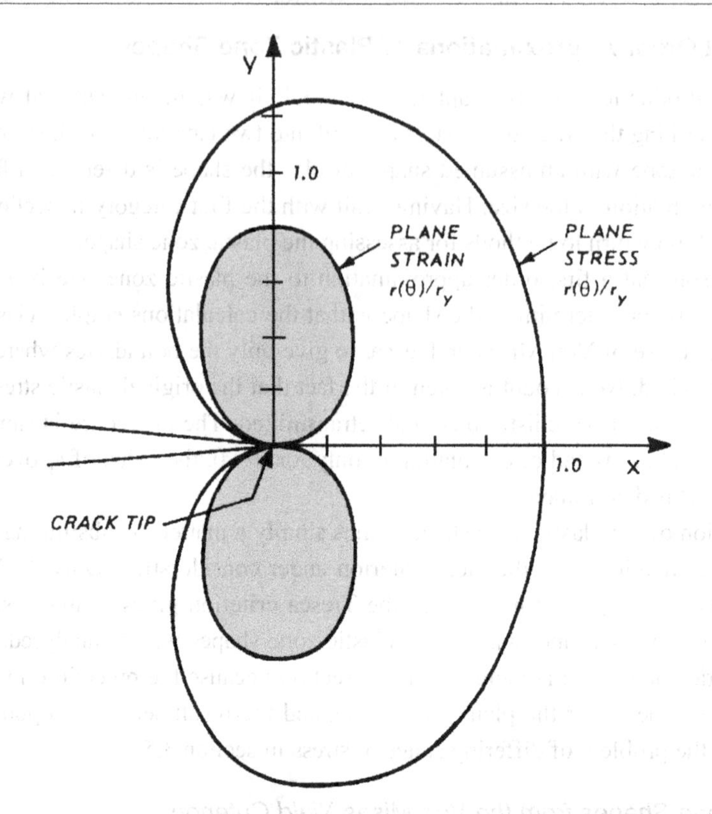

Figure 3.7. Dimensionless plastic zone shapes from the Von Mises yield criterion.

$$r(\theta)_{\text{plane stress}} = \frac{1}{4\pi}\left(\frac{K_{\text{I}}}{\sigma_{\text{ys}}}\right)^2\left(1 + \frac{3}{2}\sin^2\theta + \cos\theta\right). \tag{3.22}$$

Equation (3.22) can be made dimensionless by dividing by r_y, *i.e.* the first order approximation to the plastic zone size for $\theta = 0$ (x-axis), equation (3.2). Then

$$\frac{r(\theta)_{\text{plane stress}}}{r_y} = \frac{1}{2} + \frac{3}{4}\sin^2\theta + \frac{1}{2}\cos\theta\,. \tag{3.23}$$

Note that for $\theta = 0$ the value of $r(\theta)$ is indeed r_y and for $\theta = \pi/2$ (y-axis) the value is $5/4\, r_y$.

For plane strain, *i.e.* $\sigma_3 = \nu(\sigma_1 + \sigma_2)$

$$\frac{K_{\text{I}}^2}{2\pi r}\left\{\frac{3}{2}\sin^2\theta + (1 - 2\nu)^2(1 + \cos\theta)\right\} = 2\sigma_{\text{ys}}^2$$

and

$$\frac{r(\theta)_{\text{plane strain}}}{r_y} = \frac{3}{4}\sin^2\theta + \frac{1}{2}(1 - 2\nu)^2(1 + \cos\theta)\,. \tag{3.24}$$

Along the x-axis ($\theta = 0$) the plane strain value of $r(\theta)$ is much less than the plane stress value. Assuming $\nu = 1/3$,

$$r(\theta{=}0)_{\text{plane strain}} = \frac{1}{9}\, r(\theta{=}0)_{\text{plane stress}} = \frac{1}{9}\, r_y\,.$$

Figure 3.7 depicts the shapes of the plane stress and plane strain plastic zones in dimensionless form.

Similar derivations of plastic zone shapes can be obtained for mode II and mode III loading. Results of such derivations are given in reference 3 of the bibliography to this chapter.

3.5 The State of Stress in the Crack Tip Region

In section 3.1 it was mentioned that the state of stress, *i.e.* plane stress or plane strain, affects the plastic zone size and shape, and figure 3.7 is a good illustration of such effects. For this reason alone it is of interest to go into some detail concerning the state of stress in the crack tip region. However, there are additional important effects of stress state and these will be discussed later in the present section and in section 3.6.

Through-Thickness Plastic Zone Size and Shape

Consider a through-thickness crack in a plate. From equations (2.24) we know that there is at least a biaxial (plane stress) condition, for which the elastic stresses in the x and y directions are given by

$$\sigma_{ij} = \frac{\sigma\sqrt{\pi a}}{\sqrt{2\pi r}} \cdot f_{ij}(\theta)\,. \tag{3.25}$$

Equation (3.25) shows that for small values of r both σ_x and σ_y will exceed the material yield stress. Thus a biaxial plastic zone will form at the crack tip. Assuming in the first instance that there is a uniform state of plane stress and that the plastic zone is circular as in Irwin's analysis, then a section through the plate in the plane of the crack gives the situation shown in figure 3.8. With no strain hardening the material within the plastic zone should be able to flow freely and contract in the thickness direction: however, the adjacent (and surrounding) elastic material cannot contract to the same extent. This phenomenon, called plastic constraint, leads to tensile stresses in the thickness direction

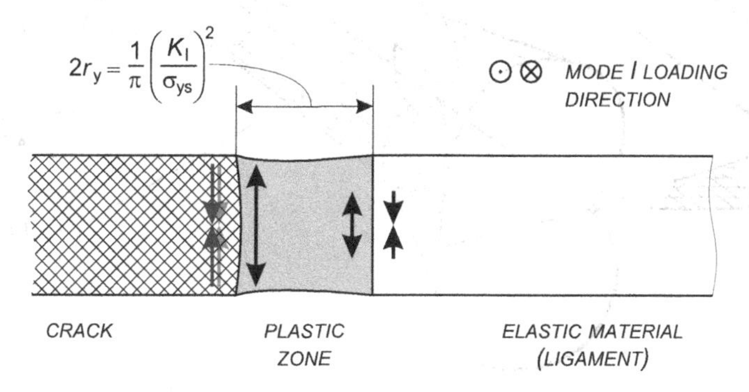

Figure 3.8. Schematic section in the crack plane.

on the plastic zone boundary, *i.e.* a triaxial stress condition which when unrelieved by deformation would correspond to plane strain.

In fact there is an interaction of stress states that can be described at best only semi-quantitatively. At the plate side surface there are no stresses in the thickness direction and so there is a biaxial condition of plane stress. Proceeding inwards there is an increasing degree of triaxiality that approaches and eventually may correspond to plane strain. Thus in a first approximation the plastic zone size and shape may be considered to vary through the thickness of the plate. For a plate of intermediate thickness that is neither fully in plane stress nor predominantly in plane strain these approximate variations are considerable, as indicated schematically in figure 3.9. However, the plane stress surface regions will be more compliant than the plane strain interior, *i.e.* the surface regions will give a larger displacement v for the same remote stress σ, *cf.* equations (2.40) and (2.41). Consequently, load shedding occurs from the surface regions to the interior. This means that the plane stress and plane strain plastic zone sizes will be respectively smaller and larger than those obtained from a first approximation. Indeed it has been found by finite element analysis that the through-thickness plastic zone size variations are much less than those indicated schematically in figure 3.9, see reference 4 of the bibliography.

Simple calculation of the stress state distribution for a certain plate thickness is not possible. However, there are *empirical* rules for estimating whether the condition is predominantly plane stress or plane strain:

1) Full plane stress may be expected if the calculated size of the plane stress plastic

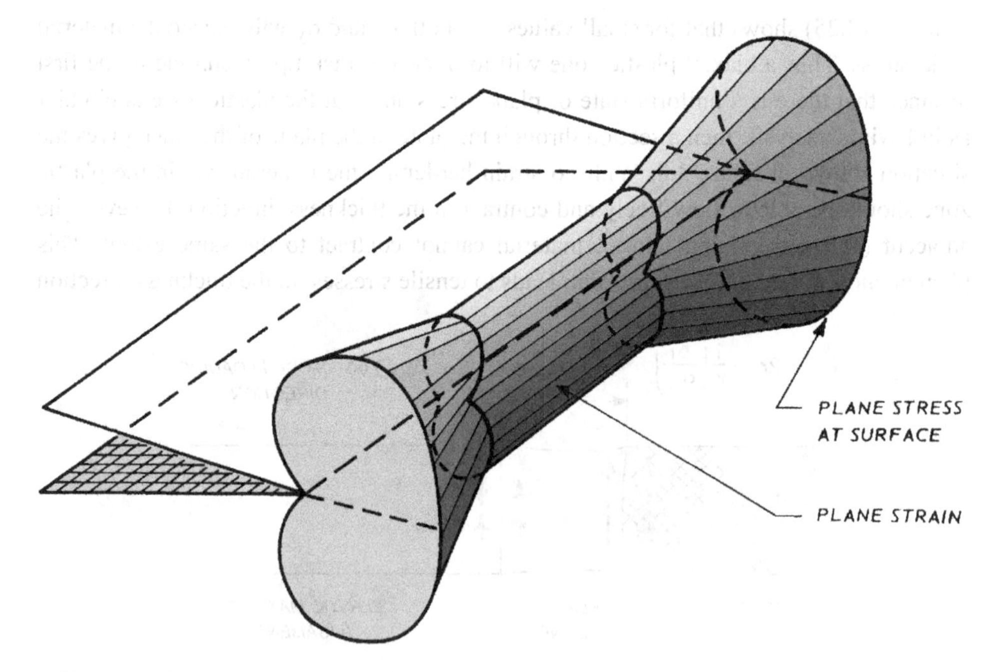

Figure 3.9. Through-thickness plastic zone in a plate of intermediate thickness.

zone, *i.e.* $2r_y$ in Irwin's analysis, is of the order of the plate thickness.

2) Predominantly plane strain may be expected when the calculated size of the plane stress plastic zone, $2r_y$ (the approximate value at the plate surfaces), is no larger than one-tenth of the plate thickness.

Through-Thickness Plastic Zone Size and the Plastic Constraint Factor

In chapter 2 the mode I stress field equations were expressed in terms of principal stresses, equations (2.33). Using these expressions we can derive the relations between σ_1, σ_2 and σ_3 for plane strain conditions:

$$\sigma_2 = \sigma_1\left(\frac{1-\sin{}^{\theta}/_2}{1+\sin{}^{\theta}/_2}\right)$$

$$\sigma_3 = \sigma_1\frac{2\nu}{(1+\sin{}^{\theta}/_2)}\,.$$

It is readily seen that if $\theta = 0$, then $\sigma_2 = \sigma_1$ and $\sigma_3 = 2\nu\sigma_1$. Assuming elastic conditions with $\nu = {}^1/_3$, the Von Mises yield criterion, equation (3.21), can be used to determine the value of σ_1 that is reached before yielding occurs:

$$(\sigma_1-\sigma_1)^2+(\sigma_1-\frac{2}{3}\sigma_1)^2+(\frac{2}{3}\sigma_1-\sigma_1)^2 = 2\sigma_{ys}^2$$

and so

$$\sigma_1 = \sigma_2 = 3\sigma_{ys} \text{ and } \sigma_3 = 2\sigma_{ys}\,.$$

This simple analysis suggests that the ratio between σ_1 and the yield stress becomes as high as 3 for plane strain. This ratio is commonly designated as the *plastic constraint factor, C*.

The first order approximation to the plane strain plastic zone size along the x-axis can now be written as

$$r_{y,\text{plane strain}} = \frac{1}{2\pi}\left(\frac{K_I}{C\sigma_{ys}}\right)^2, \tag{3.26}$$

which for $C = 3$ leads to [2]

$$r_{y,\text{plane strain}} = \frac{1}{9}r_y\,.$$

This result was already obtained in section 3.4. This plane strain value of r_y must be a considerable underestimate of the overall through-thickness plastic zone size in a plate, since at the plate surfaces there is a state of plane stress and the plastic zone size will be

---

[2] Unless stated otherwise, it is implicitly assumed that r_y is the first order approximation for the *plane stress* plastic zone size.

r_y, *i.e.* nine times as large. For this reason Irwin proposed that an intermediate value of $\sqrt{3}$ be used for C, such that the nominal plane strain value of r_y is

$$r_{y,\text{plane strain}} = \frac{1}{6\pi}\left(\frac{K_I}{\sigma_{ys}}\right)^2 = \frac{1}{3} r_y \, . \tag{3.27}$$

This value is often quoted in the literature. However, in most work pertaining to the COD concept a value of $C = 2$ is used (see also chapter 6 and 7). For this constraint factor we find $r_{y,\text{plane strain}} = \frac{1}{4} r_y$.

Planes of Maximum Shear Stress

Besides plastic zone size and shape the state of stress also influences the locations of the planes of maximum shear stress in the vicinity of the crack tip. This is shown in figure 3.10 together with Mohr's circle construction for the principal stresses in plane stress and plane strain:

1) Plane stress
   For a real crack, *i.e.* with a finite tip radius[3], $\sigma_y > \sigma_x$ for $\theta = 0$, see equations (2.28). The principal stresses σ_1 and σ_2 are σ_y and σ_x respectively and $\sigma_3 = \sigma_z = 0$ (it is customary to take $\sigma_1 > \sigma_2 > \sigma_3$). As can be seen from figure 3.10.a., the maximum shear stress, τ_{max}, acts on 45° planes along the x-axis.
2) Plane strain
   For a real crack in plane strain the situation is slightly more complicated. In this case σ_y is also always larger than σ_x in the vicinity of the crack tip. When we go from the plane stress area at the surface to the plane strain interior there is a gradual increase in σ_z from 0 (plane stress) to $\nu(\sigma_y + \sigma_x)$ for pure plane strain. Because plastic deformation signifies that there is no change in volume the 'plastic ν' must have a value of 0.5 within the plastic zone. Thus σ_z is also larger than σ_x. Consequently the orientation of the planes of maximum shear stress changes to 45° along the z-axis, figure 3.10.b.

Thus in plane stress the principal stresses σ_1, σ_2, σ_3 are equal to σ_y, σ_x, σ_z, while in plane strain they are equal to σ_y, σ_z, σ_x. Note that the situations depicted in figure 3.10 are valid only within the relatively small region of the plastic zone. The material will not shear off macroscopically along a plane of maximum shear stress, but will deform in a more complex manner as will be discussed in section 3.6.

3.6 Stress State Influences on Fracture Behaviour

In this section the effects of stress state on the macroscopic appearance of fracture and on the fracture toughness will be discussed.

[3] Assuming the crack tips to have a finite radius is legitimate since a crack will always show some blunting due to plastic deformation. If the crack is considered to be slit-shaped with zero crack tip radius then equations (2.24) will apply and in plane strain $\sigma_y = \sigma_1 = \sigma_x = \sigma_2 = \sigma_z = 0.5\,(\sigma_x + \sigma_y) = \sigma_3$. Therefore all principal stresses are equal ahead of a slit crack and there is a hydrostatic stress state in which τ_{max} is zero. Note that $\nu = 0.5$ since stresses within the plastic zone are considered.

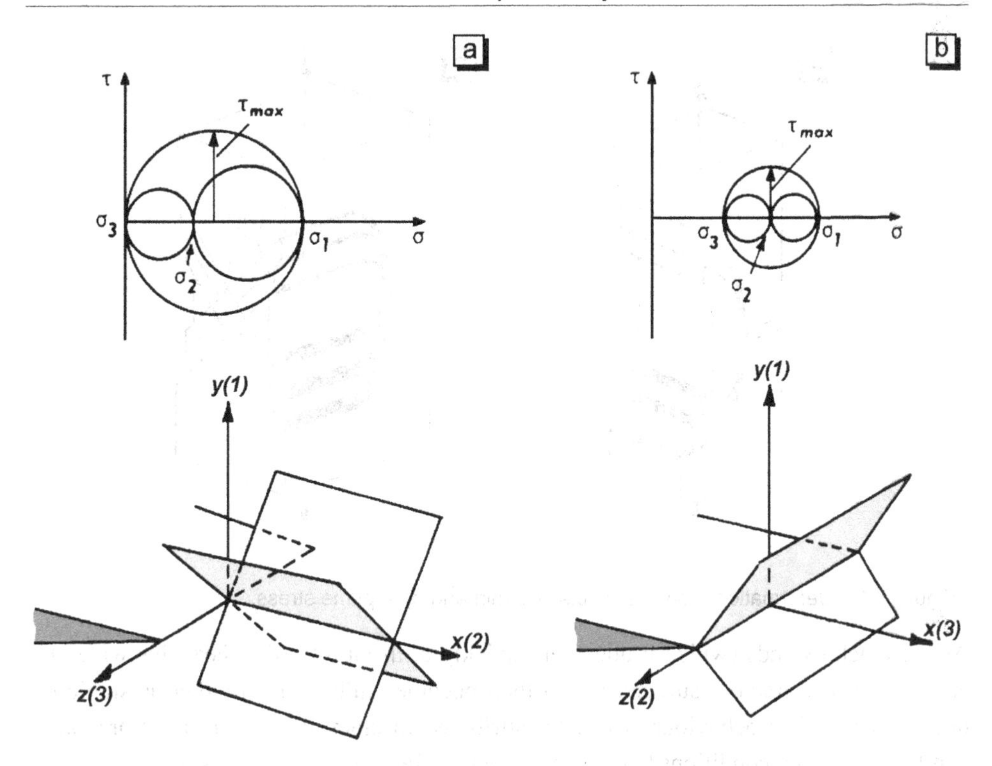

Figure 3.10. Location of the planes of maximum shear stress at the tip of a crack for a) plane stress and b) plane strain conditions.

Fracture Appearance

If a precracked specimen or component is monotonically loaded to fracture the general appearance corresponds to that sketched in figure 3.11. Crack extension begins macroscopically flat but is immediately accompanied by small 'shear lips' at the side surfaces.

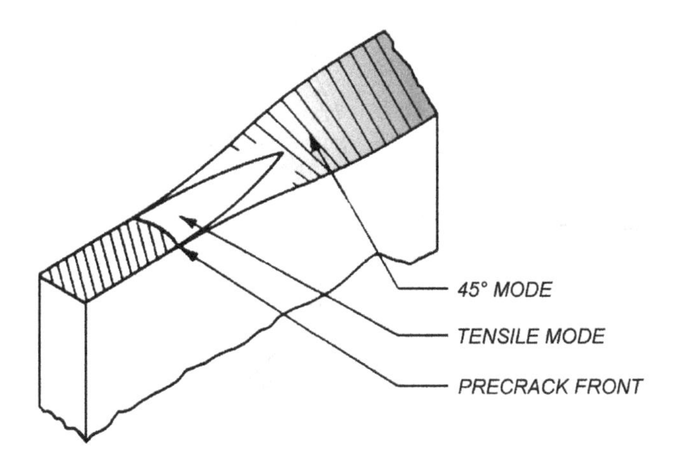

Figure 3.11. General appearance of monotonic overload fracture from a precrack.

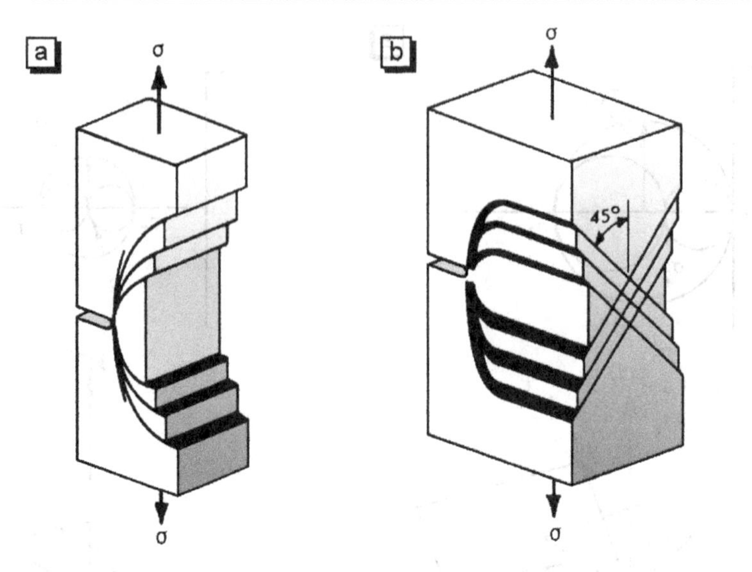

Figure 3.12. Deformation modes a) in plane strain and b) in plane stress.

As the crack extends (which it does very quickly at instability) the shear lips widen to cover the entire fracture surface, which then becomes fully slanted either as single or double shear. This behaviour is usually attributed to crack extension under predominantly plane strain conditions being superseded by fracture under plane stress.

An exact model for this flat-to-slant transition is not available, but it seems obvious that a change in the planes of maximum shear stress, see figure 3.10, plays an important role. Experimental studies by Hahn and Rosenfield (reference 5 of the bibliography to this chapter) indicate that under plane strain conditions a 'hinge' type deformation is

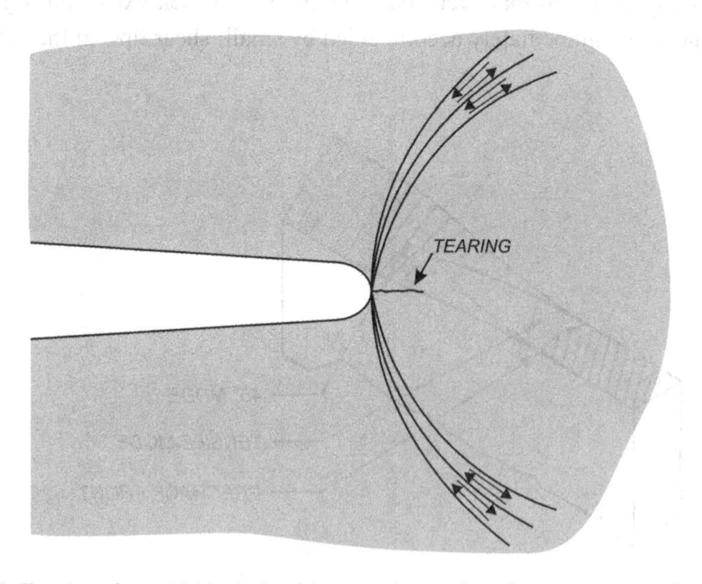

Figure 3.13. Tearing of material between hinge type shear bands.

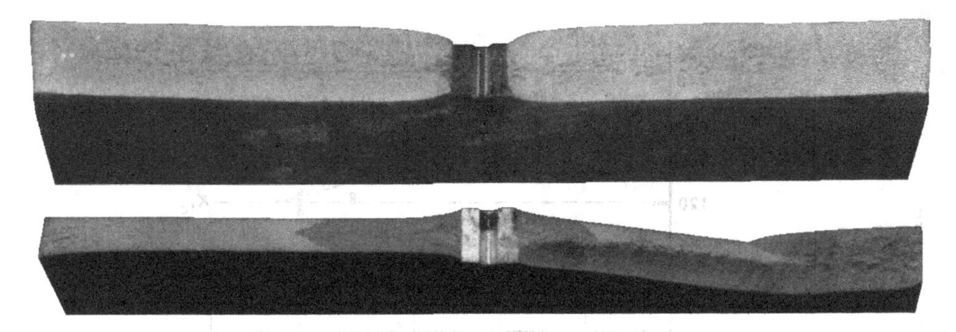

Figure 3.14. The development of a single shear fracture (top) and a transition from double to single shear fracture (bottom right-hand side) during tensile tests on centre cracked tension specimens.

followed by flat fracture, whereas under plane stress slant fracture occurs by shear after a hinge type initiation. The two deformation modes are shown in figure 3.12.

The occurrence of slant fracture by shear is reasonably clear from figure 3.12.b. However, exactly how flat fracture results from hinge type deformation requires further explanation. According to Hahn and Rosenfield flat fracture occurs by tearing of material between the extensively deformed hinge type shear bands, figure 3.13. As to the possible causes of tearing the reader is referred to the last chapter of this course, section 13.4.

As an illustration, figure 3.14 shows results of tensile tests on centre cracked tension specimens of aluminium alloy 2024. Single shear fracture occurred in a direction perpendicular to the loading direction, while double shear fracture occurred in a significantly deviating direction. In the figure a transition from double to single shear can be seen: the single shear part resumes the original crack growth direction perpendicular to the load. More about this topic can be found in reference 6.

Fracture Toughness

The critical stress intensity for fracture, K_c, depends on specimen thickness. A typical dependence is given in figure 3.15. Note that beyond a certain thickness, when the material is predominantly in plane strain and under maximum constraint, the value of K_c tends to a limiting constant value. This value is called the plane strain fracture toughness, K_{Ic}, and may be considered a material property.

The behaviour illustrated in figure 3.15 is generally ascribed to the plane stress → plane strain transition that occurs with increasing specimen thickness. However, a complete explanation for the observed effect of thickness does not exist. Also, the form of the K_c dependence for very thin specimens (less than 1 mm) is not exactly known, hence the dashed part of the plot. The most satisfactory model of the thickness effect is based on the energy balance approach and will be described in chapter 4.

Owing to the dependence of fracture toughness on specimen thickness and stress state it is evident that experimental determination of K_{Ic} will be possible only when specimens exceed a certain thickness. In turn this thickness will depend on the crack tip

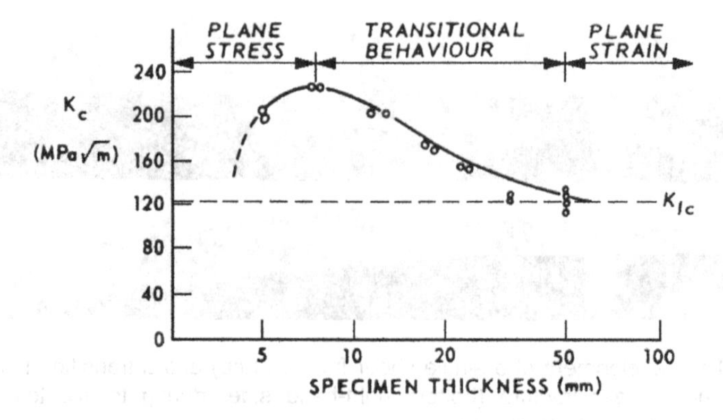

Figure 3.15. Variation in K_c with specimen thickness in a high strength maraging steel.

plastic zone size, as discussed in section 3.5, and therefore on the material yield strength. These and other aspects of plane strain fracture toughness determination and also the obtaining of K_c for thinner specimens will be dealt with in chapter 5, which concerns LEFM testing.

3.7 Some Additional Remarks on Plastic Zone Size and Shape Determination

In section 3.1 it was mentioned that it is extremely difficult to properly describe size and shape of the plastic zone at the same time. A detailed treatise on more advanced methods that attempt to do this is beyond the scope of this course, but an indication of the results is considered to be of interest here.

There are two general ways of tackling the problem: the experimental approach and finite element analysis.

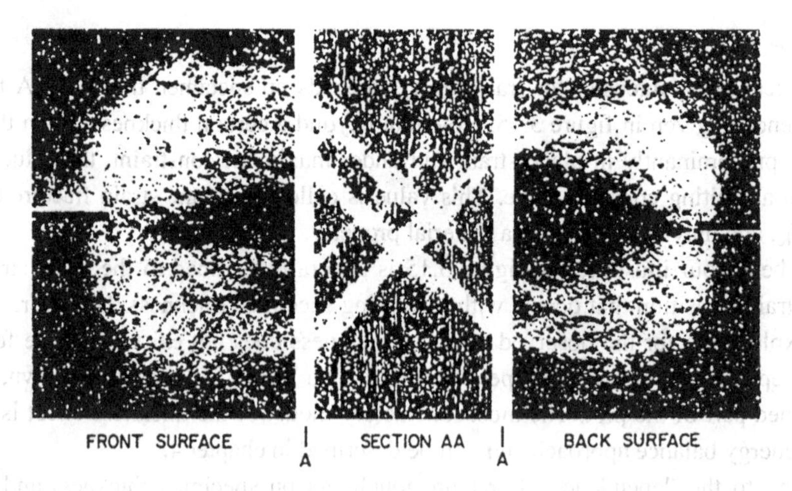

Figure 3.16. Plastic zone appearance on the front surface, back surface and a normal section of a notched silicon iron specimen in plane stress, reference 9.

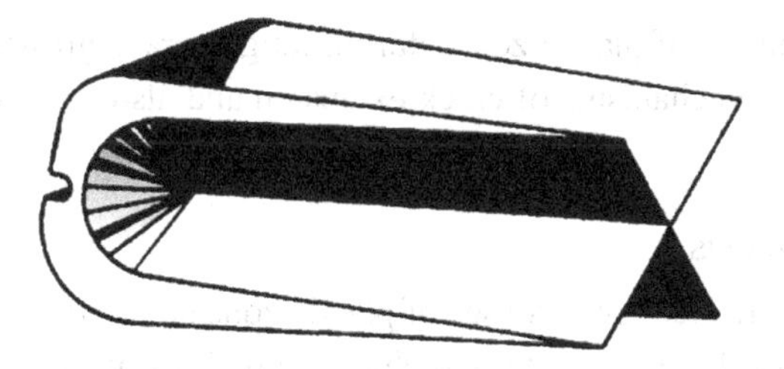

Figure 3.17. Schematic representation of the observations from figure 3.16.

The Experimental Approach

The most widely known work is that of Hahn and Rosenfield (references 5 and 9 of the bibliography to this chapter). They used specimens of silicon iron, which has the property that plastically deformed regions can be selectively etched and made visible. Some of the results are shown in figures 3.16 and 3.17. The specimen illustrated in figure 3.16 was in plane stress and its plastic zone shape is schematically represented in figure 3.17. This shape is reasonably approximated by the Dugdale strip yield model. For plane strain the plastic zones were observed to closely resemble the shape in figure 3.7, which was derived from the Von Mises yield criterion.

Other experimental techniques include the use of electron microscopy, references 7 and 8 of the bibliography, and optical interferometry. These techniques are used mostly

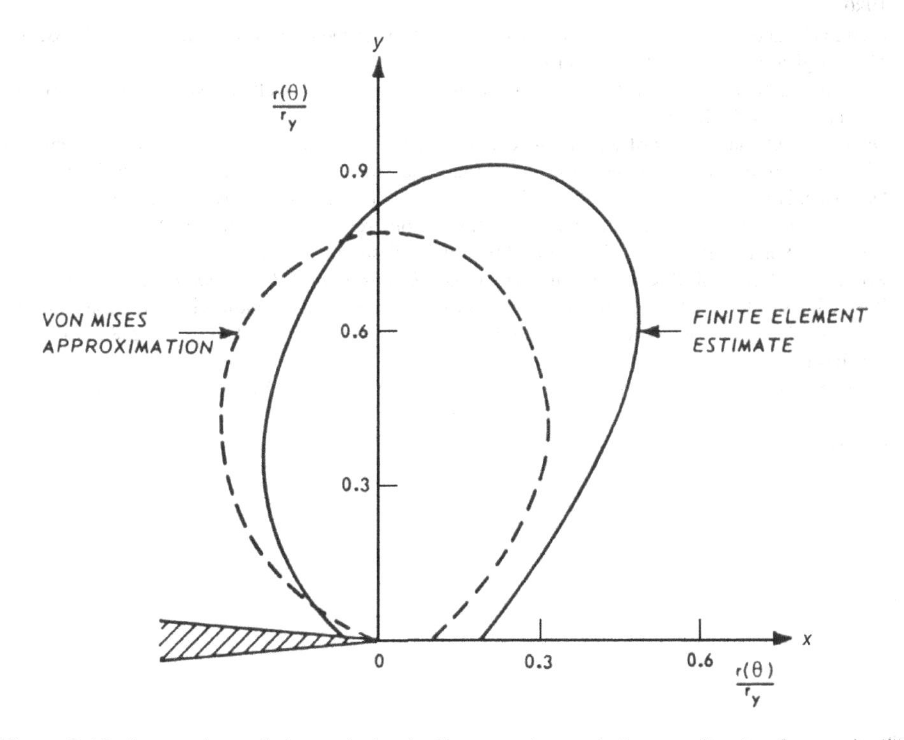

Figure 3.18. Comparison of plane strain plastic zone size and shape estimates for an elastic – perfectly plastic material.

to study the development of plastic zones during fatigue crack growth in order to obtain more insight into the mechanisms of crack extension and also to check and refine crack growth models.

Finite Element Analysis

There are various finite element analyses of plastic zone size and shape, *e.g.* the work of Levy *et al.*, reference 10 of the bibliography. Figure 3.18 depicts their estimate for a plane strain crack tip plastic zone in an elastic – perfectly plastic material together with the plastic zone derived in section 3.4 using the Von Mises yield criterion. It is seen that the latter, which is a first order approximation, is significantly smaller than the more accurate estimate provided by finite element analysis.

3.8 Bibliography

1. Tada, H., Paris, P.C. and Irwin, G.R., The Stress Analysis of Cracks Handbook, Paris Productions Incorporated (1985): St. Louis, Missouri.
2. Burdekin, F.M. and Stone, D.E.W., *The Crack Opening Displacement Approach to Fracture Mechanics in Yielding Materials*, Journal of Strain Analysis, Vol. 1, pp. 145–153 (1966).
3. McClintock, F.A. and Irwin, G.R., *Plasticity Aspects of Fracture Mechanics*, Fracture Toughness Testing and its Applications, ASTM STP 381, American Society for Testing and Materials, pp. 84–113 (1965): Philadelphia.
4. Prij, J., *Two and Three Dimensional Elasto-Plastic Finite Element Analysis of a Compact Tension Specimen*, Dutch Energy Research Centre Report ECN-80-I 78, Petten, The Netherlands, November 1980.
5. Hahn, G.T. and Rosenfield, A.R., *Local Yielding and Extension of a Crack under Plane Stress*, Acta Metallurgica, Vol. 13, pp. 293–306 (1965).
6. Zuidema, J., *Square and Slant Fatigue Crack Growth in Al 2024*, Delft University Press, ISBN 90-407-1191-7 (1995): Delft.
7. Davidson, D.L. and Lankford, J., *Fatigue Crack Tip Plasticity Resulting from Load interactions in an Aluminium Alloy*, Fatigue of Engineering Materials and Structures, Vol. 1, pp. 439–446 (1979).
8. Davidson, D.L. and Lankford, J., *The influence of Water Vapour on Fatigue Crack Plasticity in Low Carbon Steel*, Proceedings of the First International Conference on Fracture, ed. D.M.R. Taplin, University of Waterloo Press, Vol. 2, pp. 897–904 (1977): Waterloo, Ontario.
9. Rosenfield, A.R., Dai, P.K. and Hahn, G.T., *Crack Extension and Propagation under Plane Stress*, Proceedings of the First International Conference on Fracture, ed. T. Yokobori, T. Kawasaki and J.L. Swedlow, Japanese Society for Strength and Fracture of Materials, Vol. 1, pp. 223–258 (1966): Sendai, Japan.
10. Levy, N., Marcal, P.V., Ostergren, W.J. and Rice, J.R., *Small Scale Yielding near a Crack in Plane Strain: a Finite Element Analysis*, International Journal of Fracture Mechanics, Vol. 7, pp. 143–156 (1971).

4 The Energy Balance Approach

4.1 Introduction

Besides the elastic stress field (stress intensity factor) approach there is another method that can be used in LEFM, the energy balance approach mentioned in sections 1.4 and 1.5 of chapter 1. In section 4.2 the equations of energy balance and instability will be given, together with the definition of the energy release rate, *G*, which is the parameter controlling fracture.

In section 4.3 some important relations involving *G* are derived. One of these is the relationship between *G* and the change in compliance (inverse of stiffness) of a cracked specimen. This relationship has found much practical use and for that reason some applications of compliance determination are discussed in section 4.4.

In section 4.5 the energy balance interpretation of *G* is again considered in order to show its usefulness for materials exhibiting limited but significant plasticity. This leads to the concept of crack resistance, *R*, and the phenomenon of slow stable crack growth characterized by the *R*-curve, which is discussed in sections 4.6–4.8. The discussion of crack resistance is of a general nature and includes a possible explanation of *R*-curve shape and the effect of specimen thickness on fracture toughness.

The subject of *R*-curves is important and extensive information on their determination and use is given in references 1 and 2 of the bibliography at the end of this chapter. Also, the determination of *R*-curves is included in chapter 5, which concerns LEFM testing.

4.2 The Energy Balance Approach

In section 1.4 of the introduction to this course a simple energy balance derivation has been presented. This was valid only for a specimen loaded by a constant displacement, *i.e.* a fixed grip condition. Here the energy balance approach will be considered for an arbitrary loading condition, followed by a more detailed treatment of the energy release rate, *G*, as introduced by Irwin.

In the energy balance approach a cracked elastic plate and its loading system are considered. The combination of plate and loading system is assumed to be isolated from its surroundings, *i.e.* exchange of work can only take place between the two. The energy content of the plate plus the loading system, denoted as the total energy U, is written as[1]

$$U = U_o + U_a + U_\gamma - F\,, \qquad (4.1)$$

where U_o = total energy of the plate and its loading system before introducing a crack (a constant),
U_a = change in the elastic energy of the plate caused by introducing a crack,
U_γ = change in surface energy of the plate due to the introduction of a crack,
F = work performed by the loading system during the introduction of the crack = load × displacement.

In order to understand why the work F must be subtracted in equation (4.1), consider a plate placed in series with a spring between fixed grips, as shown in figure 4.1. The spring loads the plate in tension. This represents an arbitrary loading condition, since both load and displacement of the plate will change during the introduction of a crack. Furthermore, it is clear that no work can be performed from outside the combination of spring and plate, since no external displacements are allowed.

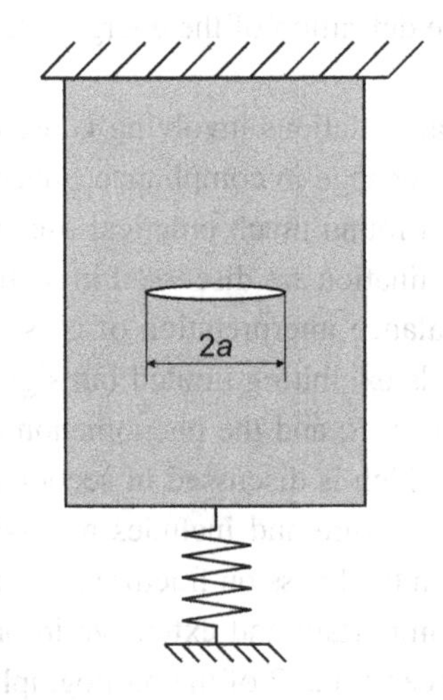

Figure 4.1. A cracked elastic plate in series with a spring between fixed grips and loaded in tension.

When a crack is introduced, the stiffness of the plate is reduced and the plate becomes somewhat longer. Consequently the spring becomes shorter by the same amount. During this process the spring performs an amount of work F on the plate. This happens at the expense of the elastic energy content of the spring, which thus decreases. Since the elastic energy content of the spring is part of the total energy U, it follows that F must be subtracted in equation (4.1).

Potential energy

At this point we define the *potential energy* of an object. Potential energy is either related to the position of an object in a conservative power field or due to its state. An evident example of the first is the potential energy of a mass which is determined by its position (height) in a gravitational field. Elastic strain energy is an example of potential

[1] As was already done in section 1.4, we consider two-dimensional geometries only and all loads and all energies are defined per unit thickness.

energy which is due to the state of an object. In all cases potential energy is characterized by its ability (potential) to perform work.

Only part of the total energy U given by equation 4.1 has the ability to perform work. This part will be designated as the potential energy, U_p, of the elastic plate and its loading system and is equal to[2]

$$U_p = U_o + U_a - F . \tag{4.2}$$

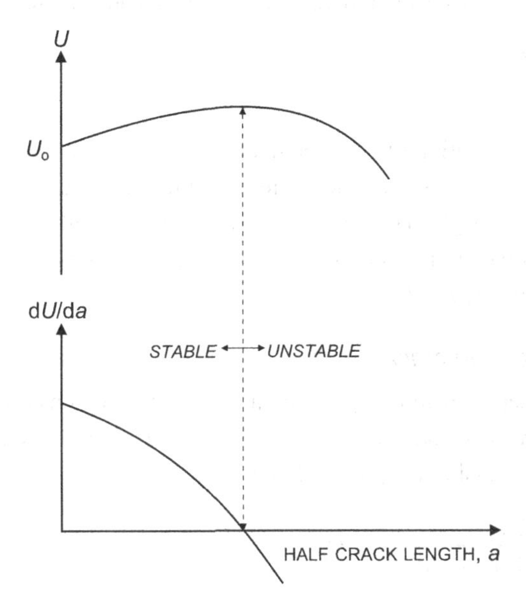

Figure 4.2. The variation of the total energy of a centre cracked plate, U, as a function of half crack length a.

Energy balance

In section 1.4 the total energy U of a large centre cracked plate loaded under fixed grip conditions was considered. In general, for a centre cracked plate under arbitrary loading conditions U will vary as a function of half crack length, a, according to the schematic plot shown in figure 4.2. Crack growth instability will occur as soon as U decreases with the crack length, *i.e.* after U has reached a maximum value. This condition is given by

$$\frac{dU}{da} < 0 \tag{4.3.a}$$

or since U_o is a constant

---

[2] In fact the term U_o need not necessarily contain potential energy only. This, however, is an academic discussion, since we will only consider changes in U_p, and U_o is defined as a constant.

$$\frac{d}{da}(U_a + U_\gamma - F) < 0 . \qquad (4.3.b)$$

Equation (4.3.b) can be rearranged to give:

$$\frac{d}{da}(F - U_a) > \frac{dU_\gamma}{da} . \qquad (4.4)$$

Using the potential energy defined in equation (4.2) we can rewrite equation (4.4) as

$$-\frac{dU_p}{da} > \frac{dU_\gamma}{da} . \qquad (4.5)$$

The left-hand side of equation (4.5) is the decrease in potential energy if the crack were to extend by da. During this same crack extension the surface energy would increase by an amount given by the right-hand side of equation (4.5). In other words, equation (4.5) states that crack growth will occur when the energy *available* for crack extension is larger than the energy *required.*

Energy release rate and crack resistance

Irwin defined the energy *available* per increment of crack extension and per unit thickness as the *energy release rate*, G. When considering a central crack with length $2a$, an increment of crack extension is $d(2a)$ and therefore:

$$G = -\frac{dU_p}{d(2a)} = \frac{d}{d(2a)}(F - U_a). \qquad (4.6)$$

Note that for a central crack, with length $2a$, G is found by differentiating U_p to $d(2a)$, while for an edge crack, with length a, differentiating to da would be sufficient. In both cases G is found as the negative derivative of the potential energy with respect to the newly formed crack area dA, where this area is defined as the projection normal to the crack plane of the newly formed surfaces. This has consequences for three-dimensional geometries: *e.g.* (i) an embedded circular crack, where $dA = d(\pi a^2) = 2\pi a da$, and (ii) an embedded elliptical crack, where $dA = d(\pi ac) = \pi(adc + cda)$. In fact $G = -dU_p/dA$, where the potential energy U_p must now be interpreted in an absolute sense, *i.e.* not per unit thickness, and dA is the decrease in the net section area.

The energy *required* per increment of crack extension is defined as the *crack resistance R*:

$$R = \frac{dU_\gamma}{d(2a)} . \qquad (4.7)$$

Thus equation (4.4) can be rewritten concisely as:

$$G > R . \qquad (4.8)$$

Elastic energy change in a remotely loaded centre cracked plate

In section 1.4 the expression $U_a = -\pi\sigma^2a^2/E$ for the change in elastic strain energy caused by introducing a central crack of length $2a$ in a plate was given without

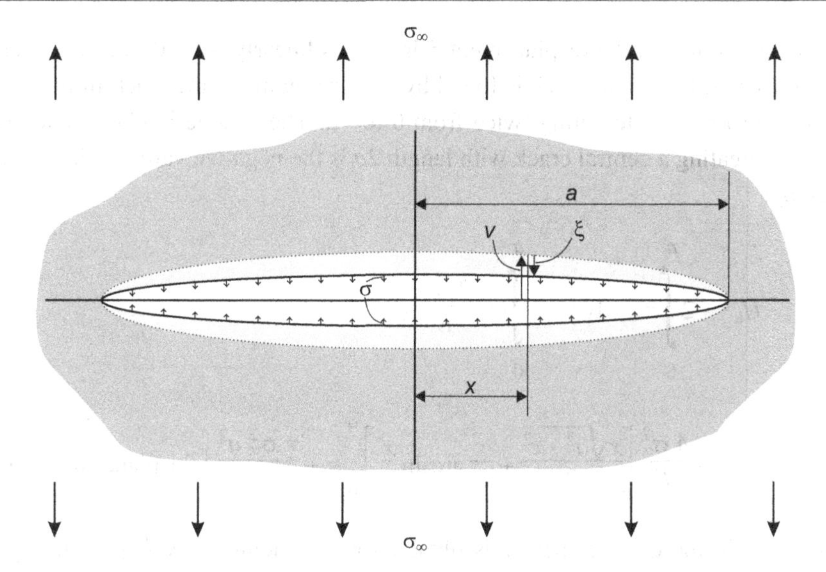

Figure 4.3. A central crack in an infinite plate. The crack flank displacement v is due to the remote load, σ_∞, while the displacement ξ is due to a closing stress, σ, applied to the flanks.

derivation. At this stage in the course it is possible to check the expression for U_a by considering crack flank displacements, figure 4.3.

In section 2.3, equation (2.40), the plane stress flank displacement of a central crack with length $2a$ in an infinite plate remotely loaded by a tensile stress, which will be temporarily denoted as σ_∞, was given as

$$v = \frac{2\sigma_\infty}{E}\sqrt{a^2 - x^2} \,. \tag{4.9}$$

Using this, we can find the change in elastic energy of the plate, U_a, by considering an open crack with stress-free crack flanks, and calculating how much work is involved in closing the crack. The situation of a closed crack resembles that without a crack, since in both cases a uniform stress field σ_∞ is present. Furthermore, closing the crack increases the elastic energy by an amount equal to the work involved.

Consider a part dx of the crack flank at a distance x from the centre of the crack. To bring together the facing crack flank parts, each must be displaced over the distance v given by equation (4.9). Application of a stress, σ, will cause a flank displacement, ξ, relative to the fully opened crack. For a flank length dx, the work involved in closing the crack is:

$$2\,\mathrm{d}x\int_0^v \sigma\,\mathrm{d}\xi = +2\,\mathrm{d}x\,\frac{1}{2}\,\sigma_\infty\,v = +\sigma_\infty\,v\,\mathrm{d}x\,.$$

In this calculation linear elastic material behaviour is assumed, *i.e.* when the stress σ

increases from 0 to σ_∞, the displacement ξ increases linearly from 0 to v. The work involved in closing the *whole* crack is found by integrating along the crack from $-a$ to $+a$, which is the same as integrating twice from 0 to $+a$. The change in elastic energy, U_a, involved in creating a central crack with length $2a$ is the negative value of this work and is given by:

$$U_a = -2\int_0^a \sigma_\infty v \,dx = -2\int_0^a \sigma_\infty \frac{2\sigma_\infty}{E}\sqrt{a^2 - x^2}\,dx$$

$$= -\frac{4\sigma_\infty^2}{E}\left[\frac{x\sqrt{a^2-x^2}}{2} + \frac{a^2}{2}\arcsin\frac{x}{a}\right]_0^a = -\frac{\pi\sigma_\infty^2 a^2}{E} \quad \text{for plane stress.} \quad (4.10)$$

For plane strain the displacement v is obtained by replacing E by $E/1-\nu^2$ in equation (4.9). Thus

$$U_a = -(1-\nu^2)\frac{\pi\sigma_\infty^2 a^2}{E} \quad \text{for plane strain.} \quad (4.11)$$

From now on we will denote the remote stress again by σ, *i.e.* without the ∞ suffix.

Equations (4.10) and (4.11) are valid for an infinite remotely loaded plate. We will now consider a plate with finite dimensions. Now it also becomes relevant what the *loading condition* is. The loading condition can be described as the dependence of the specimen load on the displacement and vice versa. Owing to crack extension the stiffness of a specimen always decreases, and so either the load or the displacement, or both, will change. Next we will consider two extreme loading conditions, namely *fixed grip* and *constant load.*

Fixed Grip and Constant Load Conditions

In a centre cracked plate we have crack extension when, according to equations (4.6) - (4.8),

$$G = \frac{d}{d(2a)}(F - U_a) > R = \frac{dU_\gamma}{d(2a)}.$$

A finite plate under *fixed grip* conditions resembles an infinite plate because no work is performed by external forces, *i.e.* F = constant during crack growth. A difference is, however, that crack extension reduces the plate stiffness and so causes the load to drop. Knowing this, it is to be expected that for a finite plate under fixed grip conditions $dU_a/d(2a)$ is not the same as for an infinite plate. It can be argued that in a finite plate loaded under fixed grip conditions the change in U_a owing to crack extension approaches that of an infinite plate if the crack size $2a$ is small compared to the plate's dimensions. For such a plate we can write

$$G = \frac{\mathrm{d}}{\mathrm{d}(2a)}(-U_a) \approx \frac{\mathrm{d}}{\mathrm{d}(2a)}\left(+\frac{\pi\sigma^2 a^2}{E}\right) = \frac{\pi\sigma^2 a}{E}.$$

The surface energy, U_γ, is equal to the product of the surface tension of the material, γ_e, and the surface area of the crack (two surfaces with length $2a$):

$$U_\gamma = 2(2a\gamma_e). \tag{4.12}$$

Therefore $R = \mathrm{d}U_\gamma/\mathrm{d}(2a) = 2\gamma_e$. Thus the criterion for crack extension is

$$G = \frac{\pi\sigma^2 a}{E} > R = 2\gamma_e. \tag{4.13}$$

Equation (4.13) is valid for a small central crack in a large plate loaded under fixed grip conditions.

If instead there is a condition of *constant load*, then crack extension results in increased displacement owing to decreased stiffness of the plate. The situation is thus more complicated than that for the fixed grip condition. To deal with this it is convenient to compare crack extension in a specimen under fixed grip and under constant load conditions. For this we refer to figure 4.4, which shows load-displacement diagrams for specimens with crack of lengths $2a$ and $2(a + \Delta a)$.

Under fixed grip conditions with a displacement v, the load on the plate will drop from P to $P + \Delta P$ (*i.e.* $\Delta P < 0$) when the crack extends by Δa at both tips. For a constant load condition with a load P, the same crack extension results in a displacement from v to $v + \Delta v$ (*i.e.* $\Delta v > 0$). We will now consider the change in the work performed by external forces, ΔF, and the change in elastic energy, ΔU_a, involved in both cases of crack extension. These changes are expressed in terms of areas in the load-displacement dia-

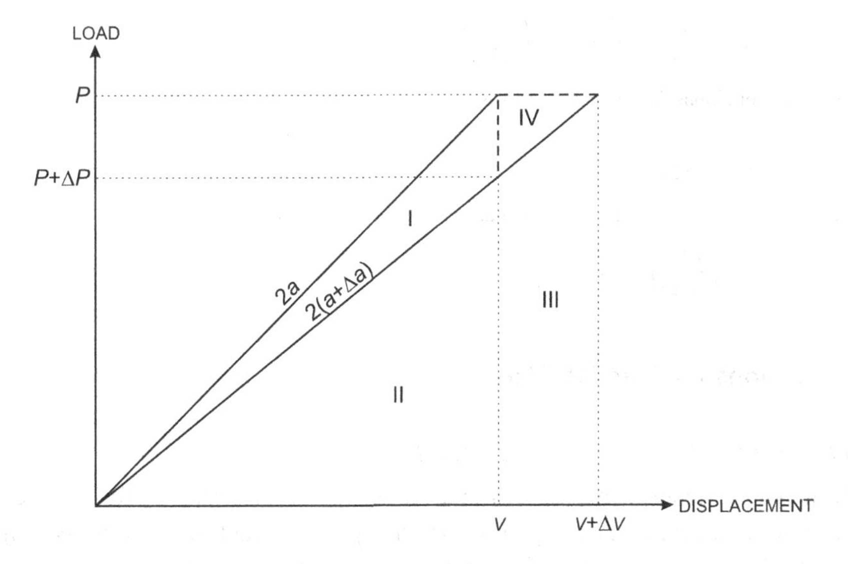

Figure 4.4. Assumed load-displacement diagram for a cracked specimen.

gram of figure 4.4, as shown in the table below.

	Fixed grip	Constant load
$\Delta F = P\Delta v$ =	0	III + IV
$\Delta U_a = \Delta(\frac{1}{2}Pv)$ =	II – (I+II) = –I	(III+II) – (I+II) = III – I
$\Delta(F - U_a)$ =	I	I + IV

If the crack extension $\Delta a \to 0$, the area IV will become negligibly small compared to area I and $\Delta(F - U_a)$ will become equal for the two loading conditions. This implies that because

$$G = \frac{\mathrm{d}}{\mathrm{d}(2a)}(F - U_a) = \lim_{\Delta a \to 0} \frac{\Delta(F - U_a)}{\Delta(2a)},$$

the energy release rate G is the same for a finite plate loaded under fixed grip conditions as well as under constant load conditions. Thus for both fixed grip and constant load conditions the criterion for crack extension in a remotely loaded large plate with a small crack is

$$G = \frac{\pi\sigma^2 a}{E} > R = 2\gamma_e. \tag{4.14}$$

It is important to note that for constant load and $\Delta a \to da$ we obtain:

$$\mathrm{d}F = P\mathrm{d}v,$$

$$\mathrm{d}U_a = \mathrm{d}(\tfrac{1}{2}Pv) = \tfrac{1}{2}P\mathrm{d}v.$$

In other words, for constant load the increase in elastic energy, $\mathrm{d}U_a$, is equal to half the work performed by external forces. The remaining other half is the energy available for crack extension. Thus G for constant load can be written as

$$G_P = \left(\frac{\partial F}{\partial(2a)} - \frac{\partial U_a}{\partial(2a)}\right)_P = +\left(\frac{\partial U_a}{\partial(2a)}\right)_P.$$

For fixed grip conditions $\mathrm{d}F = 0$, and it follows that

$$G_v = -\left(\frac{\partial U_a}{\partial(2a)}\right)_v.$$

Therefore, since G is the same for constant load and fixed grip conditions, we may write

$$G = -\left(\frac{\partial U_a}{\partial(2a)}\right)_v = +\left(\frac{\partial U_a}{\partial(2a)}\right)_P.$$

4.3 Relations for Practical Use

G in a Remotely Loaded Centre Cracked Plate

In the previous section it was shown that the energy release rate, G, for infinitesimal extension of a central crack is equal to $\mathrm{d}U_a/\mathrm{d}(2a)$ for constant load and to $-\mathrm{d}U_a/\mathrm{d}(2a)$ for fixed grips. For large plates with small central cracks G may be obtained by differentiating equations (4.10) and (4.11), which gives:

plane stress $$G=\frac{\pi\sigma^2 a}{E} \tag{4.15.a}$$

plane strain $$G=(1-\nu^2)\frac{\pi\sigma^2 a}{E} \tag{4.15.b}$$

The Relation between G and K_I

A relation of prime importance is obtained by substituting $K_I = \sigma\sqrt{\pi a}$, an expression which also is valid only for a large plate with a small central crack, in the above equations:

plane stress $$G=\frac{K_I^2}{E} \tag{4.16.a}$$

plane strain $$G=\frac{K_I^2}{E}(1-\nu^2) \tag{4.16.b}$$

This direct relation between G and K_I means that under LEFM conditions the prediction of crack growth and fracture is the same for both the energy balance and elastic stress field approaches. Irwin already demonstrated this equivalence, as was mentioned in section 1.6, and showed also that equation (4.16) is geometry-independent.

G and compliance

G has been indicated to be the controlling parameter for fracture according to the energy balance approach. It was shown that the expression for G is the same for the two most

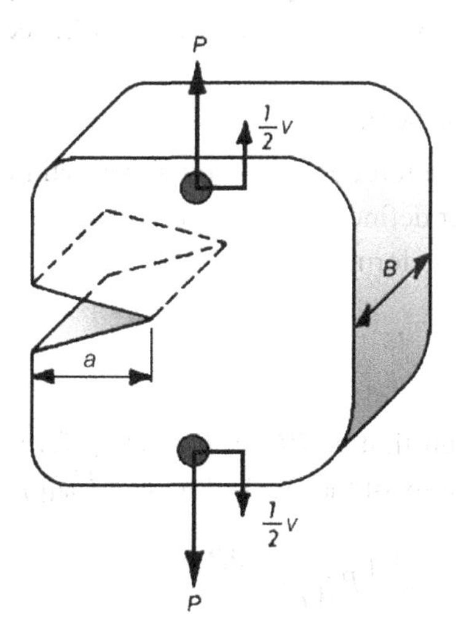

Figure 4.5. A cracked body loaded by forces P and with total displacement v at the points of load application.

extreme loading conditions, *i.e.* fixed grips and constant load. However, a more general analysis can be made, notably that of Irwin (reference 3 of the bibliography to this chapter), by considering the compliance of a cracked body.

Consider the cracked body shown in figure 4.5. For this edge crack the definition for the energy release rate, G, *i.e.* the energy available per increment of crack extension and per unit thickness, leads to:

$$G = \frac{\mathrm{d}}{\mathrm{d}a}(F - U_\mathrm{a}) . \tag{4.17}$$

The body has a thickness B, is loaded by a force P (here not defined per unit thickness) and exhibits a total displacement v. Both F and U_a are defined per unit thickness. Thus for an infinitesimal crack extension da, the change in F, dF, is equal to $(P\mathrm{d}v)/B$, while the change in U_a, dU_a, is $\mathrm{d}(\frac{1}{2}Pv)/B$. Thus G becomes

$$G = \frac{1}{B}\left(P\frac{\mathrm{d}v}{\mathrm{d}a} - \frac{\mathrm{d}(\frac{1}{2}Pv)}{\mathrm{d}a}\right) . \tag{4.18}$$

Introducing the compliance of the body, C, which is the inverse of its stiffness, *i.e.* $C = v/P$, equation (4.18) becomes

$$G = \frac{1}{B}\left(P\frac{\mathrm{d}(CP)}{\mathrm{d}a} - \frac{1}{2}\frac{\mathrm{d}(CP^2)}{\mathrm{d}a}\right) = \frac{P^2}{2B}\frac{\mathrm{d}C}{\mathrm{d}a} . \tag{4.19}$$

It should be clear that G does not depend on the loading system and is the same for fixed grip, constant load or any arbitrary loading condition. Equation (4.19) gives the explicit relation between G and the compliance, C. This important relation is the basis, together with equations (4.16), for using the compliance to determine stress intensity factors for certain specimen and crack geometries. The compliance technique is an addition to those methods discussed and listed in chapter 2, section 2.7, and will be illustrated in the next section.

Two remarks can be made here:

1) In this section we used the force P, acting on a material thickness B in the derivation of G. However we earlier defined U_a per unit thickness. If the force P were to be defined per unit thickness the formula for G would be

$$G = \frac{P^2}{2}\frac{\mathrm{d}C}{\mathrm{d}a} .$$

This is easily found by substituting PB for P and C/B for C in equation (4.19).

2) The change of elastic energy of the cracked body, $\mathrm{d}U_\mathrm{a}/\mathrm{d}a$, may be written as

$$\frac{\mathrm{d}U_\mathrm{a}}{\mathrm{d}a} = \frac{1}{B}\frac{\mathrm{d}(\frac{1}{2}Pv)}{\mathrm{d}a} = \frac{1}{2B}\left(P\frac{\mathrm{d}v}{\mathrm{d}a} + v\frac{\mathrm{d}P}{\mathrm{d}a}\right) . \tag{4.20}$$

For the constant load condition, $\mathrm{d}P/\mathrm{d}a = 0$ and thus

$$\left(\frac{\partial U_a}{\partial a}\right)_P = \frac{P}{2B}\frac{dv}{da} = \frac{P}{2B}\frac{d(PC)}{da} = \frac{P^2}{2B}\frac{dC}{da}. \tag{4.21}$$

For the other extreme loading condition, fixed grips, where displacement is kept constant, $^{dv}/_{da} = 0$ and equation (4.20) becomes

$$\left(\frac{\partial U_a}{\partial a}\right)_v = \frac{v}{2B}\frac{dP}{da} = \frac{v}{2B}\frac{d(^v/_C)}{da} = \frac{v^2}{2B}\left(\frac{dC^{-1}}{dC}\frac{dC}{da}\right) = -\frac{P^2}{2B}\frac{dC}{da}. \tag{4.22}$$

Equations (4.21) and (4.22) show that for fixed grips the change in elastic energy, $^{dU_a}/_{da}$, is the opposite of that for constant load. Comparing these equations with the expression for G, equation (4.19), it follows that

$$G = \left(\frac{\partial U_a}{\partial a}\right)_P = -\left(\frac{\partial U_a}{\partial a}\right)_v. \tag{4.23}$$

This result was already obtained at the end of section 4.2, based on the load-displacement diagram.

4.4 Determination of Stress Intensity Factors from Compliance

From equations (4.16) and (4.19) it follows that

$$K_I^2 = E'G = \frac{E'P^2}{2B}\frac{dC}{da}, \tag{4.24}$$

where $E' = E$ for plane stress and $^E/_{(1-\nu^2)}$ for plane strain. This general relation enables use of the compliance to determine stress intensity factors for certain specimen and crack geometries. A well known example is the double cantilever beam specimen (DCB) already mentioned in section 2.8 and depicted again in figure 4.6. From simple bending theory (neglecting shear displacements) the displacement v in the load line of the DCB specimen is given by

$$v = \frac{2Pa^3}{3EI} = \frac{8Pa^3}{EBh^3}.$$

Since $C = {}^v/_P$,

$$C = \frac{8a^3}{EBh^3} \quad \text{and} \quad \frac{dC}{da} = \frac{24a^2}{EBh^3}.$$

From equation (4.24)

$$K_I^2 = E'G = E'\frac{P^2}{2B}\frac{dC}{da} = \frac{E'}{E}\frac{12P^2a^2}{B^2h^3}$$

and

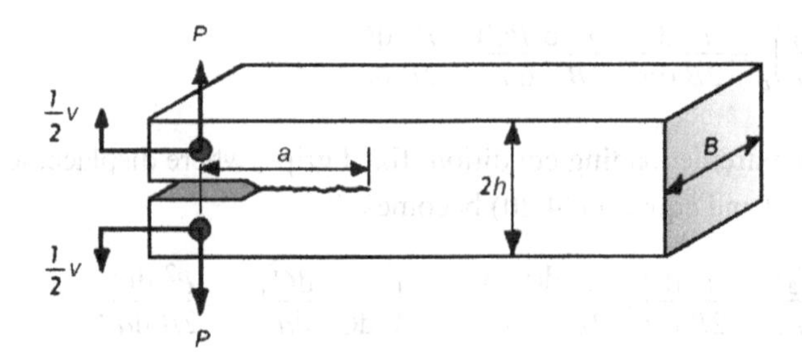

Figure 4.6. The double cantilever beam specimen (DCB).

$$K_I = 2\sqrt{3}\,\frac{Pa}{Bh^{3/2}} \quad \text{for plane stress,}$$

$$K_I = \frac{2\sqrt{3}}{\sqrt{1-\nu^2}}\,\frac{Pa}{Bh^{3/2}} \quad \text{for plane strain.} \tag{4.25}$$

These expressions were given in section 2.8. Note that under elastic conditions ($\nu \approx 0.33$) the factor $1/\sqrt{1-\nu^2}$ gives a K_I value in plane strain only 6% larger than the plane stress value.

Notes

- For beam-type loaded specimens the crack flank displacements in the load line may be used instead of the exact displacements of the points of load application (shown schematically in figures 4.5 and 4.6). This is allowed because the bending displacements are much larger then displacements in the material due to the tensile force. Another advantage is that we do not need an expensive tensile testing machine, since crack flank displacements are usually easy to measure.
- For the double cantilever beam specimen we found the crack flank displacement v proportional to a^3 and P. It follows that for fixed grip conditions the K_I *decreases* with increasing crack length, because K_I is proportional to P and a. This is an important example of a test specimen with a decreasing K_I, which has found widespread use in stress corrosion tests to find the threshold stress intensity value, $K_{I_{scc}}$ (see chapter 10).

A Constant K_I Specimen: The Tapered DCB

An interesting application of equation (4.25) is that a constant stress intensity factor may be obtained by keeping a/B or $a/h^{3/2}$ constant. The first possibility results in tapered thickness, which is not very useful since there will be a gradual change in stress state during crack growth until full plane strain is reached, equation (4.25). However, keeping $a/h^{3/2}$ constant or, still better, accounting for shear displacements by keeping $(3a^2 + h^2)/h^3$ constant, results in a constant K_I specimen, the tapered double cantilever beam (TDCB). This specimen, which is shown in figure 4.7, has been used particularly in stress corrosion testing and will be mentioned again in chapter 10. The side grooves

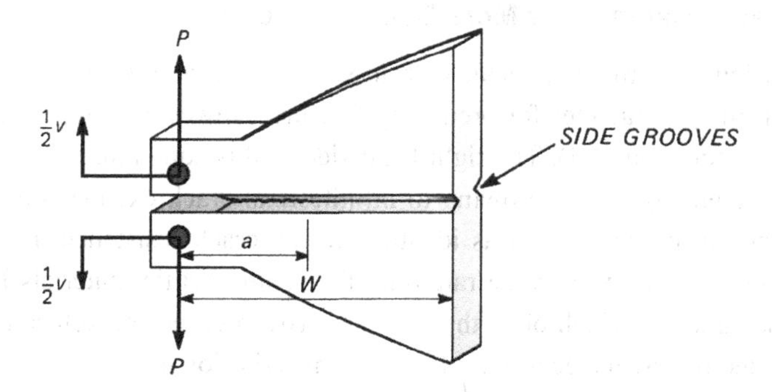

Figure 4.7. The tapered double cantilever beam specimen (TDCB).

are needed to keep crack growth perpendicular to the loading direction. A detailed discussion of the TDCB specimen is given in reference 4 of the bibliography to the present chapter.

Experimental Determination of C and K_I

Values of C and dC/da may be obtained experimentally as well as from theory. Load-displacement diagrams are made for a particular type of specimen containing cracks of different lengths, figure 4.8.a. The cracks may be extended artificially, *e.g.* a sawcut, or by fatigue loading. The compliance for different crack lengths is simply V/P, so that it is possible to construct a compliance versus crack length calibration diagram as in figure 4.8.b. From this diagram dC/da may be obtained and K_I may be calculated via equation (4.24), *i.e.*

$$K_I^2 = \frac{E'P^2}{2B}\frac{dC}{da}.$$

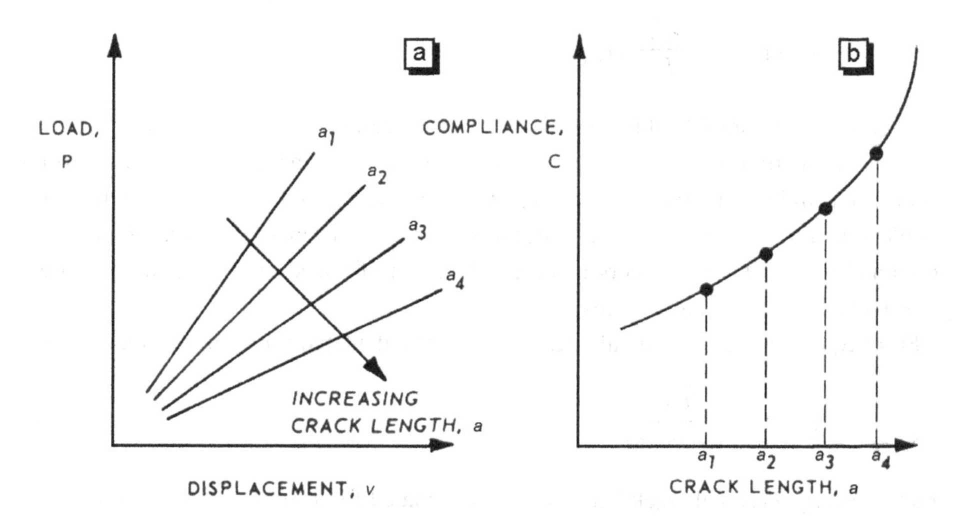

Figure 4.8. Experimental determination of compliance.

4.5 The Energy Balance for More Ductile Materials

The expressions and methods dealt with in sections 4.3 and 4.4 are all based on G being the controlling parameter for predicting fracture. However, G represents only the left-hand side of equation (4.4). The right-hand side of this equation, $R = \mathrm{d}U_\gamma/\mathrm{d}a$, represents the surface energy increase owing to infinitesimal crack extension and is thus in the first instance to be considered as identical to the crack resistance. However, R is equal to the surface energy increase rate only for ideally brittle materials like glasses, ceramics, rocks and ice, which obey the original Griffith criterion. Reformulated in the manner suggested by Irwin (see also section 1.5) this criterion is:

$$G = \frac{\pi\sigma^2 a}{E} > G_c = 2\gamma_e = R = \text{a constant}. \tag{4.26}$$

In 1948 Irwin and Orowan independently pointed out that the Griffith theory could be modified and applied to both brittle materials and metals that exhibit plastic deformation. The modification recognised that R is equal to the sum of the surface energy, γ_e, and the plastic strain work, γ_p, accompanying crack extension. Consequently, equation (4.26), being the condition for crack extension, was changed to

$$G = \frac{\pi\sigma^2 a}{E} > G_c = 2(\gamma_e + \gamma_p) = R\,. \tag{4.27}$$

For relatively ductile materials $\gamma_p >> \gamma_e$, *i.e.* R is mainly plastic energy and the surface energy can be neglected. Also, it is no longer certain that instability and fracture will occur at a constant value of G_c since R need not be a constant. In fact R, and hence G_c, are constant only for the condition of plane strain. In this case it is customary to write $R = G_{Ic}$, in an analogous way to the plane strain fracture toughness K_{Ic} discussed in section 3.6. Thus the plane strain criterion for crack extension, *i.e.* instability, is given by

$$G = (1 - \nu^2)\frac{\pi\sigma^2 a}{E} > G_{Ic}\,. \tag{4.28}$$

As is shown in figure 4.9, this condition can be represented graphically in a plot of G and R as a function of crack length. The left-hand side of inequality (4.28), G, corresponds to a straight line with a slope depending on the applied stress σ. Since the critical plane strain G value, G_{Ic}, is a constant, the crack resistance, R, is represented as a horizontal line. Instability occurs if the combination of crack length and applied stress gives rise to a G value that exceeds R.

From equation (4.28) a critical stress, σ_c, can be calculated for a certain crack length

$$\sigma_c = \sqrt{\frac{EG_{Ic}}{(1-\nu^2)\pi a}}, \tag{4.29}$$

or alternatively a critical crack length can be calculated for a certain applied stress:

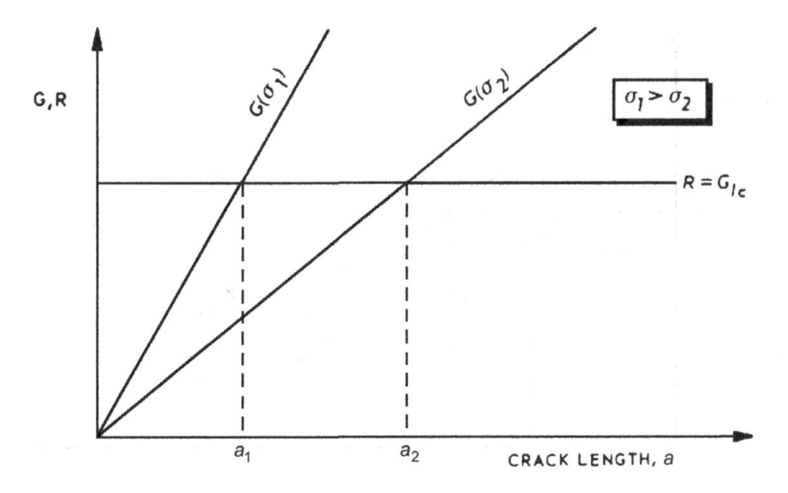

Figure 4.9. Graphical representation of the instability condition for plane strain.

$$a_c = \frac{EG_{Ic}}{(1-\nu^2)\pi\sigma^2}\,. \tag{4.30}$$

It is clear from the foregoing relations that higher stresses result in instability at shorter crack lengths or longer cracks result in instability at lower applied stresses.

4.6 Slow Stable Crack Growth and the *R*-Curve Concept

As was mentioned in section 4.5, a constant value of R, *i.e.* R independent of the crack length, is obtained only for the condition of plane strain. For plane stress and intermediate plane stress – plane strain conditions it turns out that R is no longer constant. Loading a relatively thin (predominantly in plane stress) specimen containing a crack results in the behaviour shown schematically in figure 4.10. The initial crack of length a_0 begins to extend at a certain stress σ_i. However, if the stress is maintained at σ_i no further crack growth occurs, indicating that a small increase in crack length at this stress would result in $G < R$. However, a slight increase in the stress results in additional crack extension, but the situation remains stable. The process of increasing stress accompanied by stable crack growth continues until a critical combination of stress, σ_c, and crack length, a_c, is reached, at which point instability occurs. Note that in the crack length area between a_0 and a_c G has to be equal to R, because otherwise crack arrest or crack instability would occur.

Instability is thus preceded by a certain amount of slow stable crack growth in specimens under full or predominantly plane stress. In terms of the energy balance approach this situation can be described as in figure 4.11. The value of R is depicted as a rising curve with a vertical segment corresponding to no crack extension at low stress (and G) levels. At a stress σ_i crack extension begins but R remains equal to G since the situation is stable. This is indicated by the fact that G intersects the R-curve (then $G = R$) but further crack growth cannot occur at σ_i because G_{σ_i} then becomes less than R.

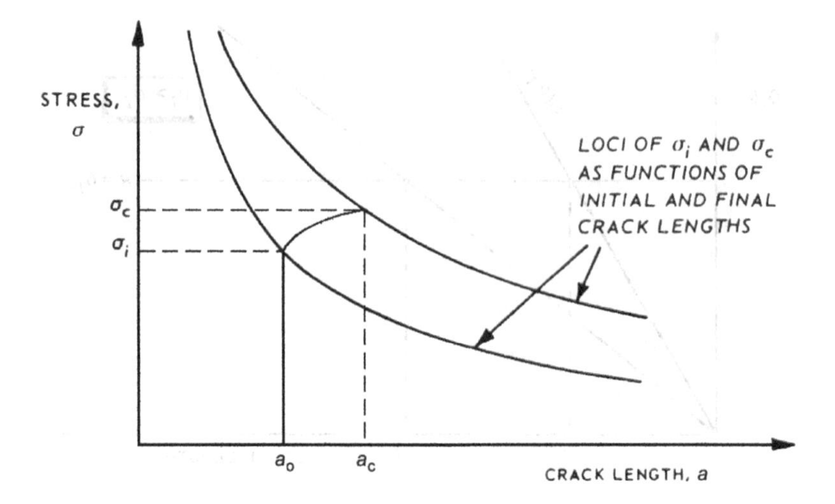

Figure 4.10. Slow stable crack growth in plane stress.

The stable condition (*i.e.* $G = R$) is maintained until σ_c and a_c are reached. Beyond this point G becomes greater than R, as indicated by the G_{σ_c} line, and instability occurs.

Figure 4.11 shows that for instability to occur in plane stress it is not only necessary to have at least a situation with $G > R$, but also the tangency condition $\partial G/\partial a > \partial R/\partial a$ should be fulfilled. This second condition is a consequence of assuming a rising R-curve. This assumption has been verified experimentally, but there is no definitive explanation as to why the R-curve rises: a possible explanation will be discussed in section 4.7. Meanwhile another important property of R-curves, their invariance with respect to initial crack length a_0, will be dealt with here.

R-Curve Invariance with Initial Crack Length

In Irwin's analysis R was considered to be independent of the total crack length a (= a_0

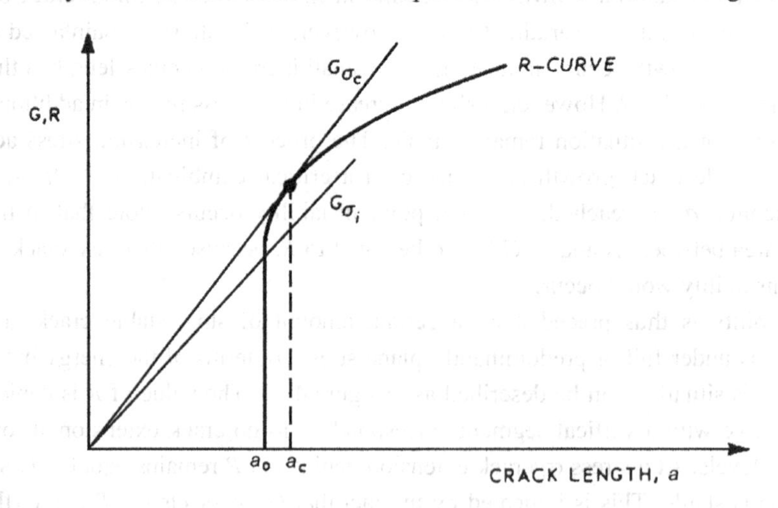

Figure 4.11. The rising *R*-curve.

+ Δa). This is true only for the plane strain condition, as illustrated in figure 4.9. If we now consider the crack resistance for a thin sheet, it is reasonable to assume that the very beginning of slow stable crack growth occurs in the middle of the specimen thickness at a relatively low G value. Therefore, this growth will be under plane strain conditions and will be independent of crack length. In addition, many tests have shown that the form of the rising part of the R-curve is also independent of crack length. Thus we may expect R-curves to be independent of the initial crack length a_0, *i.e.* an invariant R-curve may be placed anywhere along the horizontal axis of a (G,R)-crack length diagram, as in figure 4.12.

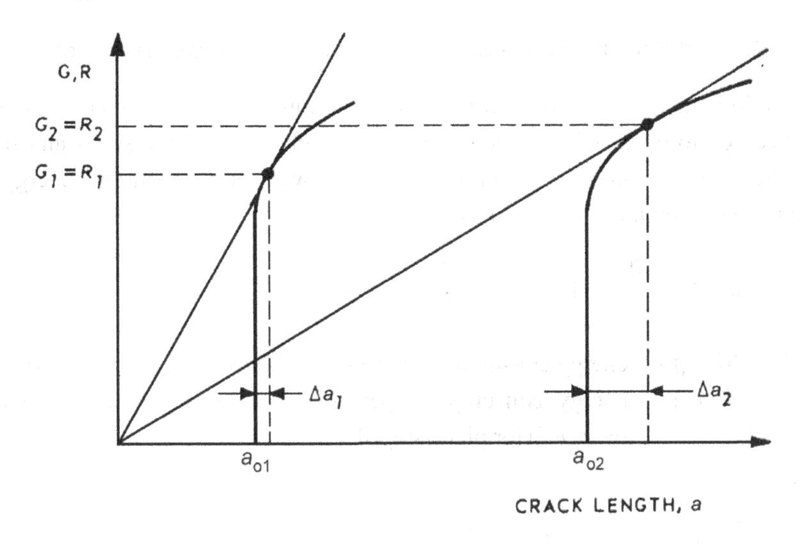

Figure 4.12. Invariant R-curves and the points of instability.

Figure 4.12 shows, however, that a shift of the complete R-curve along the crack length axis changes the point at which instability occurs, *i.e.* the tangency at which $G = R$ and $\partial G/\partial a = \partial R/\partial a$. Hence the point of instability will depend on the initial crack length a_0. In summary, an invariant R-curve has the following consequences:

1) Crack initiation is independent of initial crack length a_0.
2) Instability depends on a_0. A longer a_0 results in more stable crack growth and a higher G value at instability. In comparison with the plane strain situation it may thus be stated that instability definitely depends on total crack length a ($= a_0 + \Delta a$).

4.7 A Possible Explanation for the Rising *R*-Curve

A working hypothesis of the rising R-curve has been given by Krafft, Sullivan and Boyle (reference 5 of the bibliography). This hypothesis models the R-curve behaviour under intermediate plane stress – plane strain conditions.

Krafft *et al.* assume that in plane strain the plastic deformation energy necessary for crack extension is related to the area of newly created crack surface, but in plane stress the plastic energy is related to the volume contained by plane stress (45°) crack surfaces

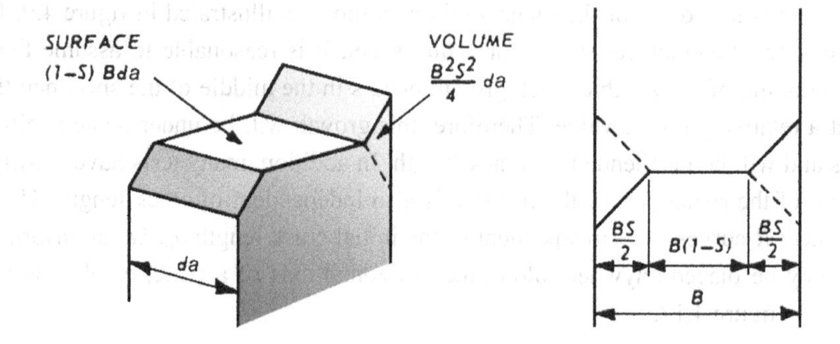

Figure 4.13. Model of Krafft *et al.* for explaining the development of crack resistance.

and their mirror images, as shown in figure 4.13. In this figure a specimen of thickness B is considered to crack with a fraction S of the thickness in plane stress and the remaining fraction $1 - S$ in plane strain. For a crack growth increment da the energy consumption depends on the crack resistance R according to

$$B\mathrm{d}a\,R = \frac{\mathrm{d}W_\mathrm{S}}{\mathrm{d}A}B(1-S)\mathrm{d}a + \frac{\mathrm{d}W_\mathrm{p}}{\mathrm{d}V}\frac{B^2S^2}{2}\mathrm{d}a\,, \tag{4.31}$$

where $\mathrm{d}W_\mathrm{S}/\mathrm{d}A$ = energy consumption per *unit crack area* (for plane strain)
$\mathrm{d}W_\mathrm{p}/\mathrm{d}V$ = energy consumption per *unit volume of plastically deformed material* (for plane stress)

Therefore

$$R = \frac{\mathrm{d}W_\mathrm{S}}{\mathrm{d}A}(1-S) + \frac{\mathrm{d}W_\mathrm{p}}{\mathrm{d}V}\frac{BS^2}{2}\,. \tag{4.32}$$

Experimentally it has been found that $B\cdot\mathrm{d}W_\mathrm{p}/\mathrm{d}V >> \mathrm{d}W_\mathrm{S}/\mathrm{d}A$, so that from equation (4.32) it is evident that as soon as shear lips (slant fracture) start to form the value of R will show a sharp increase, *i.e.* R becomes approximately proportional to S^2. In section 3.6 it was shown that monotonically loading to fracture generally results in a change from flat to slant fracture. This being so, the generality of the rising R-curve is confirmed.

An Explanation for the Effect of Specimen Thickness on Fracture Toughness

The effect of specimen thickness on fracture toughness was mentioned in section 3.6 and illustrated by figure 3.13. The model of Krafft *et al.* provides an explanation of this effect (for all but the thinnest specimens) in terms of the influence of specimen thickness on G_c, which is related to the critical stress intensity for fracture via equation (4.16), *i.e.* $K_\mathrm{c} = \sqrt{E'G_\mathrm{c}}$.

Knott (reference 6 of the bibliography) made the assumption that the absolute thickness of shear lips in a material is approximately constant, *i.e.* the fraction S in equation (4.32) decreases with increasing specimen thickness. On this basis he analysed data of Krafft *et al.* for the aluminium alloy 7075–T6, assuming a constant shear lip thickness of 2 mm. For a specimen 2 mm thick, *i.e.* $S = 1$, G_c had been found to be 200 kJ/m^2. In terms of equation (4.32) this means that

$$R = G_c = \frac{dW_p}{dV}\frac{B}{2} = 200\ kJ/m^2 .$$

On the other hand a value of 20 kJ/m^2 was estimated for dW_S/dA. Thus for this specific alloy equation (4.32) could be reduced to

$$G_c = 20(1 - S) + 200S^2 \quad kJ/m^2.$$

Figure 4.14 shows the result of using this equation as compared to experimental data for the 7075-T6 alloy. The agreement is excellent and is strong evidence for the validity of the model of Krafft *et al.*

Note that a material giving shear lips was used both in the theory and example. However, many materials show no shear lips, but do have a rising R curve. For these materials the explanation can be the same as mentioned above. When the crack grows, the plane stress plastic zone grows while the plane strain zone decreases (compare with figure 3.9). Thus the crack resistance will increase during crack growth since $B \cdot dW_p/dV >> dW_S/dA$, *cf.* equation (4.32).

4.8 Crack Resistance: a Complete Description

In sections 4.5 – 4.7 the concept of crack resistance, R, for more ductile materials has been developed. It was indicated that under plane strain conditions fracture takes place at a constant value of R and $G_c = G_{Ic}$, irrespective of the crack length. However, for plane stress and intermediate plane stress – plane strain conditions a rising R-curve de-

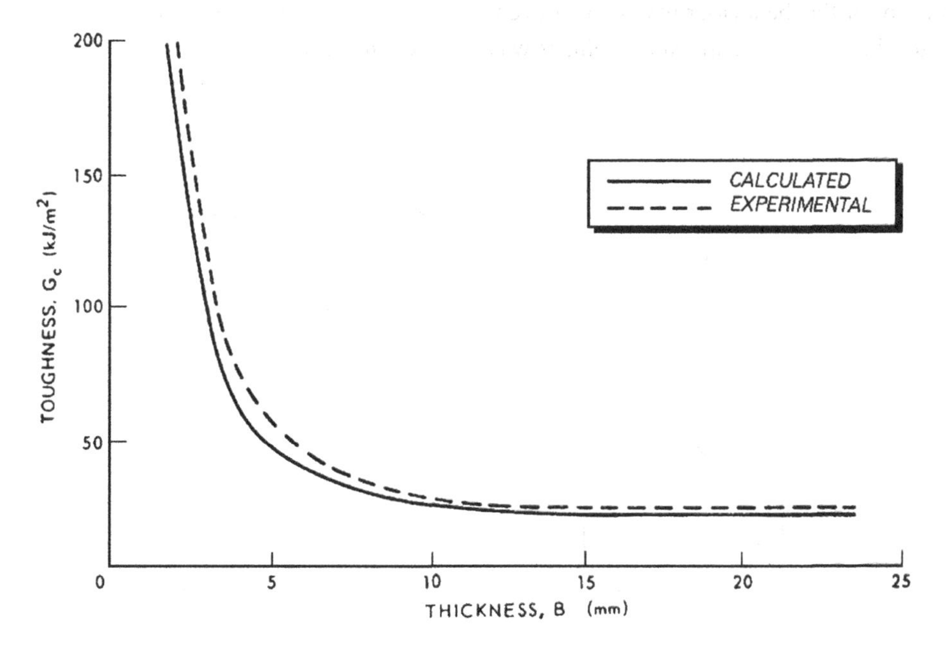

Figure 4.14. Knott's analysis of the model of Krafft *et al.* for the aluminium alloy 7075–T6.

velops and G_c is no longer constant but depends on the amount of crack extension. Finally it was shown that the *R*-curve behaviour can be fairly well explained by the model of Krafft *et al.*

At this point it is desirable to give a complete description of the *R*-curve concept, but first it is necessary to discuss a change in notation. In the literature and in practice the *R*-curve is no longer considered in terms of *G* and *R*. Instead the stress intensity factors K_G and K_R are used. This is because the stress intensity factor concept has found widespread application and the energy balance parameters *G* and *R* may be simply converted to stress intensities via the relation $K_I = \sqrt{E'G}$.

The reason for not adopting this notational conversion in previous sections of this chapter is that the *R*-curve concept is based on energy balance principles and can be explained best in those terms. However, now it is convenient to describe the crack resistance behaviour in terms compatible with current practice.

The R-Curve in Terms of Stress Intensity Factors

A schematic *R*-curve in terms of K_G and K_R is presented in figure 4.15.
In this diagram there are three important points:

1) K_i is the point of initial crack extension.
2) K_c is the critical stress intensity (instability point).
3) K_{plat} is the plateau level of the K_R-curve. (Note that in figure 4.15 the plateau is beyond instability.)

K_i has been found by means of complicated experimental techniques to be independent of specimen thickness and to have a constant value for a particular material. The reason for this behaviour may be that even in relatively thin specimens the initial crack extension takes place in plane strain, as was assumed in section 4.6.

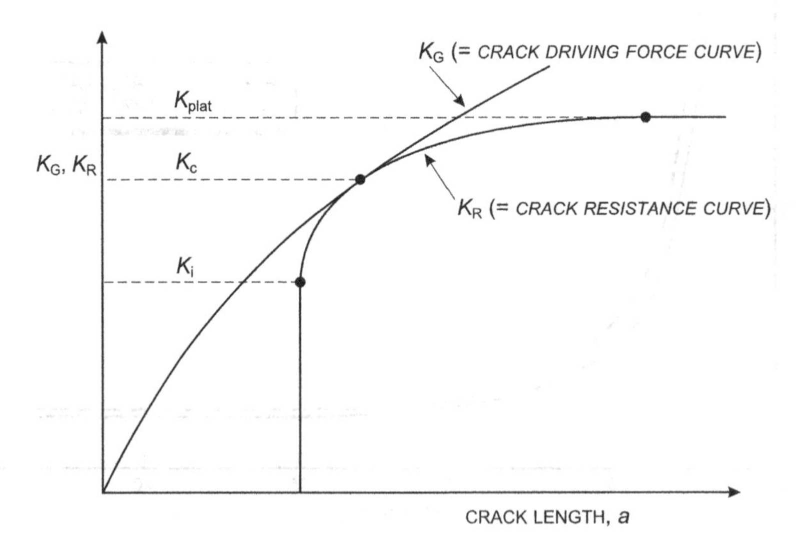

Figure 4.15. The *R*-curve in terms of stress intensity factor notation.

For K_c, however, there is a strong effect of specimen thickness: thinner specimens give higher K_c values and consequently exhibit more slow stable crack growth, since K_i remains constant. A possible explanation for this behaviour was given in section 4.7. It should be noted that a sufficiently thick specimen will result in full plane strain and K_c will then be equal to K_i.

K_{plat} also depends strongly on specimen thickness. This parameter is not a generally accepted feature of the K_R-curve. A number of authors consider the existence of K_{plat} to be due to specimen finite geometry effects and that the K_R-curve for very wide panels would attain a constant non-zero slope rather than a plateau level.

Figure 4.16 shows an example of experimentally determined values of K_i, K_c and K_{plat} as functions of specimen thickness. The data for K_{plat} (beyond instability) were obtained by a special testing technique to be described in the next chapter, section 5.4.

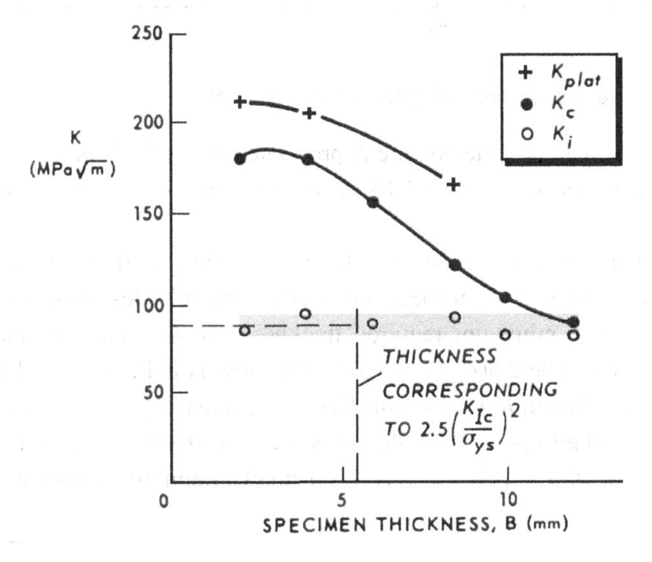

Figure 4.16. Experimental data for K_i, K_c and K_{plat} from reference 1 of the bibliography to this chapter.

Effect of Specimen Thickness on K_R-Curve Shape

From the observed dependence of K_i, K_c and K_{plat} on specimen thickness it is possible to indicate the general shapes of K_R-curves as functions of specimen thickness, figure 4.17. This figure demonstrates that a family of K_R-curves can be presented and that the curve for plane strain need not be considered as a separate case: it is simply a curve which does not show stable crack growth. In this respect the plane strain K_R-curve is analogous to the original (G,R) representation of instability, figure 4.9, since it is to be remembered that R-curves (and hence K_R-curves) are independent of the initial crack length, a_0. This analogy is schematically depicted in figure 4.18.

Crack Resistance and K_{Ic} Testing

Although the procedure for K_{Ic} testing will be fully discussed in chapter 5, some re-

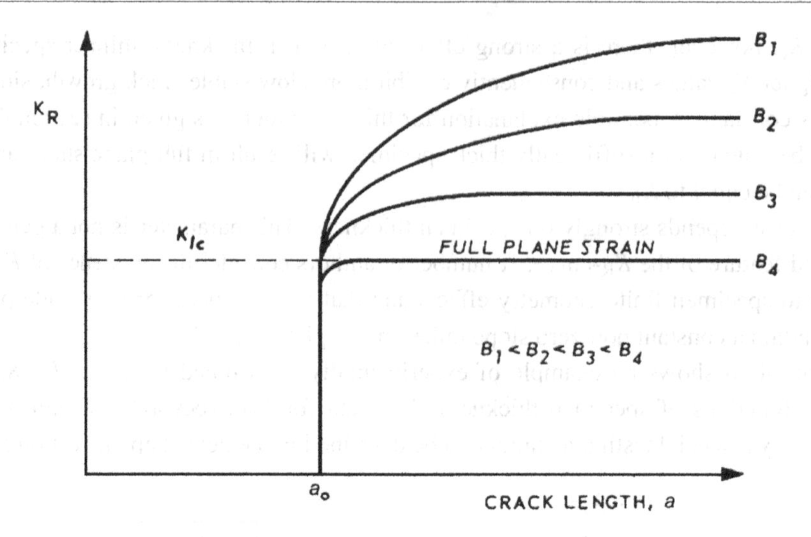

Figure 4.17. K_R-curves as functions of specimen thickness, B.

marks about the standard test method are appropriate here. This is because crack resistance test results, like those in figure 4.16, shed some light on problems in determining K_{Ic}.

It has been remarked several times that there is a thickness effect on fracture toughness, and that only above a certain specimen thickness more or less constant values (*i.e.* K_{Ic}) are obtained. The minimum required thickness for K_{Ic} determination has been found experimentally to be $2.5(K_{Ic}/\sigma_{ys})^2$. This thickness is indicated as a line in figure 4.16. It is seen that although K_i is constant, K_c is still decreasing out to greater thickness.

Note that the standard test method, described in some detail in the next chapter, defines K_{Ic} as K_I either after a 'pop-in', which is a small amount of audible unstable crack

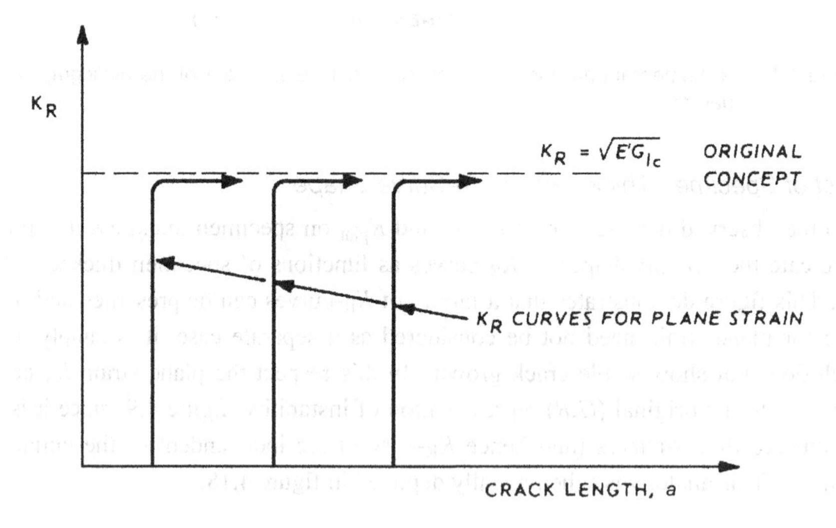

Figure 4.18. Comparison of the original and current representation of plane strain crack resistance.

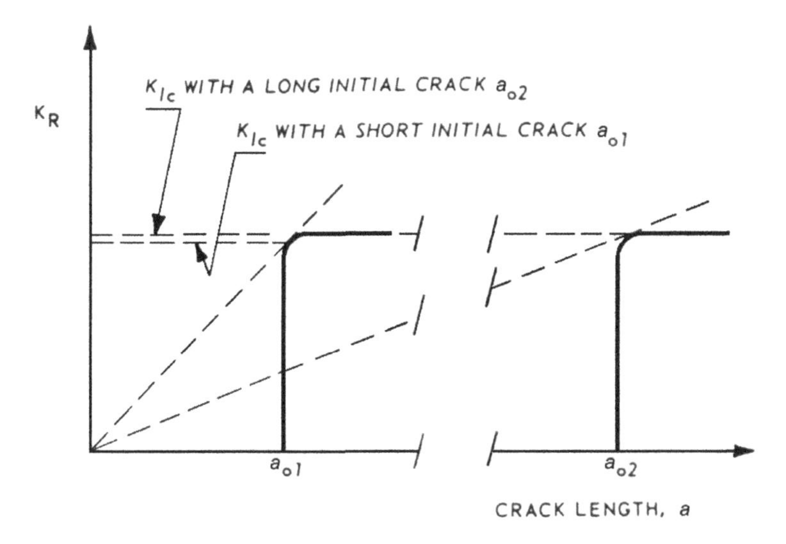

Figure 4.19. Effect on initial crack length a_o on K_{Ic}.

growth followed by crack arrest, or after 2% crack extension. From figure 4.16 it is clear that K_i does not depend on specimen thickness, while K_c does: a 2% stable crack extension can cause a significant increase in stress intensity. Now the question arises whether this definition of K_{Ic} might lead to inconsistent results, especially for thicknesses near the minimum required thickness.

Furthermore, assuming that the representation of K_R-curves as in figure 4.17 is correct, then stable crack growth should have a small but consistent effect on the value of K_{Ic} when specimens of widely differing initial crack length are used. This effect is demonstrated schematically in figure 4.19 and has indeed been observed.

It might be thought that the foregoing problems could readily be avoided by using K_i instead of K_{Ic}, but as mentioned earlier K_i has to be found by means of complicated experimental techniques. This makes its determination unsuitable for a standard test.

4.9 Bibliography

1. Schwalbe, K.-H., *Mechanik und Mechanismus des Stabilen Risswachstums in Metallischen Werkstoffen*, Fortschrittsberichten der VDI Zeitschriften Reihe 18, No. 3 (1977).
2. McCabe, D.E. (Editor), *Fracture Toughness Evaluation by R-Curve Methods*, ASTM STP 527, American Society for Testing and Materials (1973): Philadelphia.
3. Irwin, G.R., *Fracture,* Encyclopaedia of Physics, ed. S. Flüge, Springer Verlag, pp. 551-589 (1958): Berlin.
4. Mostovoy, S., Crosley, P.B. and Ripling, E.J., *Use of Crack-Line Loaded Specimens for Measuring Plane-Strain Fracture Toughness*, Journal of Materials, Vol. 2, pp. 661–681 (1967).
5. Krafft, J.M., Sullivan, A.M. and Boyle, R.W., *Effect of Dimensions on Fast Fracture Instability of Notched Sheets,* Proceedings of the Crack Propagation Symposium Cranfield, The College of Aeronautics, Vol. 1, pp. 8–28 (1962): Cranfield, England.
6. Knott, J.F., Fundamentals of Fracture Mechanics, Butterworths (1973): London.

Figure 4.19. Effect of initial crack length a_0 on K_{Ic}.

growth followed by crack arrest or after 2% crack extension. From figure 4.18 it is clear that K_R does not depend on specimen thickness, while K_c does: a 2% stable crack extension can cause a significant increase in stress intensity. Now the question arises whether this definition of K_{Ic} might lead to inconsistent results, especially for thicknesses near the minimum required thickness.

Furthermore, assuming that the representation of K_R curves as in figure 4.17 is correct, then stable crack growth should have a small but consistent effect on the value of K_{Ic} when specimens of widely differing initial crack length are used. This effect is demonstrated schematically in figure 4.19 and has indeed been observed.

It might be thought that the foregoing problems could readily be avoided by using K_R instead of K_{Ic}. But K_R is not a material constant: it has to be found as part of a complicated experimental procedure. This makes it less suitable for a standard test.

4.9 Bibliography

1. Schwalbe, K.-H., Mechanik und Mechanismen des stabilen Risswachstums in metallischen Werkstoffen, Fortschrittsberichte der VDI Zeitschriften Reihe 18, Nr. 1 (1977).
2. McCabe, D.E. (editor), Fracture Toughness Evaluation by R-Curve Methods, ASTM STP 527, American Society for Testing and Materials (1973): Philadelphia.
3. Irwin, G.R., Fracture, Encyclopaedia of Physics, ed. S. Flügge, Springer Verlag, pp. 551–590 (1958), Berlin.
4. Mostovoy, S., Crosley, P.B. and Ripling, E.J., Use of Crack-Line-Loaded Specimens for Measuring Plane-Strain Fracture Toughness, Journal of Materials, Vol. 2, pp. 661–681 (1967).
5. Krafft, J.M., Sullivan, A.M. and Boyle, R.W., Effect of Dimensions on Fast Fracture Instability of Notched Sheets, Proceedings of the Crack Propagation Symposium Cranfield, The College of Aeronautics, Vol. 1, pp. 8–28 (1962): Cranfield, England.
6. Knott, J.F., Fundamentals of Fracture Mechanics, Butterworths (1973): London.

5
LEFM Testing

5.1 Introduction

In chapter 2 a number of analytical relationships were given for the determination of stress intensity factors in elastic specimens with different crack geometries. These stress intensity factors (K_I, K_{II} or K_{III} for the opening, edge-sliding or tearing modes of crack extension, figure 2.1) are functions of load, crack size and specimen geometry.

Ideally a critical stress intensity factor, K_c, can be used to predict the fracture behaviour in an actual structure. However, K_c depends on test temperature, specimen thickness and constraint. A typical dependence is shown in figure 5.1. The form of this dependence has been discussed in section 3.6 and reasonably well explained in section 4.7.

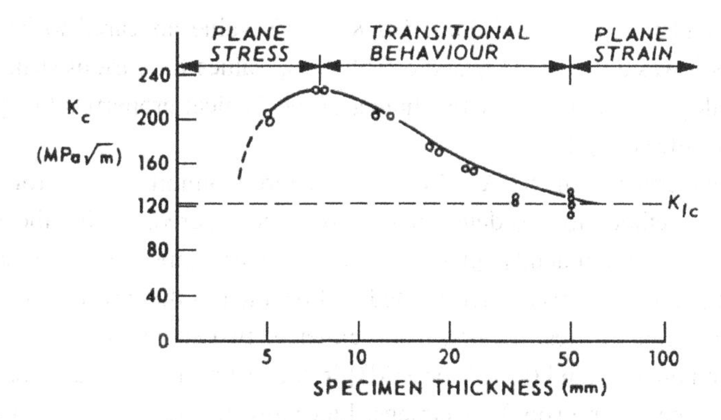

Figure 5.1. Effect of thickness on K_c behaviour for a high strength maraging steel.

Beyond a certain thickness, when the material is predominantly in plane strain and under maximum constraint, the value of K_c tends to a constant lower limit, K_{Ic}, the plane strain fracture toughness. K_{Ic} may be considered a material property, but does depend on the test temperature and loading rate. After considerable study and experimental verification the American Society for Testing and Materials (ASTM) published a standard method for K_{Ic} testing (last revision is ASTM E 399-90). This method will be described in section 5.2.

Of more general interest is the establishment of a test methodology for K_c determination, since the operating temperatures, loading rates and thicknesses of most materials used in actual structures are generally such that transitional plane strain-to-plane stress or fully plane stress conditions exist in service. Several engineering approaches to the problem of plane stress and transitional behaviour have been proposed. Only one, the Feddersen method, has the versatility required for structural design. This approach is the subject of section 5.3.

Considerable effort bas been devoted towards *R*-curve testing and analysis. A "Tentative Recommended Practice" for *R*-curve determination was issued by the ASTM in 1976 and this was followed by a standard in 1981 (last revision is ASTM E 561-94) . The determination of *R*-curves is discussed in section 5.4.

In section 5.5 a simple engineering approximation (Anderson's model) to account for the effects of yield strength and specimen thickness on fracture toughness will be given. Finally, in section 5.6 the practical use of K_c, K_{Ic} and *R*-curve data is summarised.

5.2 Plane Strain Fracture Toughness (K_{Ic}) Testing

During the period in which fracture toughness testing developed (late 1950s and the 1960s) the most suitable analyses for characterizing the resistance to unstable crack growth were those of LEFM. Although it was recognised that most structural materials do not behave in a purely elastic manner on fracturing, it was hoped that provided crack tip plasticity was very limited small specimens could be used to describe the situation of unstable crack growth occurring in a large structure.

Data like those in figure 5.1 showed that a fairly constant minimum value of K_c was obtained under plane strain conditions, *i.e.* K_{Ic}. This value appeared to be a material property. Thus it was decided to try and establish K_{Ic} values for various structural materials, in an analogous way to the establishment of mechanical properties like yield stress and ultimate tensile strength.

Under the supervision of the ASTM E-24 Fracture Committee numerous specimen designs and test methods for K_{Ic} determination were considered. During the 1960s various parameters, *e.g.* notch acuity, plate thickness, fracture appearance and stress levels during fatigue precracking were investigated and resulted in the development of a standardized, plane strain K_{Ic} test method using either of two standard specimens, namely the single edge notched bend (SENB or SE(B) in the last revision) and compact tension (CT or C(T) in the last revision) specimens. Later also the Arc-shaped Tension, Disc-shaped Compact and the Arc-shaped Bend specimens were introduced. The method was first published in 1970 and is listed in its latest (1990, ASTM E 399-90 under jurisdiction of committee E-8 on Fatigue and Fracture) version as reference 1 of the bibliography to this chapter.

The Standard K_{Ic} Specimens

The original recommended standard K_{Ic} specimens are illustrated in figures 5.2 and 5.3. For the other specimens that are allowed the reader should consult reference 1. Round robin test programmes have shown that the standard specimens enable K_{Ic} values to be reproducible to within about 15% by different laboratories.

The stress intensity factors for these standard specimens are as follows:

- Single edge notched bend specimen (SENB)

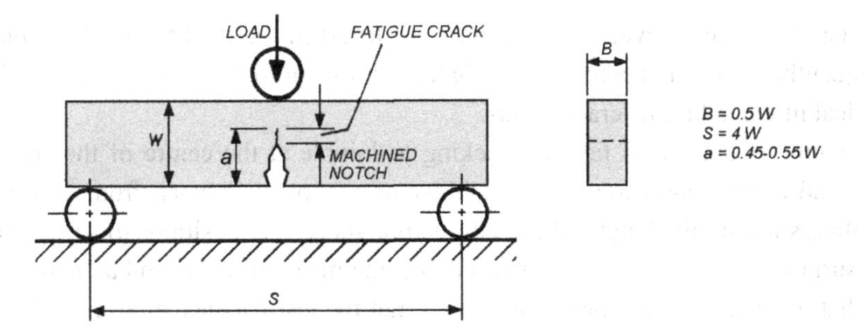

Figure 5.2. ASTM standard notched bend specimen (SENB).

$$K_{\mathrm{I}} = \frac{\mathrm{LOAD} \cdot S}{BW^{\frac{3}{2}}} \cdot f\left(\frac{a}{W}\right), \tag{5.1}$$

where

$$f\left(\frac{a}{W}\right) = \frac{3\left(\frac{a}{W}\right)^{\frac{1}{2}}\left[1.99 - \frac{a}{W}\left(1 - \frac{a}{W}\right)\left\{2.15 - 3.93\left(\frac{a}{W}\right) + 2.7\left(\frac{a}{W}\right)^{2}\right\}\right]}{2\left(1 + 2\frac{a}{W}\right)\left(1 - \frac{a}{W}\right)^{\frac{3}{2}}};$$

- Compact tension specimen (CT)

$$K_{\mathrm{I}} = \frac{\mathrm{LOAD}}{BW^{\frac{1}{2}}} \cdot f\left(\frac{a}{W}\right), \tag{5.2}$$

where

$$f\left(\frac{a}{W}\right) = \frac{\left(2 + \frac{a}{W}\right)\left\{0.886 + 4.64\left(\frac{a}{W}\right) - 13.32\left(\frac{a}{W}\right)^{2} + 14.72\left(\frac{a}{W}\right)^{3} - 5.6\left(\frac{a}{W}\right)^{4}\right\}}{\left(1 - \frac{a}{W}\right)^{\frac{3}{2}}}.$$

Experiments have shown that it is impractical to obtain a reproducible sharp, narrow machined notch that will simulate a natural crack well enough. Therefore the specimens must be fatigue precracked. To ensure that cracking occurs correctly the specimens

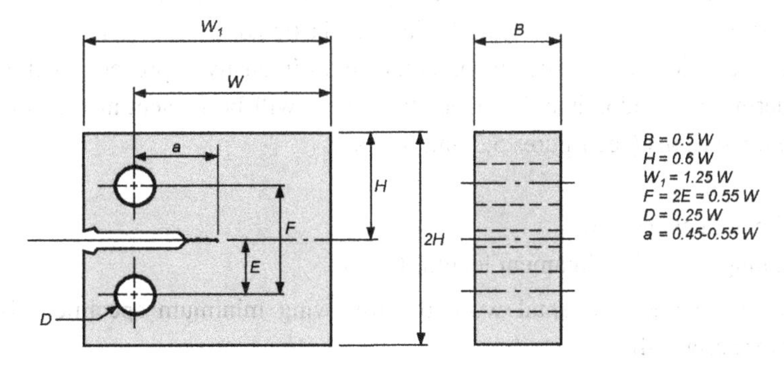

Figure 5.3. ASTM standard compact tension specimen (CT).

contain starter notches. Several possibilities are listed in the ASTM standard, but the most frequently used is a chevron notch, figure 5.4, owing to its good reproducibility of symmetrical in-plane fatigue crack fronts.

The chevron notch forces fatigue cracking to initiate at the centre of the specimen thickness and thereby increases the probability of a symmetric crack front. After fracture toughness testing the length of the fatigue precrack, a, at positions a_1, a_2, a_3 and at the side surfaces is measured. For the value of a the mean of a_1, a_2 and a_3 is used. Assuming that other test requirements have been met the test is considered a valid K_{Ic} result if the difference between any two of the three crack length measurements does not exceed 10% of the average and if the surface crack lengths a_s are within 10% of a.

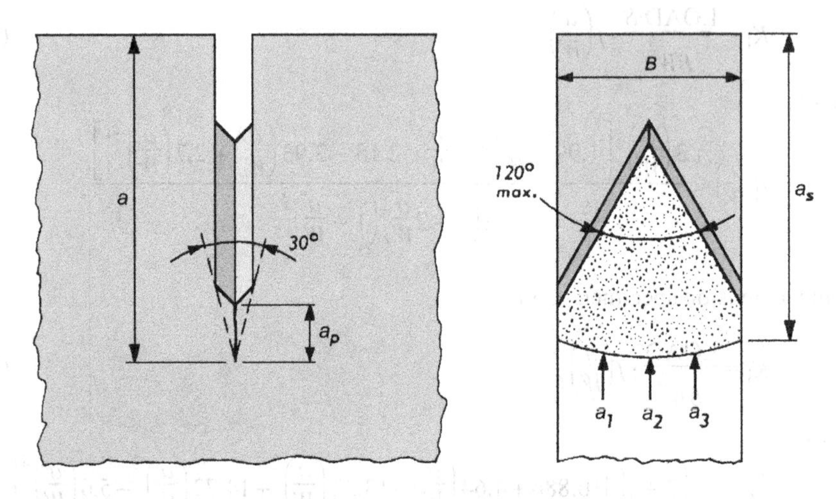

Figure 5.4. Chevron notch crack starter.

Specimen Size Requirement

The accuracy with which K_{Ic} describes the fracture behaviour depends on how well the stress intensity factor characterizes the conditions of stress and strain immediately ahead of the tip of the fatigue precrack, since it is here that unstable crack extension would originate. In establishing the specimen size requirements for K_{Ic} tests the specimen dimensions must be large enough compared with the plastic zone size, r_y. This ensures an overall elastic behaviour of the specimen, so the stress intensity approach is still appropriate. Furthermore a predominantly plane strain state will be present at the crack tip. The relevant dimensions (see figures 5.2 and 5.3) are:

1) The crack length, a.
2) The specimen thickness, B.
3) The remaining uncracked ligament length, $W - a$.

After considerable experimental work the following minimum specimen size requirements were established:

$$a \geq 2.5 \left(\frac{K_{Ic}}{\sigma_{ys}}\right)^2$$
$$B \geq 2.5 \left(\frac{K_{Ic}}{\sigma_{ys}}\right)^2, \qquad (5.3)$$

where σ_{ys} is the yield strength. Note that, since for the test specimens $0.45 < a/W < 0.55$, the size of the remaining ligament, $W - a$, will more or less satisfy the same minimum value as given in equations (5.3).

Now in chapter 3 (section 3.5) it was stated that there are empirical rules for estimating whether a specimen will be mainly in plane strain or plane stress: predominantly plane strain behaviour may be expected when the calculated size of the plane stress plastic zone, *i.e.* the diameter $2r_y$ in Irwin's analysis, is no larger than one-tenth of the specimen thickness. According to Irwin, for plane stress

$$r_y = \frac{1}{2\pi}\left(\frac{K_I}{\sigma_{ys}}\right)^2. \qquad (5.4)$$

For $K_I = K_{Ic}$ the substitution of equation (5.4) in equation (5.3) shows that the minimum thickness, B, is only about 8 times the plane stress plastic zone size, $2r_y$. The empirical rule is therefore slightly conservative (a factor 10 instead of 8).

It is important to note that the specification of a, B and W (and all the other specimen dimensions) requires that the K_{Ic} value to be obtained must already be known or at least estimated. There are three general ways of sizing test specimens before the required K_{Ic} is actually obtained:

1) Overestimate K_{Ic} on the basis of experience with similar materials and empirical correlation with other types of notch toughness test, for example the Charpy impact test.
2) Use specimens that have as large a thickness as possible.
3) For high strength materials the ratio of (σ_{ys}/E) can be used according to the following table, which was drawn up by the ASTM.

σ_{ys}/E	minimum values of a and B (mm)
0.0050 – 0.0057	75.0
0.0057 – 0.0062	63.0
0.0062 – 0.0065	50.0
0.0065 – 0.0068	44.0
0.0068 – 0.0071	38.0
0.0071 – 0.0075	32.0
0.0075 – 0.0080	25.0
0.0080 – 0.0085	20.0
0.0085 – 0.0100	12.5
≥ 0.0100	6.5

K_{Ic} Test Procedure

The steps involved in setting up and conducting a K_{Ic} test are:

1) Determine the critical dimensions of the specimen (a, B, W).
2) Select a specimen type (notch bend or compact tension) and prepare shop drawings, *e.g.* specifying a chevron notch crack starter, figure 5.4.
3) Specimen manufacture.
4) Fatigue precracking.
5) Obtain test fixtures and clip gauge for crack opening displacement measurement.
6) Testing.
7) Analysis of load-displacement records.
8) Calculation of conditional K_{Ic} (K_Q).
9) Final check for K_{Ic} validity.

At least three replicate tests should be done for every material. Steps 1 – 3 will not be discussed further. Steps 4 – 6 will now be concisely reviewed, since full details may be found in reference 1 of the bibliography. The last three steps 7 – 9 are considered under the next subheadings in this section.

The purpose of notching and fatigue precracking the test specimen is to simulate an ideal plane crack with essentially zero tip radius to agree with the assumptions made in stress intensity analyses. There are several requirements pertaining to fatigue loading. The most important is that the maximum stress intensity K_{max} during the final stage of fatigue cycling shall not exceed 60% of the subsequently determined K_Q if this is to qualify as a valid K_{Ic} result.

Recommended test fixtures for notch bend and compact tension specimen testing are described in the ASTM standard. These fixtures were developed to minimise friction and have been used successfully by many laboratories. Other fixtures may be used provided good alignment is maintained and frictional errors are minimised.

An essential part of a K_{Ic} test is accurate measurement of the crack mouth opening displacement (CMOD) as a function of applied load. The displacement is measured with a so-called clip gauge which is seated on integral or attachable knife edges on the specimen, as shown in figure 5.5. Electrical resistance strain gauges are bonded to the clip gauge arms and are connected up to form a Wheatstone Bridge circuit, as indicated.

In carrying out the test there are requirements for specimen alignment in the test fixtures, the loading rate, test records and post-test measurements on the specimen. The loading rate should be such that the rate of increase of stress intensity is within the range 0.55 – 2.75 $\mathrm{MPa}\sqrt{\mathrm{m}}/\mathrm{s}$. This is arbitrarily defined as 'static' loading. It should be noted that the ASTM standard does allow higher rates to be used. In this case the test time should be at least 1 millisecond and the test record should meet additional requirements concerning linearity.

Each test record consists of a plot of the output of a load-sensing transducer versus the clip gauge output. It is conventional to plot load along the y-axis, displacement along the x-axis. The record is continued until the specimen is no longer able to sustain a further increase in load. The maximum load must be determinable with an accuracy of

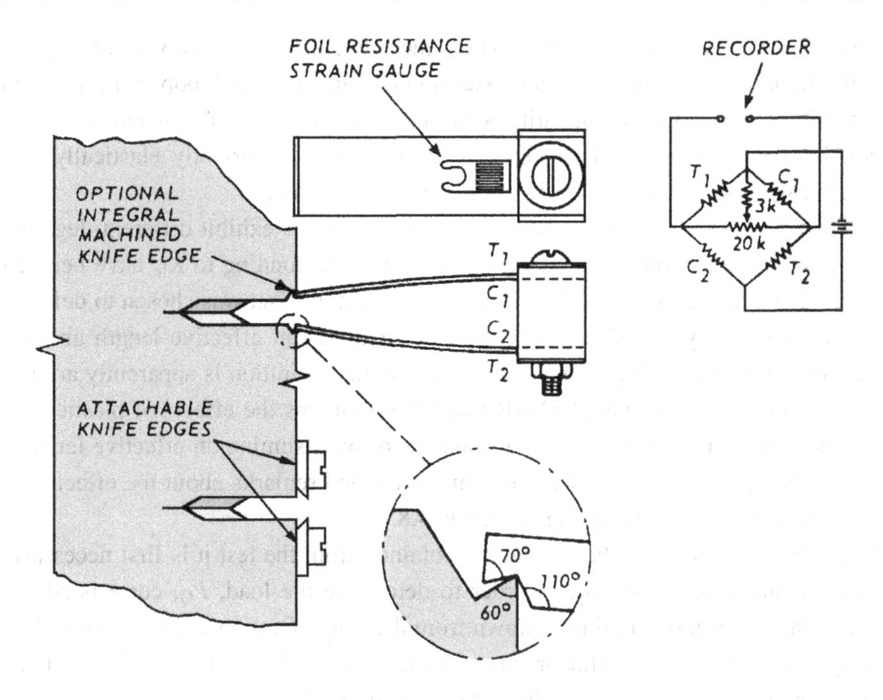

Figure 5.5. Clip gauge and its attachment to the specimen.

±1%. Post-test measurements of the specimen dimensions, B, W, S, and fatigue precrack lengths a_1, a_2, a_3, a_s (see figures 5.2 – 5.4) must be made to calculate K_Q and check its qualification as a valid K_{Ic}.

Analysis of Load-Displacement Records and Determination of Conditional K_{Ic} (K_Q)

Plots of load versus displacement may have different shapes. The principal types of diagram obtained are shown schematically in figure 5.6. Initially the displacement in-

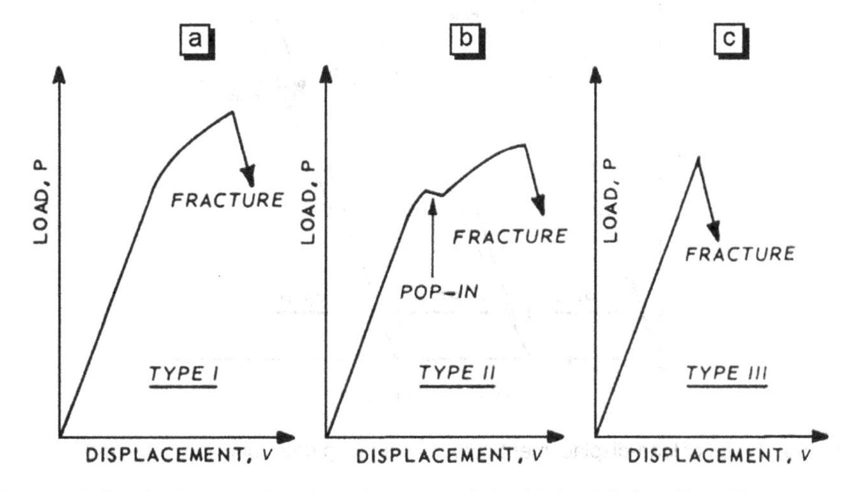

Figure 5.6. Principal types of load-displacement plots obtained during K_{Ic} testing.

creases linearly with load, P. In many cases there is either a gradually increasing nonlinearity, figure 5.6.a, or sudden crack extension and arrest (called 'pop-in') followed by nonlinearity, figure 5.6.b. Nonlinearity is caused by plastic deformation and stable crack growth before fast fracture. If a material behaves almost perfectly elastically (as is rarely the case) a diagram like that in figure 5.6.c is obtained.

As shown in figure 5.6, the general types of P-v curves exhibit different degrees of nonlinearity. Various criteria to establish the load corresponding to K_{Ic} have been considered. After considerable experimentation a 5% secant offset was chosen to define K_{Ic} as the stress intensity factor at which the crack reaches an effective length about 2% greater than at the beginning of the test. Although this definition is apparently arbitrary, it turns out that for the standard SENB and CT specimens the effects of plasticity and stable crack growth are more or less accounted for by assuming an effective length increase of 2%. (Note, however, that some more detailed remarks about the effect of stable crack growth on K_{Ic} were made in section 4.8.)

To establish whether a valid K_{Ic} can be obtained from the test it is first necessary to calculate a candidate value, K_Q. In order to determine the load, P_Q, corresponding to this candidate value a secant line is drawn from the origin O, with a slope 0.95 of that of the tangent OA to the (initial) linear part of the test record. The load P_S is the load at the intersection of the secant line with the test record, figure 5.7.

P_Q is then defined according to the following procedure. If the load at every point on the P-displacement record which precedes P_S is lower than P_S, then P_Q is P_S (Type I, figure 5.7). However, if there is a maximum load preceding P_S that is larger than P_S, then this maximum load is P_Q (Types II and III, figure 5.7). In order to prevent acceptance of a test record for a specimen in which excessive stable crack growth occurred or in which the stress state is not sufficiently plane strain, it is required at this stage of the analysis that P_{max}/P_Q be less than 1.10. The value of this ratio is based on experience.

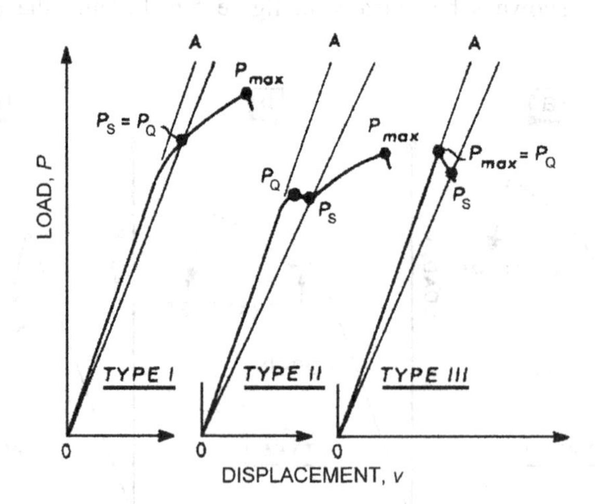

Figure 5.7. Types of load-displacement curves illustrating determination of P_S and P_Q.

Check for K_{Ic} validity

After determining P_Q the value of K_Q is calculated using the appropriate stress intensity factor expression, *i.e.* equations (5.1) or (5.2) for the SENB or CT specimens respectively. Then it is determined whether this K_Q is consistent with the specimen size and material yield strength according to equations (5.3), *i.e.* the quantity $2.5\,(K_Q/\sigma_{ys})^2$ must be less than the thickness, B, and the crack length, a, of the specimen. Finally, a check is made whether the crack front symmetry requirements mentioned in the discussion to figure 5.4 are met.

If these requirements are not all met the test must be declared invalid and the result may be used only to *estimate* the fracture toughness: it is not an ASTM standard value.

5.3 Plane Stress Fracture Toughness (K_c) Testing: the Feddersen Approach

There is no standard method of plane stress fracture toughness (K_c) testing. In what follows, the engineering approach of Feddersen, which is a good method suited to practical use besides R-curve testing, will be described. The original description is given in reference 2 of the bibliography to this chapter.

Consider a thin plate under plane stress with a central crack $2a_0$ loaded in tension, figure 5.8. On reaching a stress σ_i the crack will begin to extend by slow stable crack growth. In order to maintain crack growth the stress has to be increased further: the crack will stop growing if the load is kept constant.

Slow crack growth continues until a critical crack size $2a_c$ is reached at a stress σ_c.

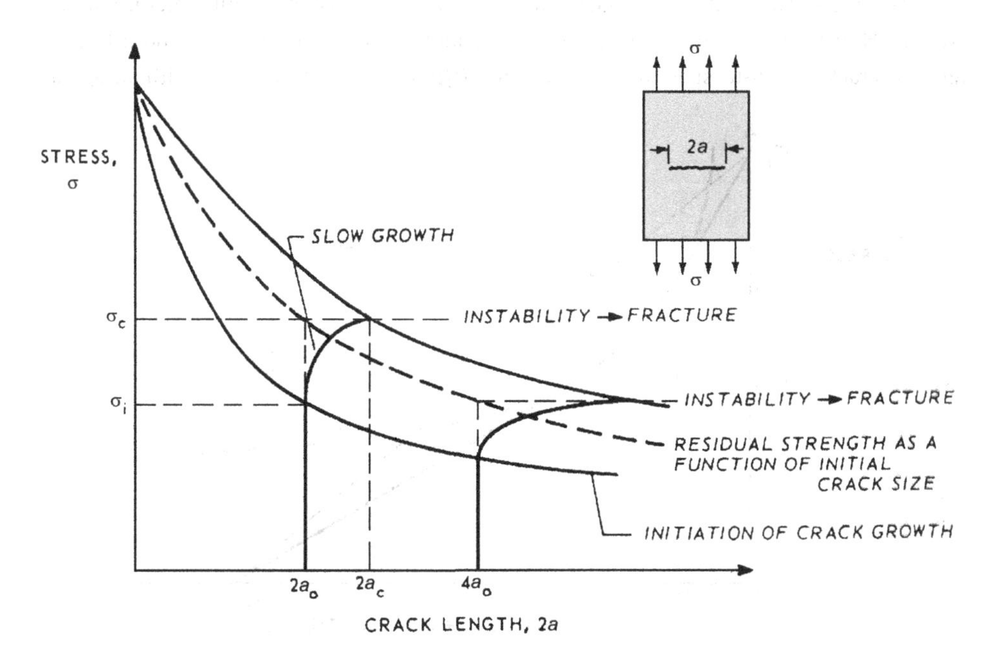

Figure 5.8. Residual strength behaviour in plane stress.

Then the crack becomes unstable and fracture occurs. If the initial crack is longer, for example $4a_o$, stable crack growth starts at a lower stress, the amount of slow crack growth is larger, but σ_c is lower.

If it is assumed that all events in crack propagation and fracture are governed by the stress intensity factor, particular stress intensity factors can be attributed to each event as follows:

$$K_i = \sigma_i \sqrt{\pi a_o} f\left(\frac{a}{W}\right)$$

$$K_c = \sigma_c \sqrt{\pi a_c} f\left(\frac{a}{W}\right) \qquad (5.5)$$

$$K_e = \sigma_c \sqrt{\pi a_o} f\left(\frac{a}{W}\right),$$

where K_i is the critical stress intensity for the onset of (stable) crack growth and K_c is the critical stress intensity for fracture. K_e is an apparent stress intensity, since it relates the initial crack size to the fracture stress: it has an engineering significance because irrespective of whether slow stable crack growth subsequently occurs the value of K_e defines the residual strength of a plate containing a crack of a given initial size.

Instead of using actual crack sizes, a, in equations (5.5) the effective crack sizes $a + r_y$ could be used, as in Irwin's analysis in section 3.2. However, this is not necessary in the Feddersen approach, since in calculating stress intensity factors to obtain the residual strength diagram and the reverse operation of using the diagram to calculate the fracture stress or initial crack size the contribution of r_y cancels out.

Tests have shown that K_i, K_c and K_e are not material constants with general validity like K_{Ic}. However, they are approximately constant for a given thickness and a limited range of crack length-to-specimen width ratio, a/W. For a given material with an appar-

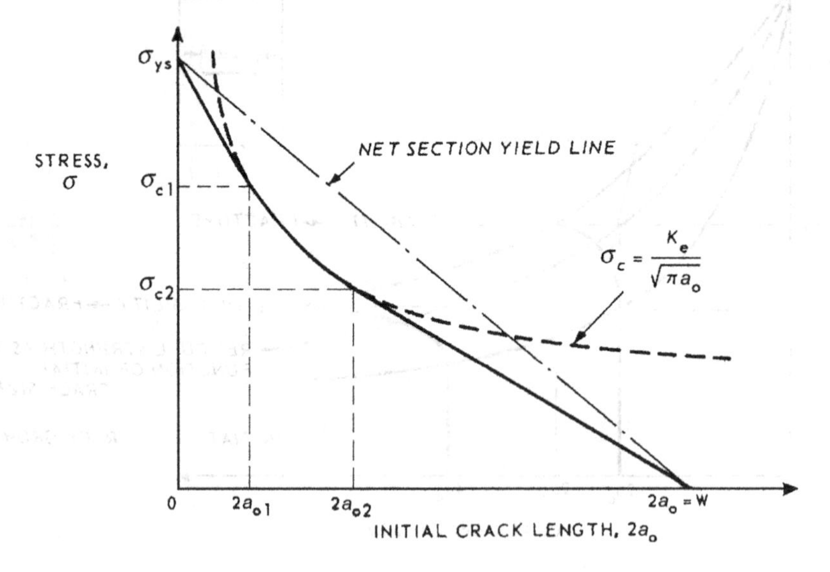

Figure 5.9. The residual strength diagram.

ent toughness, K_e, the relation between the residual strength and initial crack size of a (large) centre cracked panel is given ideally by the curve $\sigma_c = K_e/\sqrt{\pi a_o}$, as shown in figure 5.9.

However, for small crack sizes σ_c tends to infinity, but the residual strength at $a_o = 0$ cannot be larger than the yield stress. On the other hand, as $2a_o$ tends towards W, the residual strength approaches zero. At these two extremes net section yield occurs. Feddersen proposed the construction of two tangents to the curve (this was also based on experimental results), one from the stress axis at σ_{ys} and the other from the crack length axis at W. In the region between the points of tangency K_e is approximately constant: this part of the curve plus the two tangents constitute the residual strength diagram.

Now a tangent to the K_e curve at any point is

$$\frac{d\sigma}{d(2a_o)} = \frac{d}{d(2a_o)}\left(\frac{K_e}{\sqrt{\pi a_o}}\right) = -\frac{\sigma}{4a_o}. \tag{5.6}$$

For the tangent from $(0,\sigma_{ys})$ equation (5.6) gives, see figure 5.9,

$$-\frac{\sigma_{c1}}{4a_{o1}} = -\frac{\sigma_{ys} - \sigma_{c1}}{2a_{o1}}, \text{ or } \sigma_{c1} = \frac{2\sigma_{ys}}{3} \tag{5.7}$$

and so the left-hand tangency point is always at $2\sigma_{ys}/3$.

Also from figure 5.9, the tangent from $(W,0)$ is defined by

$$-\frac{\sigma_{c2}}{4a_{o2}} = -\frac{\sigma_{c2}}{W - 2a_{o2}}, \text{ or } 2a_{o2} = \frac{W}{3}. \tag{5.8}$$

Thus the right-hand tangency point is always at $W/3$.

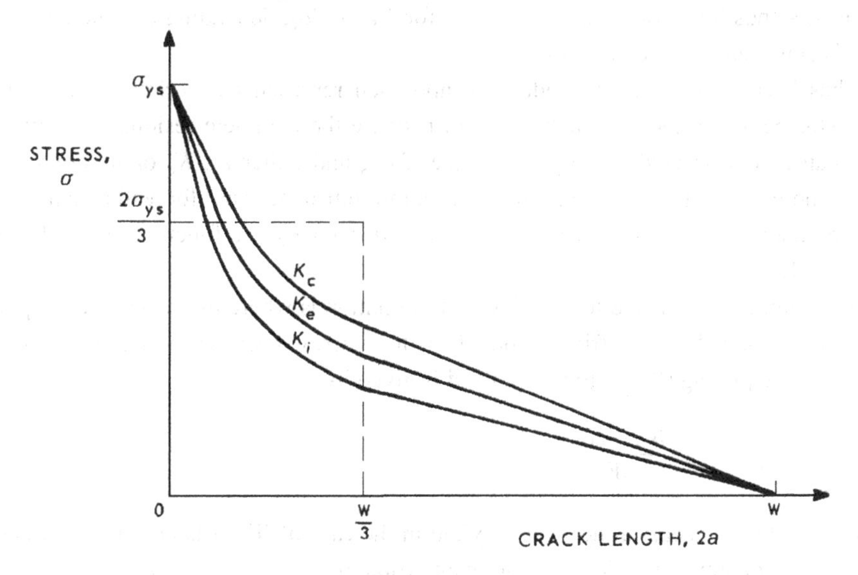

Figure 5.10. The Feddersen approach for K_i, K_e and K_c.

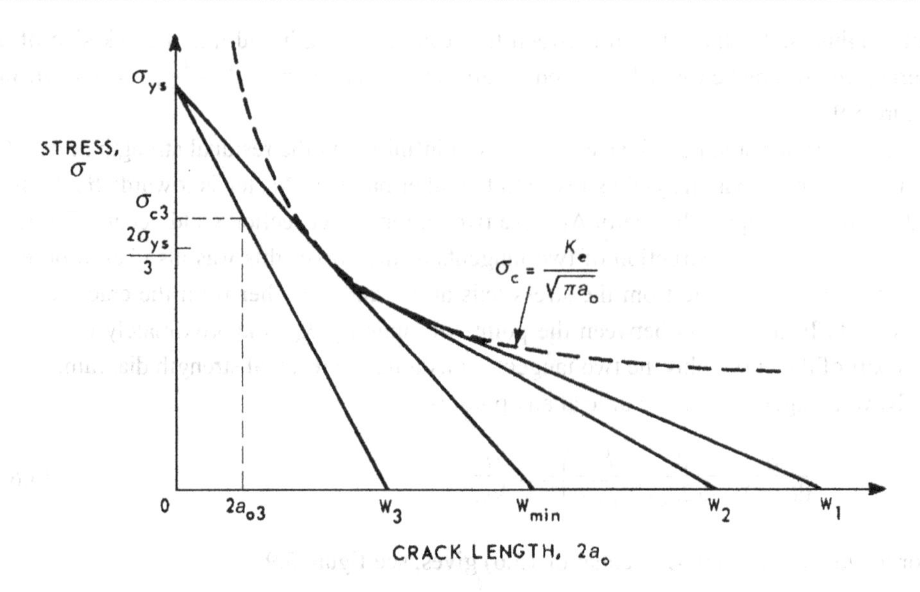

Figure 5.11. Residual strength for various panel widths.

The same construction can be made for K_i and K_c, see figure 5.10. Summarising, the screening criteria for valid plane stress fracture toughness testing are

$$\sigma_x < \frac{2\sigma_{ys}}{3}$$

and (5.9)

$$2a_x < \frac{W}{3},$$

where σ_x stands for σ_i or σ_c and $2a_x$ stands for $2a_o$ or $2a_c$, depending on whether K_i, K_c or K_e is considered, *cf.* equations (5.5).

It has been found that the Feddersen approach represents experimental data quite well. This being so, the approach is useful because the complete residual strength diagram can be constructed for any panel size if σ_{ys} and either K_i, K_e or K_c are known. Furthermore, one can easily determine the minimum panel size for valid plane stress fracture toughness testing. Consider the construction for K_e as a function of panel width, figure 5.11.

The minimum panel size for valid K_e determination is where the two tangency points coincide. W_1 and W_2 are sufficient, but W_3 is too narrow because the failure stress lies along the line joining $(0,\sigma_{ys})$ to $(W_3,0)$ and is given by

$$\sigma_{c3} = \sigma_{ys}\left(\frac{W_3 - 2a_{o3}}{W_3}\right), \tag{5.10}$$

i.e. failure always occurs at net section yield in the case of W_3: it is no longer a fracture mechanics problem, but a yielding dominant problem.

From $K_e = \sigma\sqrt{\pi a_o}$ the left-hand point of tangency is given by

$$K_e = \frac{2\sigma_{ys}}{3}\sqrt{\pi a_o}\,, \quad \text{or} \quad 2a_o = \frac{9}{2\pi}\left(\frac{K_e}{\sigma_{ys}}\right)^2 . \tag{5.11}$$

The two tangency points coincide when $W_{min}/3 = 2a_o$. Thus from equation (5.11)

$$W_{min} = \frac{27}{2\pi}\left(\frac{K_e}{\sigma_{ys}}\right)^2 . \tag{5.12}$$

Similarly, the value of W_{min} can be found for K_i or for K_c.

Note that equation (5.12) cannot be used as a screening criterion. This is because violation of one of the criteria of equation (5.9) would lead to a lower apparent K value (on a tangent rather than the residual strength curve) and substitution of this K value in equation (5.11) would give too small a value of W_{min}, *i.e.* an invalid test could be declared valid.

Some Experimental Considerations for K_c Testing

Actual test conditions for K_c determination are not rigorously defined as in standard tests. However, there are a number of guidelines and remarks:

1) It is obvious from what has previously been discussed that K_c tests should be conducted for material thicknesses representative of actual applications.
2) The tests should preferably be done with centre cracked panels, for which the stress intensity factors are well defined and the problems of secondary bending and buckling are minimised (but see point 5 below).
3) Fatigue precracking is advisable, but not necessary if the notch is sharp enough to start slow crack growth well before fracture.
4) The maximum load in the test should be taken as the fracture load. Only if K_c is to be determined should slow crack growth be measured, and then a load-time record synchronised with the crack growth record must be made.
5) Under the subheading "Edge Notched Specimens" in section 2.4 it was argued that compressive stresses $\sigma_x = -\sigma$ are present along the flanks of a central crack in a uniaxially loaded plate. Especially in thin sheets these compressive stresses can cause local buckling near the crack, and for long cracks buckling occurs well before the specimen is ready to fail and may significantly affect the residual strength. If such buckling would be restrained in service (by structural reinforcements) then anti-buckling guides must be used in the specimen test. On the other hand, anti-buckling guides should not be used to prevent buckling that could occur in service.
6) In figure 5.9 the ideal fracture mechanics curve ($\sigma_c = K_e/\sqrt{\pi a_o}$) is shown dotted. This formula is valid only for a large plate with a small crack, *i.e.* the right-hand part of the curve where $2a \rightarrow W$ is incorrect. Instead the $f(a/W)$ corrections given in sections (2.4) or (2.8) should be used.

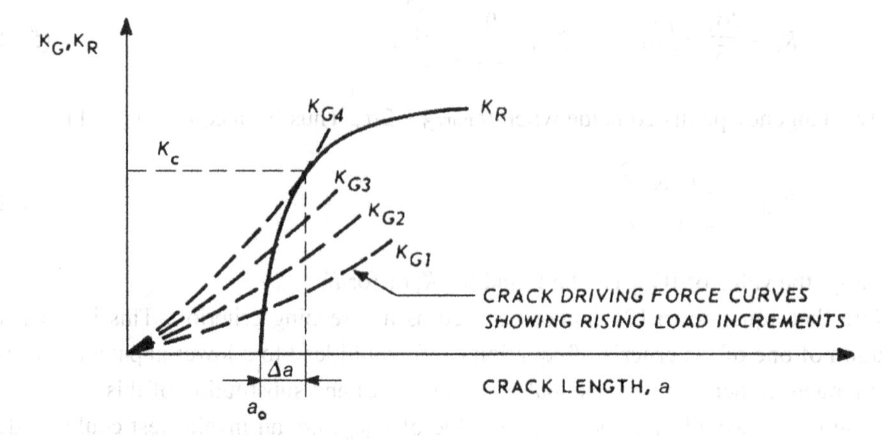

Figure 5.12. Crack growth resistance and crack driving force curves for a load controlled test.

5.4 The Determination of *R*-Curves

A limitation of K_c testing is that the effect of slow stable crack growth is not properly characterized. If we restrict ourselves to the apparent fracture toughness K_e this is not a problem, because then only the initial crack length is used. However if the K_c value is needed, it is obtainable from *R*-curves, as has been discussed in sections 4.6 – 4.8.

An *R*-curve is a plot of crack-extension resistance as a function of stable crack extension, Δa. The crack growth resistance may be expressed in the same units as G or, as is now customary, in terms of stress intensity factors, *i.e.* $K_G = \sqrt{E'G}$, $K_R = \sqrt{E'R}$.

R-curves can be determined by either of two experimental techniques: load control or displacement control. The crack grows owing to increments of increased load or displacement. The load control method involves rising load tests with crack driving force (K_G) curves like those shown in figure 5.12. Under rising load conditions ($P_1 < P_2 < P_3 < P_4$) the crack extends gradually to a maximum of Δa, where unstable crack growth occurs at $K_G = K_c$. This point is determined as the tangency point between the K_R-curve and one of the lines representing a crack driving force curve, $K_G = f(P,\sqrt{a},a/W)$, in this case the K_G-curve corresponding to load P_4. During the slow stable fracturing, the developing crack growth resistance K_R is equal to the applied K_G. Clearly, this testing method is capable only of obtaining that portion of the *R*-curve up to $K_R = K_c$, when instability occurs.

Under displacement control a suitable specimen results in negatively sloped crack driving force curves, as shown in figure 5.13.

In order to understand the conditions for which the slope of the crack driving force becomes negative it is convenient to consider K_I as a function of displacement, v, and crack length, a, *i.e.* $K_I = K_I(v,a)$. We can write

$$\frac{dK_I}{da} = \left(\frac{\partial K_I}{\partial a}\right)_v + \left(\frac{\partial K_I}{\partial v}\right)_a \frac{dv}{da}.$$

The derivative $(\partial K_I/\partial v)_a$ is always positive, while dv/da is either zero (fixed grip) or positive (constant

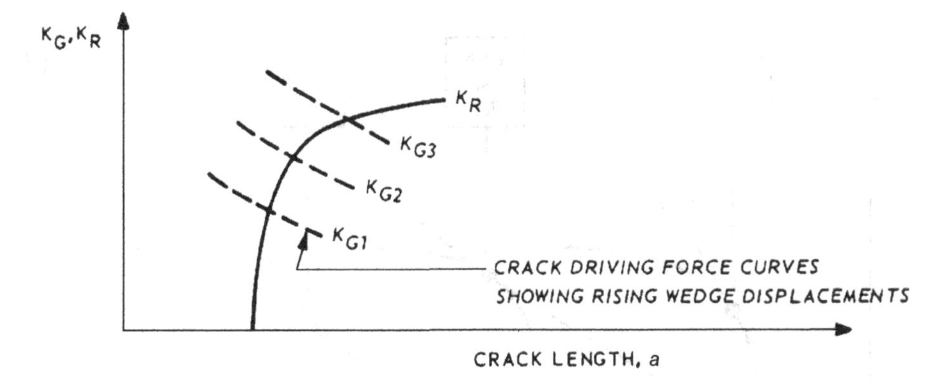

Figure 5.13. Crack growth resistance and crack driving force curves for a displacement controlled test.

load). A negatively sloped crack driving force curve, *i.e.* $dK_I/da < 0$, can be obtained by using fixed grip conditions *and* a specimen configuration such that $(\partial K_I/\partial a)_v$ is negative. Specimen configurations of this kind are widely available and will be discussed later on in the present section. In fact the only case for which a fixed grip condition does not ensure a negative dK_I/da is a remotely loaded centre cracked plate. It should be noted that even under constant load conditions dK_I/da can be negative, albeit for a limited number of specimen configurations. One example is the crack-line loading case for a central crack mentioned in section 2.5.

Under displacement control the specimen can be loaded by a wedge, which must be progressively further inserted in order to obtain greater displacements ($v_1 < v_2 < v_3$) and further crack growth. For each displacement the crack arrests when the crack driving force curve intersects the *R*-curve. Because there can be no tangency to the developing crack growth resistance, K_R, the crack tends to remain stable up to a plateau level, *i.e.* the entire *R*-curve can be obtained.

Relation Between R-Curve and K_c Testing

In section 4.6 it was indicated that *R*-curves are invariant, *i.e.* independent of initial crack length, a_o. However, K_c is approximately constant for only a limited range of crack lengths. The relation between *R*-curve and K_c testing is summarised schematically in figure 5.14. The shape of the K_c-a_o curve in figure 5.14.b is due to two effects which partly oppose each other. First, moving the *R*-curve along the *a* axis tends to raise the (G,R) and hence (K_G,K_R) tangency points. Second, the K_G line becomes markedly curved for longer initial crack lengths, figure 5.14.a, owing to the influence of finite specimen width on the stress intensity factor, as discussed in section 2.4. Increasing curvature of the K_G lines tends to lower the (K_G,K_R) tangency points.

For given test and material conditions a K_c value (obtained by rising load testing) represents only a single point on an *R*-curve. But since *R*-curves are invariant, with the *R*-curve approach the complete variation of K_c with changes in initial crack length can be described. Thus an *R*-curve is equivalent to a large number of direct K_c tests conducted with various initial crack lengths. In practice, however, this is not too important in view of the success of the Feddersen approach, *i.e.* the tangency constructions to obtain the complete residual strength diagram. It is only when the estimation of the

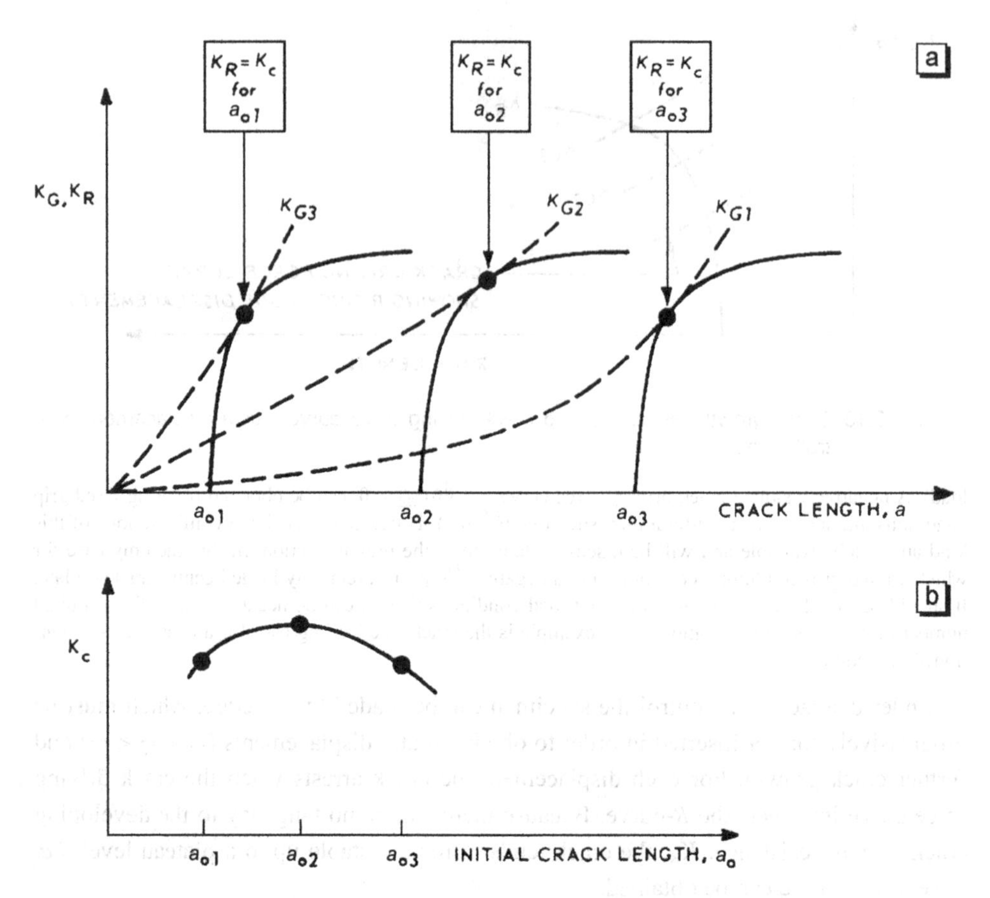

Figure 5.14. Use of *R*-curves for determining K_c as a function of initial crack length.

amount of stable crack growth is important that *R*-curve testing must be done.

Recommended Specimens for R-Curve Testing

In 1976 the ASTM published a recommended practice for *R*-curve determination, followed in 1981 by a standard (the last revision of the standard was made in 1994: ASTM E 561-94, reference 3 of the bibliography). The ASTM method will be concisely discussed in what follows. In general the specimens will have a thickness representative of plates considered for actual service. The ASTM recommends three types of specimens:

1) The centre cracked tension specimen (CCT or also called M(T), the middle-cracked tension specimen in the last revision of the standard).
2) The compact specimen (CS or C(T) in the last revision). These are the same as compact tension specimens (CT) used for K_{Ic} testing except that they may be of any thickness.
3) The crack-line wedge-loaded specimen (CLWL or C(W) in the last revision).

The first two types of specimen are tested under load control (rising K_G curves), while the CLWL specimen may be used for displacement control tests. The specimens are illustrated in figures 5.15 – 5.17. In figure 5.17 V_1 and V_2 refer to locations at which displacements are measured in order to determine compliance and hence effective crack length. The same locations can be used for the CS specimen.

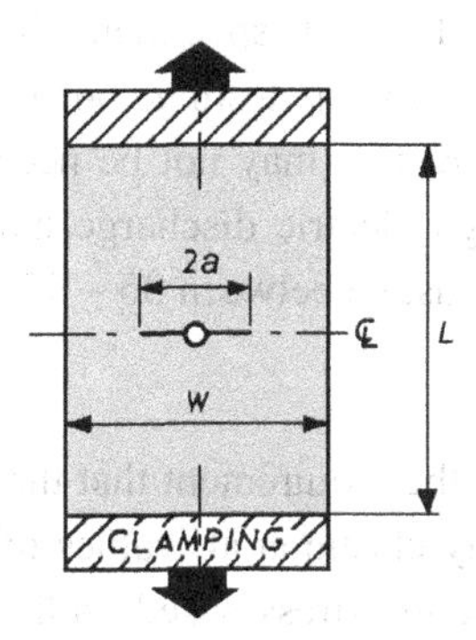

Figure 5.15. Centre cracked tension specimen (CCT or M(T)).

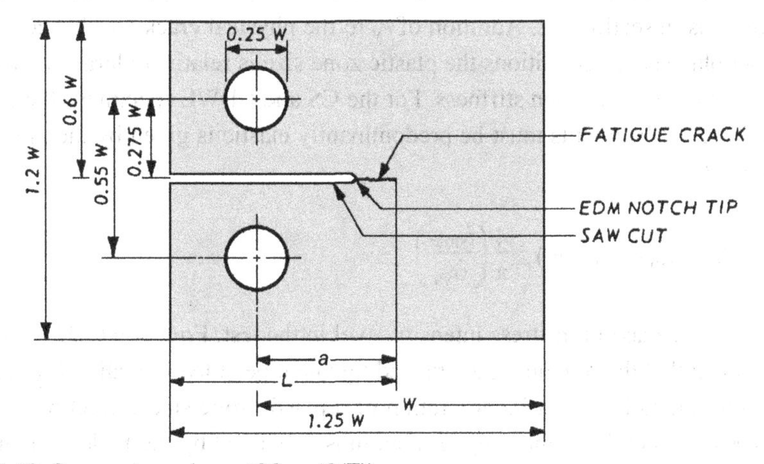

Figure 5.16. Compact specimen (CS or (C(T)).

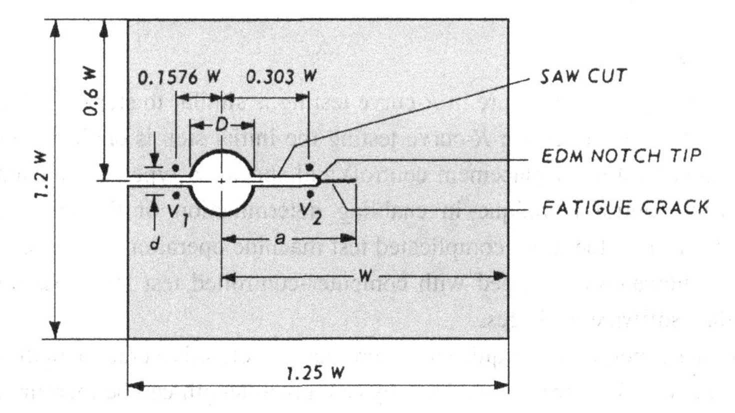

Figure 5.17. Crack-line wedge-loaded specimen (CLWL or C(W)).

The specimens must be fatigue precracked unless it can be shown that the machined notch root radius effectively simulates the sharpness of a fatigue precrack. For the CCT specimen the machined notch must be 30 – 35% of W with fatigue cracks not less than 1.3 mm in length. For the CS and CLWL specimens the starter notch configuration is basically similar to that required for K_{Ic} testing, but owing to the lesser thickness a chevron notch crack starter (figure 5.4) may not be necessary to obtain a symmetrical crack front, *i.e.* a straight through electric discharge machined (EDM) slot will often suffice. The initial crack length must be between 35 – 45% of W.

Specimen Size

Specimen size is based solely on the requirement that the uncracked ligaments ($W - 2a$) or ($W - a$) must be predominantly elastic at all values of applied load. More precisely, for the CCT specimen the net section stress based on the effective crack size (which is the physical crack size augmented for the effects of crack-tip plastic deformation, $2(a_o + \Delta a + r_y)$) must be less than the yield stress. The radius r_y of the plastic zone is given by Irwin's analysis in section 3.2. Addition of r_y to the physical crack size is necessary because under plane stress conditions the plastic zone size is relatively large and has a significant effect on the specimen stiffness. For the CS and CLWL specimens the condition that the uncracked ligaments must be predominantly elastic is given by the more or less empirical relation

$$W - (a_o + \Delta a + r_y) \geq \frac{4}{\pi}\left(\frac{K_{max}}{\sigma_{ys}}\right)^2, \tag{5.13}$$

where K_{max} is the maximum stress intensity level in the test. Equation (5.13) amounts to the requirement that the remaining uncracked ligament be at least equal to $8r_{y\,max}$.

It is worth noting here that incorporation of a plastic zone size correction will result in K_c values consistently slightly higher than those obtained by the Feddersen approach (although there is no fundamental objection to using a plastic zone size correction for the latter, see section 5.3).

R-Curve Test Procedure

Broadly speaking, the procedure in R-curve testing is similar to steps 1 – 7 for K_{Ic} testing, section 5.2. However, for R-curve testing the initial step is choice of testing technique (load control or displacement control) and specimen type. The advantage of the displacement control technique in enabling determination of the entire R-curve is somewhat offset by the more complicated test machine operation. However, most laboratories are nowadays equipped with computer-controlled test machines and more or less standard software packages.

Additional experimental requirements are that the effective crack length must be determined and buckling prevented. The physical crack length can be measured using *e.g.* optical microscopy or the electrical potential method, and subsequently the measured crack length can be adjusted by the addition of r_y. Alternatively, the effective crack length can be determined directly by means of compliance measurements: the procedure for this is fully described in the ASTM standard practice. Use of compliance instrumentation also makes it possible to determine whether the specimen develops undesirable buckling despite the presence of anti-buckling guides.

All types of specimen must be loaded incrementally, allowing time between steps for the crack to stabilise before measuring load and crack length, except if autographic instrumentation is used. In the latter case the load versus crack extension can be monitored continuously, but the loading rate must be slow enough not to introduce strain rate effects into the R-curve. To develop an R-curve the load versus crack extension data ($a_o + \Delta a + r_y$) can be used to calculate the crack driving force K_G, and hence K_R, using one of the two following expressions for the stress intensity factor:

- CCT specimen

$$K_I = \frac{\text{LOAD}}{BW}\sqrt{\pi a \sec\left(\frac{\pi a}{W}\right)} \tag{5.14}$$

or

$$K_I = \frac{\text{LOAD}}{BW}\sqrt{a}\left\{1.77 - 0.177\left(\frac{2a}{W}\right) + 1.77\left(\frac{2a}{W}\right)^2\right\} \tag{5.15}$$

- CS and CLWL specimens

$$K_I = \frac{\text{LOAD}}{B\sqrt{W}}\frac{\left(2+\frac{a}{W}\right)\left\{0.886+4.64\left(\frac{a}{W}\right) - 13.32\left(\frac{a}{W}\right)^2 + 14.72\left(\frac{a}{W}\right)^3 - 5.6\left(\frac{a}{W}\right)^4\right\}}{\left(1-\frac{a}{W}\right)^{\frac{3}{2}}}, \tag{5.16}$$

where B is the material thickness and W the specimen width measured from the load line. Note further that for the CLWL specimen the load is indirectly obtained from a load-displacement calibration curve. The procedure for obtaining such curves is given in the ASTM standard practice.

5.5 An Engineering Approximation to Account for the Effects of Yield Strength and Specimen Thickness on Fracture Toughness: Anderson's Model

Figure 5.1. shows that the value of K_c depends on thickness, decreasing gradually to a limiting lower value of K_{Ic}. The effect of sheet thickness is related to the gradual transition from plane stress to plane strain, and this transition is strongly influenced by the yield strength, as shown schematically in figure 5.18. A higher yield strength signifies a smaller plastic zone, so that there is more material in plane strain and the fracture toughness in the transition region is lower. It is also found that K_c and K_{Ic} generally decrease with increasing yield strength.

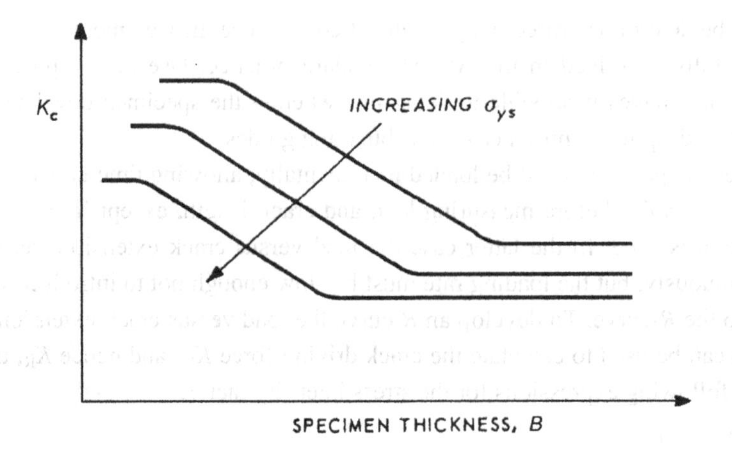

Figure 5.18. Schematic of the effects of yield strength and specimen thickness on K_c.

Although the qualitative trend shown in figure 5.18 is well established, there is no generally accepted quantitative model of the thickness effect. The simplest and also the most readily usable model is that of Anderson, figure 5.19.

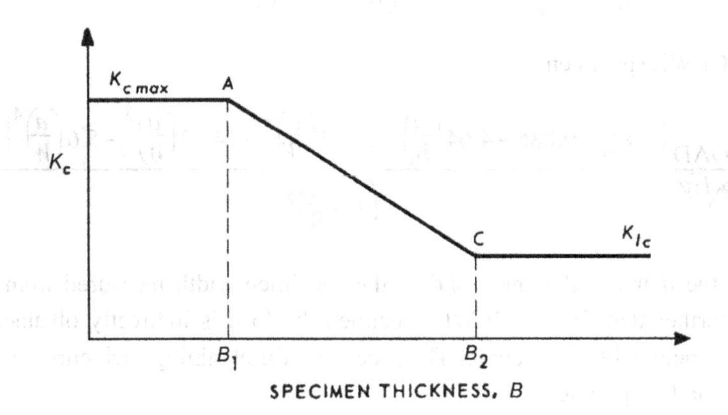

Figure 5.19. The thickness effect according to Anderson (reference 4 of the bibliography to this chapter).

The model is empirical. With knowledge of the two 'basic' fracture toughness values $K_{c,max}$ and K_{Ic} a line is drawn between the points A and C, which can be obtained from the following empirical relations:

1) Point A is given by

$$B_1 = \frac{1}{3\pi}\left(\frac{K_{c,max}}{\sigma_{ys}}\right)^2. \tag{5.17}$$

2) Point C is obtained from the limit of the ASTM condition for nominal plane strain behaviour (see section 5.2), *i.e.*

$$B_2 = 2.5\left(\frac{K_{Ic}}{\sigma_{ys}}\right)^2 . \tag{5.18}$$

The simplicity of Anderson's model makes it useful in an engineering sense, *i.e.* estimating fracture toughness for a practical sheet thickness when a full range of data are not available.

5.6 Practical Use of K_{Ic}, K_c and R-Curve Data

K_{Ic} data can be useful in two general ways. First, they may be used directly for choosing between materials for a particular application, especially high strength aerospace materials. More generally, since it is desirable (if possible) to avoid plane strain/low energy fracture, K_{Ic} values may be used as a basis for a screening criterion to ensure plane stress/high energy fracture. Several criteria have been proposed. One of the simplest is the through-thickness yielding criterion

$$K_{Ic} \geq \sigma_{ys}\sqrt{B} , \tag{5.19}$$

which gives the desired increase in toughness with increasing yield strength and sheet thickness in order to obtain plane stress fracture. This criterion is useful when σ_{ys} and K_{Ic} of a material are known. In this case the thickness B has to be less than or equal to $(K_{Ic}/\sigma_{ys})^2$ in order to avoid low energy fracture. A full derivation of this criterion is given in reference 5 of the bibliography.

Use of the through-thickness yield criterion can be demonstrated with the help of

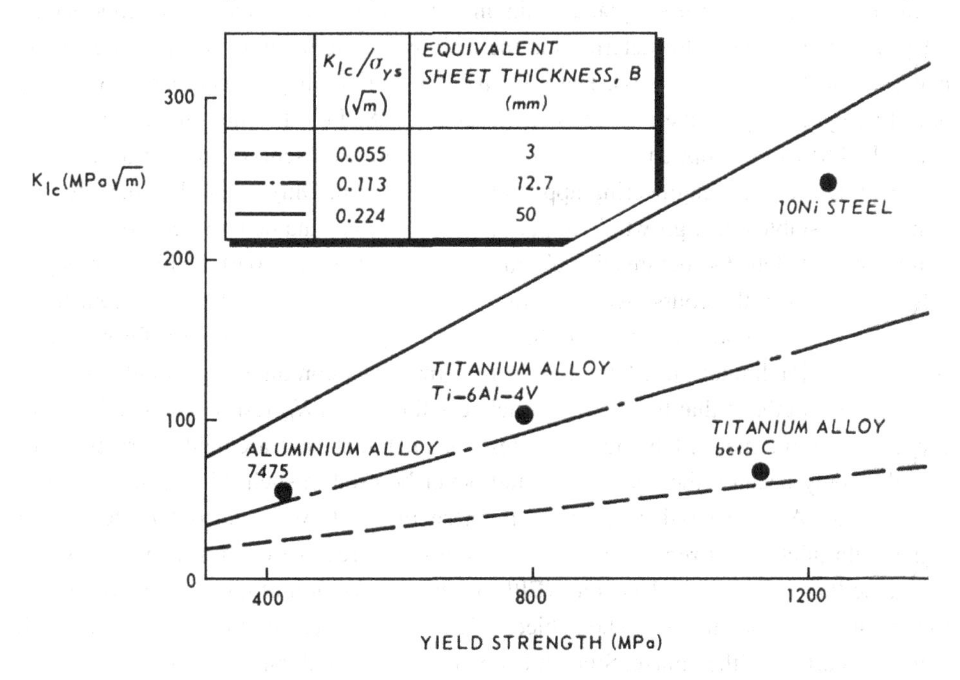

Figure 5.20. Illustration of the use of the through-thickness yielding criterion.

figure 5.20, which shows $K_{Ic} - \sigma_{ys}$ data for various materials, together with lines corresponding to different ratios of K_{Ic}/σ_{ys}. From equation (5.19) it is seen that these ratio lines have the dimension $\sqrt{\text{LENGTH}}$. Thus each line may be considered to correspond to a particular sheet thickness, B, as indicated in the table in figure 5.20.

The lines give minimum values of K_{Ic}/σ_{ys} necessary for through-thickness yielding to occur in a sheet of given thickness, *e.g.* for a sheet 3 mm thick the minimum value of K_{Ic}/σ_{ys} is 0.055. Comparison of actual $K_{Ic} - \sigma_{ys}$ data with the ratio lines in figure 5.20 shows the following.

1) The titanium alloy beta C has high strength but low toughness. However, the alloy can still meet the through-thickness yielding criterion, equation (5.19), for sheet thicknesses up to 3 mm.
2) 7475 aluminium alloy and Ti-6Al-4V titanium alloy meet the through-thickness yielding criterion for sheet thicknesses up to at least 12.7 mm.
3) 10 Ni steel combines high toughness with high strength and meets the through-thickness yielding criterion for heavy sections approaching 50 mm thickness.

Clearly, if it is required to meet the through-thickness yielding criterion in practice, then 10 Ni steel would be selected for heavy sections. On the other hand it would be possible to use beta C titanium alloy in thin sheet applications, thereby taking advantage of the lower density of titanium as compared to steel. Also, it should be noted that other materials can be plotted on diagrams like figure 5.20, and that actual selection of materials has to take into account many other factors, some or most of which are unrelated to fracture mechanics considerations.

Intermediate plane stress – plane strain and fully plane stress fracture toughness data are primarily of interest for determining the residual strengths of actual structures using materials of the same thickness. Here again the materials may be compared, as in the introductory example given in chapter 1, section 1.8. For the majority of situations where LEFM can be applied there is little incentive to use R-curves instead of the relatively straightforward engineering approach of Feddersen. Only when the characterization of slow stable crack growth is important will R-curve data be required, for example in thin sheet stiffened structures like aircraft fuselages, reference 6 of the bibliography.

At this point in the course we come to the end of Part II, which has been concerned with LEFM. It is appropriate to note that the inability to account properly for plasticity is often a major limitation. Many engineering materials combine high toughness with low yield strength, so that the required thickness for a valid K_{Ic} test may reach the order of magnitude of a metre! Obviously, K_{Ic} tests on such materials are neither practical nor useful, if only because the materials would never be used in such thicknesses (see also section 7.5). Also, excessive plasticity in these materials will rule out K_c testing of plates with thicknesses representative for actual structures. Resort has then to be made to Elastic Plastic Fracture Mechanics (EPFM) characterization of the crack resistance or to a plastic collapse analysis. The subject of EPFM is treated in chapters 6 – 8, which comprise Part III of the course. Since it is not a fracture-dominant failure mode (see also section 1.2), plastic collapse will only be mentioned shortly in chapter 6. In chapter 8 some attention is paid to failure due to the combination of fracture and plasticity.

5.7 Bibliography

1. ASTM Standard E 399-90, *Standard Test Method for Plane-Strain Fracture Toughness of Metallic Materials*, ASTM Standards on Disc, Vol. 03.01 (2001): West Conshohocken, Philadelphia.
2. Feddersen, C.E., *Evaluation and Prediction of the Residual Strength of Center Cracked Tension Panels*, Damage Tolerance in Aircraft Structures, ASTM STP 486, American Society for Testing and Materials, pp. 50–78 (1971): Philadelphia.
3. ASTM Standard E 561-98, *Standard Practice for R-Curve Determination*, ASTM Standards on Disc, Vol. 03.01 (2001): West Conshohocken, Philadelphia.
4. Anderson, W.E., *Some Designer-Oriented Views on Brittle Fracture*, Battelle Memorial Institute Pacific Northwest Laboratory Report 5A 2290, February 11 (1969): Richland, Washington.
5. Rolfe, S.T. and Barsom, J.M., *Fracture and Fatigue Control in Structures*, Prentice-Hall, Inc. (1977): Englewood Cliffs, New Jersey.
6. Vlieger, H. *Damage Tolerance of Skin-Stiffened Structures: Prediction and Experimental Verification*, Fracture Mechanics: 19th Symposium, ASTM STP 969, American Society for Testing and Materials, pp. 169-219 (1988): Philadelphia.

Part III
Elastic-Plastic Fracture Mechanics

6 Basic Aspects of Elastic-Plastic Fracture Mechanics

6.1 Introduction

Linear Elastic Fracture Mechanics (LEFM) was originally developed to describe crack growth and fracture under essentially elastic conditions, as the name implies. In this case plasticity is confined to a very small region surrounding the crack tip. However, such conditions are met only for plane strain fracture of high strength metallic materials and for fracture of intrinsically brittle materials like glasses, ceramics, rocks and ice.

Later it was shown that LEFM concepts could be slightly altered in order to cope with limited plasticity in the crack tip region. In this category falls the treatment of fracture problems in plane stress, *e.g.* the *R*-curve concept discussed in chapter 4. Nev-

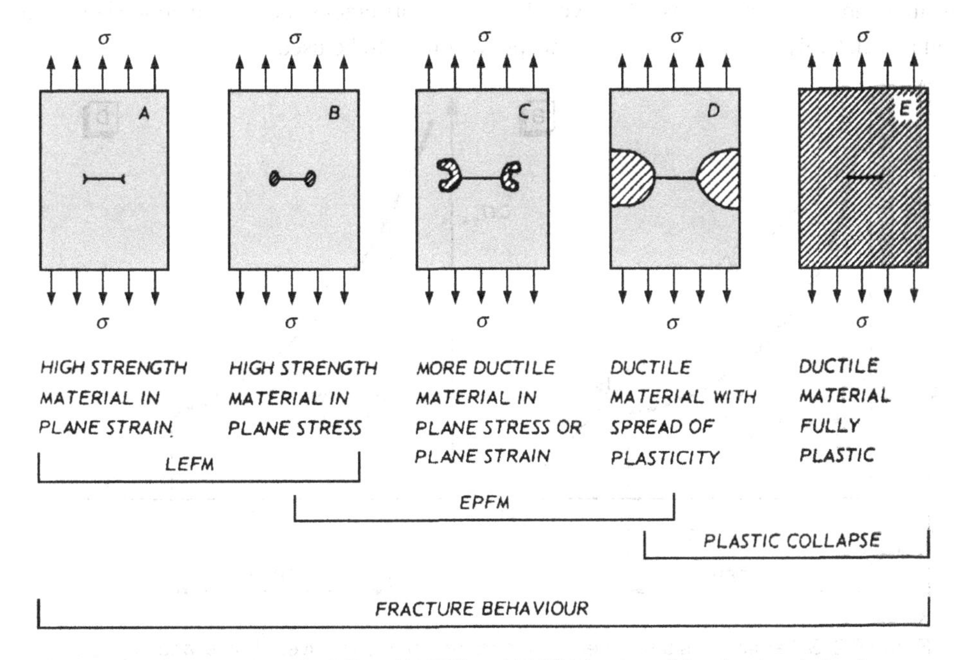

Figure 6.1. Ranges of applicability of LEFM and EPFM for describing fracture behaviour.

ertheless, there are many important classes of materials that are too ductile to permit description of their behaviour by LEFM: the crack tip plastic zone is simply too large. For these cases other methods must be found.

In this part of the course we shall discuss methods in the category of Elastic-Plastic Fracture Mechanics (EPFM). These methods significantly extend the description of fracture behaviour beyond the elastic regime, but they too are limited. Thus EPFM cannot treat the occurrence of general yield leading to so-called plastic collapse. Figure 6.1 gives a schematic indication of the ranges of applicability of LEFM and EPFM in the various regimes of fracture behaviour.

Since this course concerns fracture mechanics concepts, no further attention will be paid to plastic collapse, which is a yielding-dominant failure mode (see section 1.2). Discussion is confined to cases A, B, C and sometimes D of figure 6.1, *i.e.* the fracture-dominant failure modes. The LEFM concepts applicable to cases A and B have been treated in the previous chapters 2 – 5. Here and in chapters 7 and 8 the principles of EPFM, which are applicable to cases B, C and D, will be given.

Note that the ranges of applicability of LEFM and EPFM overlap in figure 6.1. Before proceeding to the development of EPFM it is worthwhile to discuss these ranges of applicability in some more detail. This will be done with the help of figure 6.2.

Figure 6.2.a gives a schematic residual strength diagram for a relatively brittle material in terms of the dimensionless crack length, $2a/W$ (W = panel width), of a centre cracked panel. Except for very short cracks the residual strength is determined by the stress intensity factor, since the K_c curve lies well below the line representing net section yield (and hence plasticity induced failure) of the uncracked ligaments. Thus LEFM is applicable for most cases. However, for very short cracks the plastic zone size is no longer relatively small, and EPFM concepts will have to be used.

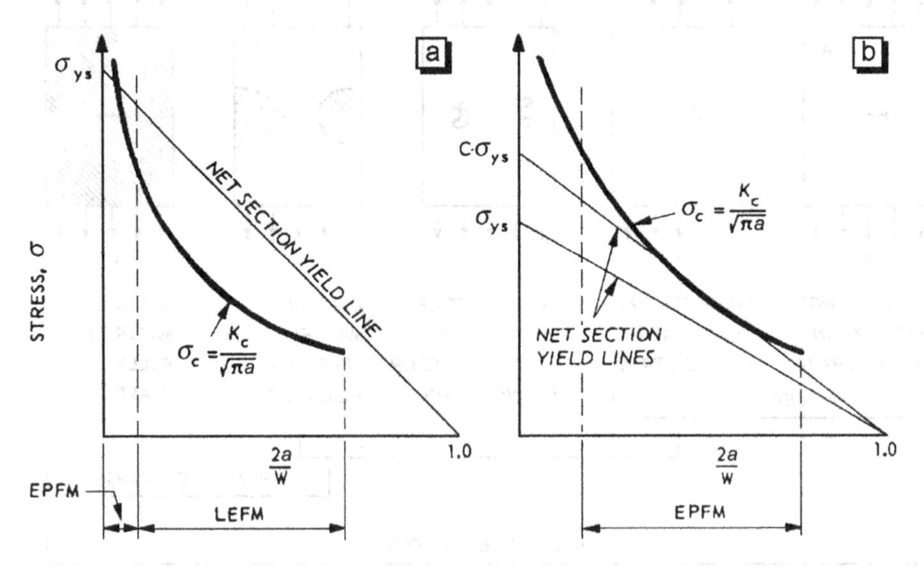

Figure 6.2. Schematic residual strength diagrams for (a) relatively brittle and (b) relatively ductile materials.

Figure 6.2.b gives a residual strength diagram for a relatively ductile material. Clearly the unconstrained yield stress, σ_{ys}, will be reached in the uncracked ligaments well before the critical stress $\sigma_c = K_c/\sqrt{\pi a_c}$ because the critical stress intensity factor, K_c, is high in ductile materials. The fracture behaviour is therefore likely to be controlled by general yielding, *i.e.* neither LEFM or EPFM are applicable. However, in situations of high constraint, *e.g.* cracks in thick sections, the effective yield stress will increase to $C\sigma_{ys,}$ where C is the plastic constraint factor discussed in section 3.5. The K_c curve may then predict a failure stress, σ_c, of the same order of magnitude as that given by the net section yield line: this is shown for a fairly wide range of $2a/W$ in figure 6.2.b. In such situations EPFM can be used to predict fracture behaviour. LEFM cannot be applied because σ_c will be too large a fraction of the effective yield stress and the plastic zone size will be too large.

It might be thought that situations of high constraint are rather special cases. In fact they are of prime importance with respect to practical applications. In the power generating and chemical processing industries most cracks occur in high pressure parts, which are of course thick-walled vessels and pipes. Also, the offshore industry has to cope with cracks in very large thick-sectioned welded structures. Seen in this light it is therefore not surprising that most contributions to the development of EPFM have come from these industries. In contrast LEFM is principally applied in the aerospace industry, where weight savings are at a premium and high strength, relatively brittle materials must be used.

6.2 Development of Elastic-Plastic Fracture Mechanics

Within the context of EPFM two general ways of trying to solve fracture problems can be identified:

1) A search for characterizing parameters (*cf.* K, G, R in LEFM).
2) Attempts to describe the elastic-plastic deformation field in detail, in order to find a criterion for local failure.

It is now generally accepted that a proper description of elastic-plastic fracture behaviour, which usually involves stable crack growth, is not possible by means of a straightforward, single parameter concept. Numerous detailed studies are being made of local failure criteria and elastic-plastic crack tip stress fields, but these are unlikely to give results suitable for practical use in the near future.

So far a notable success of EPFM for practical applications, however, is the ability to describe the *initiation of crack growth* and also a *limited amount of actual growth* using one or two parameters. Of the concepts developed for this purpose two have found a fairly general acceptance: the J integral and the Crack Opening Displacement (COD) approaches. Besides these concepts a number of others exist, but none have received widespread recognition.

Since this course is intended to provide a basic knowledge of fracture mechanics the discussion of EPFM concepts will be limited to the generally accepted J integral and COD approaches, which are treated in sections 6.3 – 6.5 and in section 6.6 respectively.

In section 6.8 the relation between J and COD is discussed. This is of itself interesting, but it is also helpful for illustrating the equivalence of J and G in the LEFM regime and that J is compatible with LEFM principles.

6.3 The *J* Integral

The J integral concept was first introduced by Rice, references 1 and 2 of the bibliography. Based on an energy approach Rice formulated J as a path-independent line integral with a value equal to the decrease in potential energy per increment of crack extension in linear or nonlinear elastic material. Its path independence implies that J can be seen as a measure for the intensity of stresses and strains at the tips of notches and cracks (see also section 6.5). Therefore the J integral can be viewed both as an energy parameter, comparable to G, and as a stress intensity parameter comparable to K.

In this section the energy description of J, *i.e.* as a nonlinear elastic energy release rate, will be considered first. Then a derivation is given that expresses this energy release rate as a line integral and it will be shown that this integral is path independent. Finally the usefulness of the J integral concept is discussed.

Energy Description of J

In section 4.2 the total energy of an elastic cracked plate *and* its loading system was given as[1]

$$U = U_o + U_a + U_\gamma - F\,. \tag{4.1}$$

In chapter 4 we have considered only linear elastic behaviour. However, there is no reason why equation (4.1) should not be valid for elastic material behaviour that is nonlinear: the essence is that the behaviour is elastic. A load-displacement diagram for a nonlinear elastic body is shown schematically in figure 6.3.a.

An important consequence of the extended validity of equation (4.1) is that under certain restrictions nonlinear elastic behaviour can be used to model plastic behaviour of a material. This is known as the deformation theory of plasticity. The main restriction is that no unloading may occur in any part of a body since for actual plastic behaviour the plastic part of the deformation is irreversible. The difference in unloading behaviour of bodies made of nonlinear elastic or plastic material is illustrated by comparing figures 6.3.a and 6.3.b.

In section 4.2 a part U_p of the total energy U of a cracked plate and its loading system was regarded as *potential energy*, *i.e.*

$$U_p = U_o + U_a - F \tag{4.2}$$

The energy available in linear elastic material per unit of new crack area was derived and designated as the energy release rate G. In equation (4.6) G was defined for a cen-

---

[1] As before, we consider two-dimensional geometries only, and all loads and energies are defined per unit thickness.

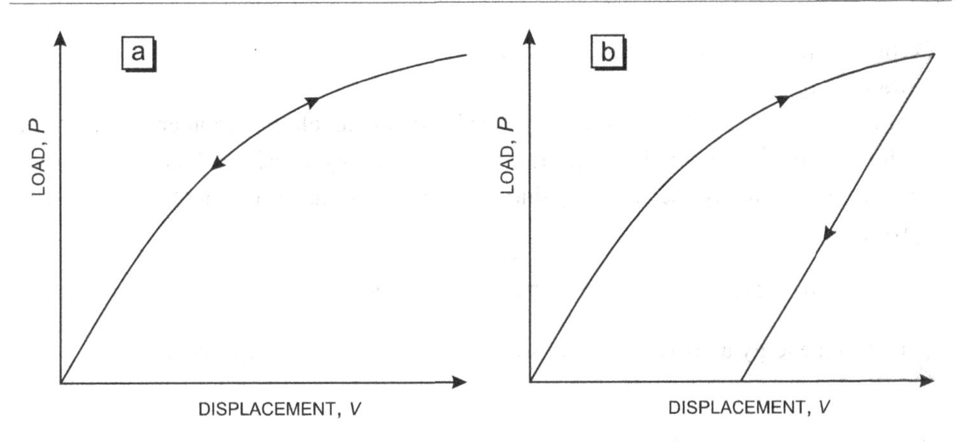

Figure 6.3. Load-displacement diagrams for bodies of (a) nonlinear elastic material and (b) plastically deformable material.

tral crack. Here we will consider an edge crack and define a nonlinear elastic equivalent accordingly:

$$J = -\frac{dU_p}{da} = \frac{d}{da}(F - U_a) . \tag{6.1}$$

Equation (6.1) gives the *energy definition* of J. Note that for linear elastic material behaviour $J = G$ by definition (see also remark 4 in section 6.8).

Concepts Necessary for Deriving J as a Line Integral

For the derivation of an expression for J as a line integral a two-dimensional situation is assumed, *i.e.* there is no dependence of relevant quantities upon the thickness coordinate. This simplifies the analysis but does have consequences for its generality, as will be mentioned in section 6.4.

To understand the derivation the reader should be familiar with the following:

- Index notation

  In this derivation it is more convenient to use an index notation instead of the engineering notation used until now. In the index notation the coordinate axes are no longer x, y and z, but x_i with the index i ranging from 1 to 3. The components of an arbitrary vector $\underline{v}$ are v_i and the components of the stress and strain tensors are σ_{ij} and ε_{ij}.

  For the current two-dimensional analysis the indices take only the values 1 or 2. The matrix representations of stress and strain are

$$[\sigma_{ij}] = \begin{bmatrix} \sigma_x & \tau_{xy} \\ \tau_{yx} & \sigma_y \end{bmatrix} = \begin{bmatrix} \sigma_{11} & \sigma_{12} \\ \sigma_{21} & \sigma_{22} \end{bmatrix} \quad \text{and} \quad [\varepsilon_{ij}] = \begin{bmatrix} \varepsilon_x & \varepsilon_{xy} \\ \varepsilon_{yx} & \varepsilon_y \end{bmatrix} = \begin{bmatrix} \varepsilon_{11} & \varepsilon_{12} \\ \varepsilon_{21} & \varepsilon_{22} \end{bmatrix} .$$

  As is customary with the index notation, the summation convention will be used, implying a summation with respect to symbol indices that are the same for the components of a term, see for example equation (6.2).

- Strain energy density

For an *elastic* material a strain energy density W, *i.e.* an elastic strain energy per unit volume, can be defined. The infinitesimal strain energy density $\mathrm{d}W$ is the work per unit volume done by the stress σ_{ij} during an infinitesimal strain increment $\mathrm{d}\varepsilon_{ij}$. It is given by

$$\mathrm{d}W = \sigma_{11}\,\mathrm{d}\varepsilon_{11} + \sigma_{21}\,\mathrm{d}\varepsilon_{21} + \sigma_{12}\,\mathrm{d}\varepsilon_{12} + \sigma_{22}\,\mathrm{d}\varepsilon_{22} = \sigma_{ij}\,\mathrm{d}\varepsilon_{ij}\,. \tag{6.2}$$

The strain energy density for a total strain ε_{kl} is obtained by integration, *i.e.*

$$W = W(\varepsilon_{kl}) = \int_0^{\varepsilon_{kl}} \mathrm{d}W = \int_0^{\varepsilon_{kl}} \sigma_{ij}\,\mathrm{d}\varepsilon_{ij}\,. \tag{6.3}$$

This means that the strain energy density W can be calculated if the strain, ε_{kl}, is known as well as the (linear or nonlinear) elastic relation between stress and strain, *i.e.* σ_{ij} as a function of ε_{ij}.

- Traction

The traction vector $\underline{T}$ is a force per unit area acting on some plane in a stressed material. It can be expressed in terms of the stress tensor $\underline{\sigma}$ according to:

$$T_i = \sigma_{ij}\, n_j \;\Rightarrow\; \begin{bmatrix} T_1 \\ T_2 \end{bmatrix} = \begin{bmatrix} \sigma_{11} & \sigma_{12} \\ \sigma_{21} & \sigma_{22} \end{bmatrix} \begin{bmatrix} n_1 \\ n_2 \end{bmatrix} = \begin{bmatrix} \sigma_{11}n_1 + \sigma_{12}n_2 \\ \sigma_{21}n_1 + \sigma_{22}n_2 \end{bmatrix}, \tag{6.4}$$

where n_1 and n_2 are the components of the unit vector $\underline{n}$ normal to the plane on which $\underline{T}$ acts. Note that the dimension of $\underline{T}$ is force per unit area.

- Principle of virtual work

When solving problems of elasticity it is often convenient to use the *principle of virtual work* (see reference 3 of the bibliography). In the case of a particle this principle states that if such a particle is in equilibrium, the total work of all real forces acting on the particle in any *virtual displacement* vanishes. For a deformable body we have to deal with internal stresses and strains too. The principle now states that the body is in a state of equilibrium if for any virtual displacement field the total virtual work done by all real external forces is equal to the total virtual work done by all real internal stresses, *i.e.*

total external virtual work = total internal virtual work.

Equilibrium in a two-dimensional body, or even part of a body, can therefore also be expressed in terms of the *virtual work equation*

$$\int_{\Gamma} T_i \, \delta u_i \, ds = \int_{A} \sigma_{ij} \, \delta\varepsilon_{ij} \, dA \,, \tag{6.5}$$

where Γ, A = perimeter and area of the two-dimensional body respectively,
ds = increment along perimeter Γ,
T_i = traction acting on perimeter,
δu_i = any suitable virtual displacement field,
$\delta\varepsilon_{ij}$ = virtual strain field corresponding to δu_i.

A suitable virtual displacement field is any field of small displacements that is *kinematically admissible*, *i.e.* it must be (i) differentiable and (ii) compatible with imposed conditions at the surface of the body. Virtual strain is related to virtual displacement according to

$$\delta\varepsilon_{ij} = \tfrac{1}{2}\left(\frac{\partial}{\partial x_j}\delta u_i + \frac{\partial}{\partial x_i}\delta u_j\right). \tag{6.6}$$

For an elastic material, *i.e.* stress and strain are uniquely related, the virtual work equation can also be written in terms of the virtual strain energy density $\delta W = \sigma_{ij} \, \delta\varepsilon_{ij}$:

$$\int_{\Gamma} T_i \, \delta u_i \, ds = \int_{A} \delta W \, dA \,. \tag{6.7}$$

Potential Energy of an Elastic Body and its Loading System

Consider the two-dimensional cracked body consisting of nonlinear elastic material shown in figure 6.4. The body has a surface A^* and a perimeter Γ^*. The crack flanks are not considered to be part of the perimeter. Tractions T_i are prescribed along a part Γ_T^* of the perimeter, while along another part displacements may be prescribed. It is assumed that the crack flanks are traction free, *i.e.* they are not in any way mechanically loaded. Consequently the loading system can perform work only along Γ_T^*.

Recall that the potential energy of an elastic body and its loading system was found as

$$U_p = U_o + U_a - F \,. \tag{4.2}$$

The purpose of this equation was to express the potential energy change owing to the introduction of a crack. The term U_o represents the potential energy for the uncracked configuration, and therefore can be considered as the elastic strain energy of the body minus the work performed by the loading system, both before a crack is present. Therefore the terms in equation (4.2) can be rearranged by combining the strain energy parts and the work parts. Expressing the strain energy as an integral of the strain energy den-

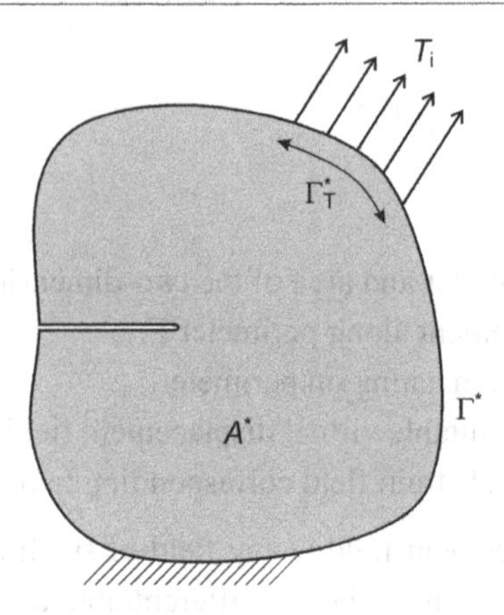

Figure 6.4. A two-dimensional cracked body consisting of nonlinear elastic material.

sity, W, over the whole area of the body and expressing the work as the traction T_i times the displacement u_i integrated along the part of the body perimeter Γ_T^*, we obtain the following expression for the potential energy, U_p:

$$U_p = \int_{A^*} W \, dA - \int_{\Gamma_T^*} T_i \, u_i \, ds \,. \tag{6.8}$$

Potential Energy Change Owing to Crack Growth

Consider the difference in potential energy, ΔU_p, between two cases where the crack length in the body differs by an amount Δa. The prescribed loads, *i.e.* tractions along Γ_T^* and possibly displacements along some other part of Γ^*, are identical for the two cases considered. Using equation (6.8), while denoting the displacement and strain energy density differences between the two cases by Δu_i and ΔW respectively, we obtain

$$\Delta U_p = \int_{A^*} \Delta W \, dA - \int_{\Gamma_T^*} T_i \, \Delta u_i \, ds = \int_{A^*} \Delta W \, dA - \int_{\Gamma^*} T_i \, \Delta u_i \, ds \,. \tag{6.9}$$

Note that in the latter expression the line integral is conveniently evaluated along the whole body perimeter Γ^* instead of along Γ_T^* only. This is permitted because either T_i or Δu_i vanishes along the part of Γ^* not belonging to Γ_T^*.

Using equation (6.9), the decrease in potential energy per increment of crack growth, $-dU_p/da$, can be expressed in terms of the following limit:

$$-\frac{dU_p}{da} = \lim_{\Delta a \to 0} -\frac{\Delta U_p}{\Delta a} = \lim_{\Delta a \to 0} \frac{1}{\Delta a} \left(\int_{\Gamma^*} T_i \, \Delta u_i \, ds - \int_{A^*} \Delta W \, dA \right) . \tag{6.10}$$

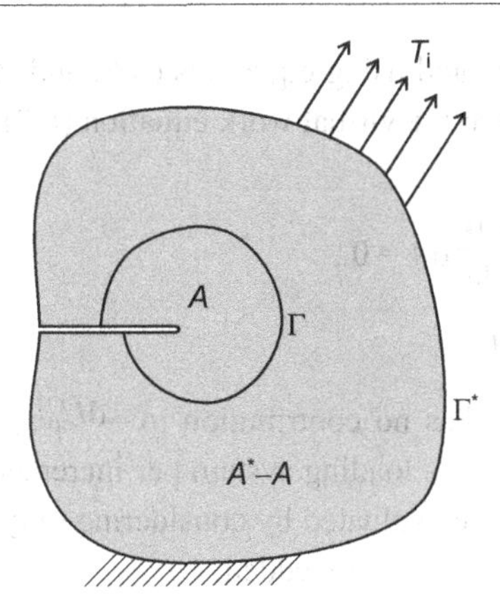

Figure 6.5. Elastic body in which an arbitrary area A is defined embedding the crack tip.

The integration domains used in equation (6.10) refer to the elastic body as a whole. To show that this is not strictly necessary, consider an arbitrary part of the body which however still includes the crack tip, as shown in figure 6.5. This part of the body has an area A and a perimeter Γ. As before, the crack flanks are not assumed to be part of the perimeter, so in fact Γ is a contour surrounding the crack tip starting and ending somewhere on the respective crack flanks.

For the part of the body A^*-A the decrease in potential energy per increment of crack growth can be expressed as

$$\left(-\frac{dU_p}{da}\right)_{A^*-A} = \int_{\Gamma^o} T_i \frac{du_i}{da} ds - \int_{A^*-A} \frac{dW}{da} dA , \qquad (6.11)$$

where the limit of equation (6.10) is now expressed in terms of derivatives, and Γ^o is the curve bounding the area A^*-A. We will now apply the principle of virtual work to the area A^*-A. The virtual work equation (6.7) must hold for any kinematically admissible virtual displacement field δu_i. Here we choose the displacement field resulting from a virtual crack extension δa, *i.e.*

$$\delta u_i = \frac{du_i}{da} \delta a . \qquad (6.12)$$

This displacement field is kinematically admissible because the area A^*-A does not contain the crack tip singularity caused by crack growth.

Although for an infinitely sharp crack both displacement and stress can be expected to become singular at the tip, these singularities will disappear in the case of a blunted crack tip (see for example equation (2.28) derived for the linear elastic case). When considering crack growth, however, the situation is inherently different. The reason is that new traction-free surface is created at material points which previously were loaded to some finite extent. Therefore the displacement field involved in crack growth (equation 6.12) will be singular at the tip, even if the tip is blunted.

The strain energy density field, δW, corresponding to the virtual displacement δu_i of equation (6.12) can be calculated using equations (6.6) and (6.3), and can be written as $\mathrm{d}W/\mathrm{d}a \cdot \delta a$. By substitution into the virtual work equation (6.7), it follows that

$$\int_{\Gamma^o} T_i \frac{\mathrm{d}u_i}{\mathrm{d}a}\,\mathrm{d}s - \int_{A^*-A} \frac{\mathrm{d}W}{\mathrm{d}a}\,\mathrm{d}A = 0\,, \tag{6.13}$$

and therefore the area A^*-A has no contribution to $-\mathrm{d}U_p/\mathrm{d}a$, the decrease in potential energy of the whole body and its loading system per increment of crack growth. Consequently, equation (6.10) can be evaluated by considering only an arbitrary but finite part of the body in which the crack tip is embedded, *i.e.*[2]

$$-\frac{\mathrm{d}U_p}{\mathrm{d}a} = \lim_{\Delta a \to 0} \frac{1}{\Delta a}\left(\int_{\Gamma} T_i\,\Delta u_i\,\mathrm{d}s - \int_{A} \Delta W\,\mathrm{d}A \right). \tag{6.14}$$

A Moving Coordinate System

Until now the usual coordinate system has been used, *i.e.* the coordinates of a given material point were fixed. At this stage it is convenient to introduce a 'moving' coordinate system with its origin at the crack tip irrespective of the crack length a. Relative to a fixed system x_1,x_2, of which the x_1 axis is chosen parallel to the crack, the moving coordinates X_1,X_2 are

$$X_1 = x_1 - a\,,$$
$$X_2 = x_2\,. \tag{6.15}$$

Consider an arbitrary quantity f that depends on position as well as crack length. This quantity can be expressed as a function of either the fixed coordinate x_i or the moving coordinate X_i, *i.e.* $f = f(x_1,x_2,a)$ or $f = \tilde{f}(X_1,X_2,a)$. For a given material point the total derivative of f with respect to the crack length a is

$$\frac{\mathrm{d}f}{\mathrm{d}a} = \frac{\mathrm{d}\tilde{f}}{\mathrm{d}a} = \frac{\partial \tilde{f}}{\partial X_1}\frac{\mathrm{d}X_1}{\mathrm{d}a} + \frac{\partial \tilde{f}}{\partial X_2}\frac{\mathrm{d}X_2}{\mathrm{d}a} + \frac{\partial \tilde{f}}{\partial a}\,. \tag{6.16}$$

The coordinates x_1 and x_2 of the point considered are independent of the crack length. Thus

$$\frac{\mathrm{d}X_1}{\mathrm{d}a} = \frac{\mathrm{d}(x_1 - a)}{\mathrm{d}a} = -1\,, \tag{6.17}$$

---

[2] In the derivation of the J integral it will prove more convenient to use the limit form rather than writing this expression in terms of derivatives.

$$\frac{dX_2}{da} = \frac{dx_2}{da} = 0\,, \tag{6.18}$$

and furthermore, because of equation (6.15), we may write

$$\frac{\partial \tilde{f}}{\partial X_1} = \frac{\partial f}{\partial x_1}\,. \tag{6.19}$$

Applying equations (6.17) – (6.19) to equation (6.16) gives

$$\frac{df}{da} = \frac{\partial \tilde{f}}{\partial a} - \frac{\partial f}{\partial x_1}\,. \tag{6.20}$$

The first term of the right-hand part of this expression is the change in the quantity f for a point with constant moving coordinates X_i, thus having a fixed position relative to the crack tip. The second term represents the correction that becomes necessary because the X_1 coordinate of a material point decreases (becomes more negative) due to the crack growth.

J as a Line Integral

Using equation (6.20) the line integral term in equation (6.14) can be written as

$$\lim_{\Delta a \to 0} \frac{1}{\Delta a} \int_{\Gamma} T_i\, \Delta u_i\, ds = \int_{\Gamma} T_i \frac{du_i}{da} ds = \int_{\Gamma} T_i \left(\frac{\partial \tilde{u}_i}{\partial a} - \frac{\partial u_i}{\partial x_1} \right) ds\,. \tag{6.21}$$

where $\tilde{u}_i$ is the displacement expressed as a function of the moving coordinates X_i.

The area integral of the strain energy density difference in equation (6.14) could also be expanded in an analogous manner. In order to express J in the form of a line integral, a conversion would then be required of the area integral of $\partial W/\partial x_1$ to a line integral. However, in the case of an infinitely sharp crack such a conversion is not possible since the integrand is singular at the tip. The way in which Rice deals with this problem, reference 2 of the bibliography, is described in the following.

The area integral of equation (6.14) should be evaluated over an area A. Obviously the position of this area is not affected by crack growth. However, straightforward evaluation of an area integral in terms of the moving coordinate system X_i would involve an integration area that moves with the growing crack. Therefore a correction of the integration area becomes necessary. Figure 6.6 schematically shows the principle. Denoting the strain energy density for the initial crack length, a, as W^o, the area integral of equation (6.14) can be expanded as

$$\lim_{\Delta a \to 0} \frac{1}{\Delta a} \int_{A} \Delta W\, dA = \lim_{\Delta a \to 0} \frac{1}{\Delta a} \left\{ \iint_{A} (W^o + \Delta W)\, dx_1 dx_2 - \iint_{A} W^o\, dx_1 dx_2 \right\}. \tag{6.22}$$

The second integral refers to the initial crack length a. In this case the relation between moving coordinates X_i and fixed coordinates x_i (equation 6.15) is independent of the

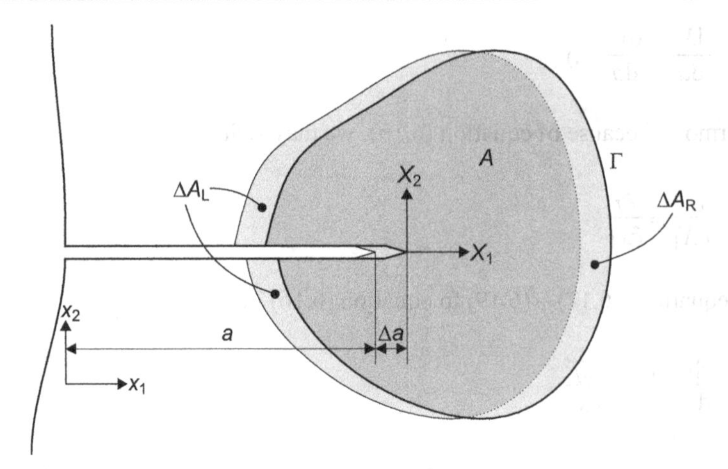

Figure 6.6. Correction of the integration area necessary because of use of the moving coordinate system X_i.

crack length difference Δa. Therefore, the integration area for this integral is not affected by the use of moving coordinates and remains equal to area A. For the first integral, however, the crack length is $a+\Delta a$. If we were to integrate over the area A while using moving coordinates, we would refer to the original area A moved in the positive X_1 direction over a distance Δa. To correct for this an integration area ΔA_L, with a width Δa, must be added on the left side of area A and likewise an area ΔA_R must be subtracted on the right side. Thus, using moving coordinates X_i and denoting the strain energy density as a function of the moving coordinates as $\widetilde{W}$, equation (6.22) can be rewritten and further expanded as

$$\begin{aligned}
\lim_{\Delta a\to 0}\frac{1}{\Delta a}\int_A \Delta W\,\mathrm{d}A &= \lim_{\Delta a\to 0}\frac{1}{\Delta a}\left\{\iint_{A+\Delta A_L-\Delta A_R}(\widetilde{W}^{\mathrm{o}}+\Delta\widetilde{W})\,\mathrm{d}X_1\mathrm{d}X_2-\iint_A \widetilde{W}^{\mathrm{o}}\,\mathrm{d}X_1\mathrm{d}X_2\right\}\\
&= \lim_{\Delta a\to 0}\frac{1}{\Delta a}\iint_A \Delta\widetilde{W}\,\mathrm{d}X_1\mathrm{d}X_2+\lim_{\Delta a\to 0}\frac{1}{\Delta a}\iint_{\Delta A_L-\Delta A_R}(\widetilde{W}^{\mathrm{o}}+\Delta\widetilde{W})\,\mathrm{d}X_1\mathrm{d}X_2\\
&\overset{(1)}{=} \int_A \frac{\partial\widetilde{W}}{\partial a}\mathrm{d}A+\lim_{\Delta a\to 0}\frac{1}{\Delta a}\left\{\iint_{\Delta A_L}(\widetilde{W}^{\mathrm{o}}+\Delta\widetilde{W})\,\mathrm{d}X_1\mathrm{d}X_2-\iint_{\Delta A_R}(\widetilde{W}^{\mathrm{o}}+\Delta\widetilde{W})\,\mathrm{d}X_1\mathrm{d}X_2\right\}\\
&\overset{(2)}{=} \int_A \frac{\partial\widetilde{W}}{\partial a}\mathrm{d}A+\int_{\Gamma_L} W\,\mathrm{d}x_2-\int_{\Gamma_R} W\,\mathrm{d}x_2\\
&\overset{(3)}{=} \int_A \frac{\partial\widetilde{W}}{\partial a}\mathrm{d}A-\int_{\Gamma} W\,\mathrm{d}x_2\,.
\end{aligned}\tag{6.23}$$

The steps (1) – (3) in this expansion require some additional explanation:

1) The first limit is to be evaluated for a point having a fixed position relative to the crack tip. Therefore this expression is converted to the partial derivative $\partial\widetilde{W}/\partial a$, *i.e.* keeping the moving coordinates X_i constant.

   Note that until now this type of limit expression was converted to a total derivative (*cf.* equations (6.11) and (6.21)). However, in this conversion it was implicitly assumed that material points were considered, *i.e.* the fixed coordinates x_i remained constant. For this case the total derivative with respect to crack length is equivalent to the partial derivative.

2) The areas ΔA_L and ΔA_R both have a width (in the X_1 direction) of Δa. Therefore, in the limit of $\Delta a \to 0$, the integral of $\widetilde{W}^o+\Delta\widetilde{W}$ over this width is equal to $\Delta a \times \widetilde{W}^o$, since $\Delta\widetilde{W} \to 0$. Thus in the limit the area integrals are converted to line integrals along the contour parts corresponding to the areas ΔA_L and ΔA_R. These contour parts are denoted as Γ_L and Γ_R respectively. At the same time Δa in the denominator cancels out. Furthermore, a transition is made from moving coordinate X_2 to fixed coordinate x_2 and $\widetilde{W}^o$ is written as W.
3) Since the line integrals along Γ_L and Γ_R should yield a positive value, they must both be evaluated in the positive x_2 direction. Consequently, the two integrals can be replaced by a single line integral along Γ, evaluated in a counterclockwise direction.

Substituting equations (6.21) and (6.23) in equation (6.14), the decrease in potential energy per increment of crack growth can be expressed as

$$-\frac{dU_p}{da} = \int_\Gamma T_i \frac{\partial \widetilde{u}_i}{\partial a} ds - \int_\Gamma T_i \frac{\partial u_i}{\partial x_1} ds - \int_A \frac{\partial \widetilde{W}}{\partial a} dA + \int_\Gamma W dx_2 . \quad (6.24)$$

The virtual work equation (6.7) will be applied once more using the virtual displacement field

$$\delta u_i = \frac{\partial \widetilde{u}_i}{\partial a} \delta a . \quad (6.25)$$

Note that this displacement field is different from the one described by equation (6.12). The latter field described the displacement of points with constant fixed coordinates x_i, *i.e.* actual material points. The current field, however, describes the displacement of points with *constant moving coordinates* X_i, *i.e.* points with a fixed position relative to the crack tip. The consequence is that this field is kinematically admissible throughout the entire body, since it represents the infinitesimal displacement change owing to an infinitesimal increase in crack length, a value which is not singular even at the (moving) crack tip.

The strain energy density field, δW, corresponding to the virtual displacement field of equation (6.25) can be written as $\partial\widetilde{W}/\partial a \cdot \delta a$. Substituting δu_i and δW in the virtual work equation (6.7), we obtain[3]

---

[3] Note that for evaluating the line integral in the virtual work equation (6.7) the crack flanks need not be considered since they are assumed traction-free. Therefore the current definition of Γ is compatible with that used in equation (6.7).

$$\int_{\Gamma} T_{\mathrm{i}} \frac{\partial \widetilde{u}_{\mathrm{i}}}{\partial a} \mathrm{d}s = \int_{A} \frac{\partial \widetilde{W}}{\partial a} \mathrm{d}A . \tag{6.26}$$

It turns out that the first and the third term in equation (6.24) cancel each other. Thus we can finally write $-\mathrm{d}U_{\mathrm{p}}/\mathrm{d}a$ in terms of a single line integral. In view of the energy definition of J given in equation (6.1), this also leads to the expression for the J integral, *i.e.*

$$J = -\frac{\mathrm{d}U_{\mathrm{p}}}{\mathrm{d}a} = \int_{\Gamma} \left(W \mathrm{d}x_2 - T_{\mathrm{i}} \frac{\partial u_{\mathrm{i}}}{\partial x_1} \mathrm{d}s \right) . \tag{6.27}$$

Thus for a crack with the tip pointing in the positive x_1 direction, this expression enables J, or the decrease in potential energy per increment of crack growth, to be evaluated as a line integral along an arbitrary path surrounding the crack tip, starting somewhere on the lower crack flank and ending somewhere on the upper crack flank. The integration should be performed in a counterclockwise direction.

Note that the fact that an arbitrary integration path may be used implies that the J integral is path independent.

Alternative Expression for the J Integral

Equation (6.27) is often written differently. Consider an increment ds along the contour Γ. If $\mathrm{d}\underline{s} = (\mathrm{d}x_1,\mathrm{d}x_2)$ is the vector coinciding with this part of the contour, then

$$\mathrm{d}x_1 = -n_2 \mathrm{d}s \quad \text{and} \quad \mathrm{d}x_2 = n_1 \mathrm{d}s , \tag{6.28}$$

where $\underline{n} = (n_1,n_2)$ is the outward-directed unit vector normal to the contour.

Equation (6.28) can be derived as follows. The scalar product of the vectors $\mathrm{d}\underline{s} = (\mathrm{d}x_1,\mathrm{d}x_2)$ and $\underline{n} = (n_1,n_2)$ is zero because $\underline{n}$ is perpendicular to $\mathrm{d}\underline{s}$. Thus $n_1\mathrm{d}x_1+n_2\mathrm{d}x_2 = 0$. The vector product $\underline{n} \times \mathrm{d}\underline{s}$ is perpendicular to both $\underline{n}$ and $\mathrm{d}\underline{s}$ and has the value $(0,0,n_1\mathrm{d}x_2-n_2\mathrm{d}x_1)$. The absolute value (length) of this vector, $\underline{n} \times \mathrm{d}\underline{s}$, is $|\underline{n}||\mathrm{d}\underline{s}|\sin(90°) = \mathrm{d}s$, because the absolute value of the unit vector $|\underline{n}| = 1$, while the length of vector $\mathrm{d}\underline{s}$ is defined as ds. The length of $\underline{n} \times \mathrm{d}\underline{s}$ is also equal to $n_1\mathrm{d}x_2-n_2\mathrm{d}x_1$. We now have two equations for the two unknown variables dx_1 and dx_2: $\mathrm{d}s = n_1\mathrm{d}x_2-n_2\mathrm{d}x_1$ and $n_1\mathrm{d}x_1+n_2\mathrm{d}x_2 = 0$. Solving these two equations for dx_1 and dx_2 leads to the result of equation (6.28).

Substituting equation (6.28) for dx_2 in equation (6.27) we obtain:

$$J = \int_{\Gamma} \left(W n_1 - T_{\mathrm{i}} \frac{\partial u_{\mathrm{i}}}{\partial x_1} \right) \mathrm{d}s . \tag{6.29}$$

Usefulness of the J Integral Concept

In this section a path-independent integral expression has been derived for J, representing a nonlinear elastic energy release rate. Two observations are made:

- Under certain restrictions this nonlinear elastic energy release rate can be used as an elastic-plastic energy release rate. This is interesting because in LEFM there is a critical value for the energy release rate, G_c, which predicts the onset of crack extension.
- Path independence of the J integral allows the contour Γ to be chosen just as small as the crack tip area. This illustrates that J is in fact a measure for the stresses and strains at the very tip (see also section 6.5). It seems reasonable to assume that the onset of crack extension is determined by these stresses and strains.

Both observations suggest that there is a critical J value, J_c, at which crack growth is initiated. By analogy with G_c in LEFM, for a loaded cracked component calculated J values can be compared to the critical value, J_c, characteristic for a given material. Thus a fracture mechanics analysis can be carried out: J must remain less than J_c.

Note that path independence of J also allows calculation along a contour remote from the crack tip. Such a contour can be chosen to contain only elastic loads and displacements. Thus an elastic-plastic energy release rate can be obtained from an elastic calculation along a contour for which loads and displacements are known.

6.4 Remarks Concerning the *J* Integral Concept

The J integral concept is not easy to understand. However, the concept is undeniably useful, and so at this point it is also worthwhile to direct some remarks to the derivation, applications and restrictions of J:

1) At the beginning of section 6.3 it was stated that J would be derived assuming the deformation behaviour to be nonlinear elastic and therefore reversible. But plastic deformation is not reversible, and the energy dissipated cannot be transformed into other kinds of energy. Thus strictly speaking, concepts such as strain energy density and potential energy are not legitimately usable with true plastic deformation.
2) It was also stated that the assumption of nonlinear elasticity is compatible with actual deformation behaviour only if no unloading occurs in any part of the material. But during crack growth the newly formed crack flanks are completely unloaded from stresses as high as σ_{ys} (or even higher in the case of plane strain and/or work hardening). Therefore J is in principle applicable only up to the beginning of crack extension and not for crack growth. However, under certain restrictions J can be used to characterize crack growth. This will be discussed further in section 8.3.
3) The controlling parameters in the derivation of J are the stress and strain fields in the cracked body. In order to simplify the analysis, the derivation was limited to a two-dimensional configuration. Stresses and strains in the thickness direction were not taken into account, although they obviously affect the assumed nonlinear relation between the in-plane stresses and strains. Full plane strain or plane stress conditions can indeed be described two-dimensionally. For intermediate conditions the effect of thickness stress and strain will vary as a function of the thickness coordinate and the question arises whether J can be used in such cases. However the J integral concept can be fully extended to three-dimensional crack geometries, see reference 4. J then

has a local value that varies along the crack front, but loses its meaning as a path-independent line integral.

4) By definition, J for the linear elastic case is equal to G and we may write

$$J = G = \frac{K^2}{E'}, \tag{6.30}$$

where $E' = E$ for plane stress and $E' = E/(1 - \nu^2)$ for plane strain. Thus the J integral concept is compatible with LEFM.

Note that by analogy with G the dimensions of J are [ENERGY]/[LENGTH] per unit thickness of material, *i.e.* Joules/m^2 or N/m.

5) As stated before, it is to be expected that there is a critical material parameter, J_c, which predicts the onset of crack extension. Methods for measuring J_c will be presented in chapter 7.

6) Obtaining solutions for the J integral in actual specimens or components turns out to be difficult. It is generally necessary to use finite element techniques. However, some simple expressions have been developed for standard specimens. They will also be presented in chapter 7.

7) The J integral concept has been developed mainly in the USA as a fracture criterion for materials used in the power generating industry, particularly nuclear installations. In this area of application high level technology and costly production techniques are generally used and no large differences are to be expected for material behaviour in laboratory specimens and actual structures. For instance: local differences in behaviour of welded joints are not normally accounted for. This contrasts with the COD approach, which will be discussed in section 6.6. This concept was developed in the UK at the Welding Institute and is obviously more directed to the design of welded structures (see section 6.7).

6.5 *J* as a Stress Intensity Parameter

From work done by Hutchinson (references 5 and 6) and independently by Rice and Rosengren (reference 7) the crack tip stresses and strains can be expressed in terms of J according to the so-called HRR solution:

$$\sigma_{ij} = \sigma_o \left(\frac{E}{\alpha\, \sigma_o^2\, I_n} \frac{J}{r}\right)^{\frac{1}{n+1}} \bar{\sigma}_{ij}(\theta,n),$$

$$\varepsilon_{ij} = \alpha \frac{\sigma_o}{E} \left(\frac{E}{\alpha\, \sigma_o^2\, I_n} \frac{J}{r}\right)^{\frac{n}{n+1}} \bar{\varepsilon}_{ij}(\theta,n). \tag{6.31}$$

They assumed a power-law hardening material, *i.e.* the relation between the uniaxial stress σ and strain ε is given by the so-called Ramberg-Osgood relation

$$\frac{\varepsilon}{\varepsilon_o} = \frac{\sigma}{\sigma_o} + \alpha \left(\frac{\sigma}{\sigma_o}\right)^n, \tag{6.32}$$

where α is a dimensionless constant, $\varepsilon_o = \sigma_o/E$ with σ_o usually equal to the yield stress, and n is the strain hardening exponent. I_n in equations (6.31) is a dimensionless constant depending on the strain hardening exponent n and the stress state (plane stress or plane strain), and $\bar{\sigma}_{ij}$ and $\bar{\varepsilon}_{ij}$ are dimensionless functions of n, the angle θ and the stress state. Values of I_n, $\bar{\sigma}_{ij}$ and $\bar{\varepsilon}_{ij}$ are given in tabular form by Shih (reference 8).

Note that for $n = 1$ (linear elastic material behaviour), equations (6.31) show a $1/\sqrt{r}$ singularity, which is consistent with LEFM. In fact it can be shown that for $n = 1$ equations (6.31) become identical to the elastic solution given by equations (2.24). On the other hand, for $n = \infty$, *i.e.* ideal plastic material behaviour, the solution becomes equal to the so-called Prandtl slip-line field solution (reference 1).

Equations (6.31) imply that the stress/strain field in the direct vicinity of a crack tip is completely characterized by a single parameter J. Different geometries with identical J values can be expected to have the same stresses and strains near the crack tip, and thus show identical responses. Therefore J can be considered as a single fracture mechanics parameter for the elastic-plastic regime (with the restriction of no unloading), analogous to K for the linear elastic regime.

Note that the HRR singularity contains an anomaly similar to the LEFM singularity, namely that both predict infinite stresses for $r \rightarrow 0$. However the large plastic strains at the crack tip cause the crack to blunt, which reduces the stresses locally. The HRR solution is thus not valid all the way to the tip. At a distance less than twice the CTOD value the HRR singularity becomes invalid (reference 9).

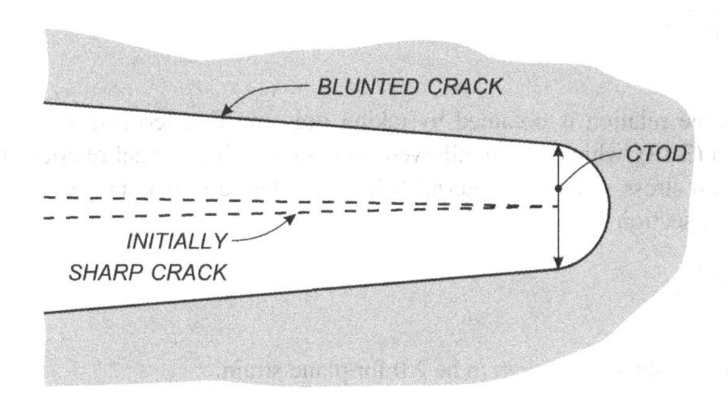

Figure 6.7. Crack tip opening displacement (CTOD or δ_t).

6.6 The Crack Opening Displacement (COD) Approach

The COD approach was introduced by Wells, reference 10 of the bibliography, as long ago as 1961. The background philosophy to the approach is as follows. In the regimes of fracture-dominant failure, cases A, B, C and partly D in figure 6.1, the stresses and strains in the vicinity of a crack or defect are responsible for failure. At crack tips the stresses will always exceed the yield strength and plastic deformation will occur. Thus failure is brought about by stresses and hence plastic strains exceeding certain respective limits. Wells argued that the stress at a crack tip always reaches the critical value (in the purely elastic case $\sigma \rightarrow \infty$). If this is so then it is the plastic strain in the crack tip region that controls fracture.

A measure of the amount of crack tip plastic strain is the displacement of the crack flanks, especially at or very close to the tip. The Crack Opening Displacement (COD) at the *original* crack tip is called Crack Tip Opening Displacement (CTOD) or δ_t (see figure 6.7): an initially sharp crack blunts by plastic deformation, resulting in a finite displacement at the original tip.[4]

Thus it might be expected that at the onset of fracture the crack tip opening displacement, δ_t, has a characteristic critical value, $\delta_{t_{crit}}$, for a particular material and therefore could be used as a fracture criterion.

In 1966 Burdekin and Stone provided an improved basis for the COD concept. They used the Dugdale strip yield model to find an expression for CTOD. Their analysis has already been reviewed in section 3.3 of the course and is given in full in reference 11 of the bibliography to this chapter. The strip yield model assumes plane stress conditions and ideal plastic (*i.e.* non-hardening) material behaviour. The result is

$$\delta_t = \frac{8\sigma_{ys}a}{\pi E} \ln \sec \frac{\pi\sigma}{2\sigma_{ys}} . \tag{3.19}$$

In sections 3.2 and 3.3 it was also shown that under LEFM conditions there are direct relations between δ_t and K_I. Thus, for the Dugdale analysis

$$\delta_t = \frac{K_I^2}{E\sigma_{ys}} . \tag{3.20}$$

Note that this simple relation is obtained by taking only the first term of a series expansion of the ln sec part of equation (3.19), which is only allowed for $\sigma << \sigma_{ys}$. The actual relationship between CTOD and K_I also depends on stress state and material behaviour. These effects are represented by the plastic constraint factor C (see section 3.5), *i.e.*

$$\delta_t = \frac{K_I^2}{EC\sigma_{ys}} .$$

C is equal to 1.0 for plane stress and taken to be 2.0 for plane strain.

For the Irwin plastic zone analysis an analogous relation was found:

$$\delta_t = \frac{4}{\pi} \frac{K_I^2}{E\sigma_{ys}} . \tag{3.8}$$

The foregoing relations between δ_t and K_I are important because they show that in the linear elastic regime the COD approach is compatible with LEFM concepts. However, the COD approach is not basically limited to the LEFM range of applicability, since occurrence of crack tip plasticity is inherent to it.

The major disadvantage of the COD approach is that equation (3.19) is valid only for an infinite plate with a central crack with length $2a$, and it is very difficult to derive similar formulae for practical geometries. This contrasts with the stress intensity factor

[4] Strictly speaking, a crack in a purely elastic material will have no CTOD. This is of course a hypothetical limiting case.

and J integral concepts. Thus in the first instance a characteristic value of CTOD at the onset of fracture can be used to compare the crack resistance of materials. It cannot be used, however, to calculate a critical crack length in a structure. In an attempt to overcome this disadvantage the COD design curve has been developed (see chapter 8).

6.7 Remarks on the COD Approach

Some additional remarks are given here concerning use of the COD approach:

1) When comparing the crack resistance of materials it is necessary to obtain $\delta_{t_{crit}}$. It has been shown experimentally that the COD depends on specimen size, geometry and plastic constraint, and so a standard COD test has been developed. This standard does not, however, precisely define the event at which δ_t is to be considered critical. There are three possibilities: δ_c, the value at instability without prior crack extension; δ_u, the value at the point of instability after stable crack extension; and δ_m, the value at maximum load (which is not necessarily identical to the other values, *e.g.* for specimens that still exhibit stable crack extension beyond the point of maximum load). The standard test is discussed further in chapter 7.
2) The COD approach has been developed mainly in the UK: more specifically, at the Welding Institute. The chief purpose was to find a characterizing parameter for welds and welded components of structural steels, which are difficult to simulate on a laboratory scale. Thus the COD approach is more strongly directed towards use in design of welded structures. (This, of course, does not mean that COD values cannot be used to compare and select materials.)
3) In welded steel structures the welds are most liable to fracture, not the material itself. At present the COD approach is a reliable way of accounting for the crack resistance of welds, since several weld quality specifications incorporating COD exist.

6.8 Relation Between J and CTOD

The J integral and COD concepts have been developed mainly in the USA and UK respectively. In the first instance the two concepts seem to be unrelated. In the late 1970s a number of expressions relating J and CTOD were published. They all take the form

$$J = M\sigma_{ys}\delta_t, \tag{6.33}$$

where M varies between 1.15 and 2.95. General acceptance of equation (6.33) is indicated by the use of a similar expression $J = \delta_t\sigma_o$ in the blunting line procedure for J_{Ic} testing, see section 7.4.

Hutchinson (reference 12 of the bibliography to this chapter) showed that derivation of equation (6.33) is relatively simple when the Dugdale strip yield model is used, although since it uses a model it does not constitute a definite proof of the relation between J and CTOD.

Consider a Dugdale type crack as shown in figure 6.8. The Dugdale plastic zone is

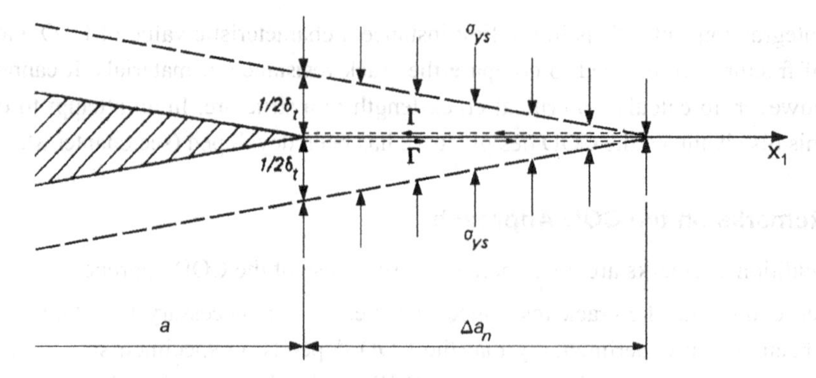

Figure 6.8. Dugdale model with contour Γ around the yielded strip.

assumed to be a strip of plastically deformed material along the x_1 axis carrying the yield stress, *i.e.* $\sigma_{22} = \sigma_{ys}$. Furthermore, the model assumes a state of plane stress, *i.e.* $\sigma_{33} = 0$. Given the symmetry of both geometry and load, the x_1 and x_2 axes must be principal stress directions, meaning that $\sigma_{12} = 0$.

Now the J integral formula (equation 6.27) is applied to a contour Γ that proceeds counterclockwise around the yielded strip at an infinitesimal distance from the x_1 axis. The contour starts and ends on the lower and upper crack flanks, respectively, at an infinitesimal distance from the crack tip. Thus

$$J = \int_\Gamma \left(W\,dx_2 - T_i \frac{\partial u_i}{\partial x_1} ds \right) \overset{(1)}{=} -\int_\Gamma \left(T_1 \frac{\partial u_1}{\partial x_1} ds + T_2 \frac{\partial u_2}{\partial x_1} ds \right)$$

$$\overset{(2)}{=} -\int_\Gamma \sigma_{ys}\, n_2 \frac{\partial u_2}{\partial x_1} ds \overset{(3)}{=} -\sigma_{ys} \int_\Gamma n_2 \frac{\partial u_2}{\partial x_1} \frac{dx_1}{-n_2}$$

$$\overset{(4)}{=} \sigma_{ys} \int_\Gamma du_2 = \sigma_{ys} \left[u_2 \right]_{-½\delta_t}^{+½\delta_t} = \sigma_{ys}\, \delta_t \qquad (6.34)$$

The steps (1) - (4) in this derivation are explained here:

1) Along the contour x_2 = constant, and thus $dx_2 = 0$. Note that there is no contribution to J from the right-hand end of the contour, since the vertical dimension (in the x_2 direction) of the contour is assumed infinitesimally small.
2) Along the contour $\underline{n} = (0,n_2)$. Using equation (6.4) in combination with the stress tensor for the yielded strip leads to a traction

$$\begin{bmatrix} T_1 \\ T_2 \end{bmatrix} = \begin{bmatrix} \sigma_{11} & 0 \\ 0 & \sigma_{ys} \end{bmatrix} \begin{bmatrix} 0 \\ n_2 \end{bmatrix} = \begin{bmatrix} 0 \\ \sigma_{ys}\, n_2 \end{bmatrix}.$$

3) According to equation (6.28) $ds = dx_1/{-n_2}$.

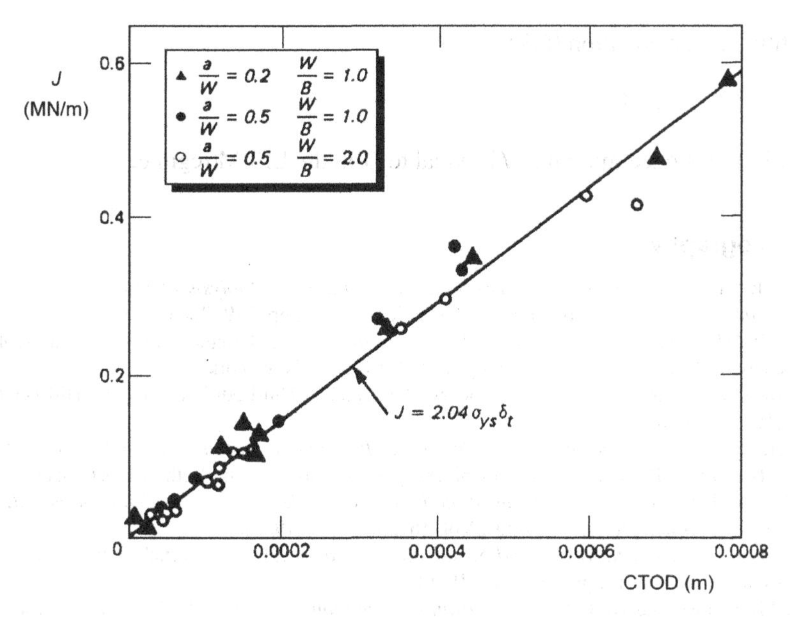

Figure 6.9. Experimental relation between J and CTOD for notched bend (SENB) tests of a steel with σ_{ys} = 370 MPa. After reference 15.

4) Since $dx_2 = 0$, the following holds:

$$du_2 = \frac{\partial u_2}{\partial x_1} dx_1 + \frac{\partial u_2}{\partial x_2} dx_2 = \frac{\partial u_2}{\partial x_1} dx_1 \,.$$

Remarks

1) Since only the definition of the strip yield model is used, and not the derivation of Δa_n, there is no restriction on the length of the plastic zone (Δa_n). Thus Hutchinson's analysis is valid under both LEFM and EPFM conditions. As stated earlier, the analysis is not a definite proof since it uses a model. However, it clearly suggests the existence of a relation between J and CTOD of the type given in equation (6.33).
2) More complex analyses, often using finite element calculations, also give results of the form suggested by equation (6.33). The multiplication factor M is found to be a function of σ_{ys}/E and the strain hardening exponent n, see references 13 and 14 of the bibliography.
3) Experimental values of M are often ≈ 2, *e.g.* figure 6.9, in contrast to equation (6.34), where $M = 1$. The higher values of M found experimentally are most probably due to the real plastic zone behaving differently from that assumed by the Dugdale approach.
4) Equation (3.20), relating the CTOD to K_I for LEFM conditions, can be written as

$$\delta_t = \frac{K_I^2}{E\sigma_{ys}} = \frac{G}{\sigma_{ys}} \,. \qquad (6.35)$$

Substitution into equation (6.34) gives

$$J = \delta_t \sigma_{ys} = G\,. \tag{6.36}$$

This illustrates once more that J is equal to G in the LEFM regime.

6.9 Bibliography

1. Rice, J.R., *A Path Independent Integral and the Approximate Analysis of Strain Concentration by Notches and Cracks*, Journal of Applied Mechanics, Vol. 35, pp. 379–386 (1968).
2. Rice, J.R., *Mathematical Analysis in the Mechanics of Fracture*, Fracture, An Advanced Treatise, ed. H. Liebowitz, Academic Press, Vol. 2, pp. 192–311 (1968): New York.
3. Timoshenko, S.P. and Goodier, J.N., Theory of Elasticity, Third edition, McGraw-Hill Book Company (1970): Tokyo.
4. Bakker, A., *The Three-Dimensional J-Integral. An Investigation into its use for Post-Yield Fracture Safety Assessment*, Ph.D. Thesis, Delft University of Technology, The Netherlands (1984).
5. Hutchinson, J.W., *Singular Behaviour at the End of a Tensile Crack in Hardening Material*, Journal of the Mechanics and Physics of Solids, Vol. 16, pp. 13–31 (1968).
6. Hutchinson, J.W., *Plastic Stress and Strain Fields at a Crack Tip*, Journal of the Mechanics and Physics of Solids, Vol. 16, pp. 337–347 (1968).
7. Rice, J.R. and Rosengren, G.F., *Plane Strain Deformation near a Crack Tip in Power Law Hardening Material*, Journal of the Mechanics and Physics of Solids, Vol. 16, pp. 1–12 (1968).
8. Shih, C.F. *Tables of Hutchinson-Rice-Rosengren Singular Field Quantities*, Brown University Report MRL E-147, Division of Engineering (1983): Providence, Rhode Island.
9. McMeeking, R.M. and Parks, D.M., *On Criteria for J-dominance of Crack Tip Fields in Large Scale Yielding*, Elastic-Plastic Fracture, ASTM STP 668, American Society for Testing and Materials, pp. 175–194 (1979): Philadelphia.
10. Wells, A.A., *Unstable Crack Propagation in Metals: Damage and Fast Fracture*, Proceedings of the Crack Propagation Symposium Cranfield, The College of Aeronautics, Vol. 1, pp. 210–230 (1962): Cranfield, England.
11. Burdekin, F.M. and Stone, D.E.W., *The Crack Opening Displacement Approach to Fracture Mechanics in Yielding*, Journal of Strain Analysis, Vol. 1, pp. 145–153 (1966).
12. Hutchinson, J.W., Nonlinear Fracture Mechanics, Technical University of Denmark, Department of Solid Mechanics (1979): Copenhagen.
13. Tracey, D.M., *Finite Element Solutions for Crack Tip Behaviour in Small Scale Yielding*, Transactions ASME, Journal of Engineering Materials and Technology, 98, pp. 146-151 (1976).
14. Rice, J.R. and Sorensen, E.P., *Continuing Crack Tip Deformation and Fracture for Plane Strain Crack Growth in Elastic-Plastic Solids*, Journal of the Mechanics and Physics of Solids, 26, pp. 163-186 (1978).
15. Dawes, M.G., *The COD Design Curve*, Advances in Elasto-Plastic Fracture Mechanics, ed. L.H. Larsson, Applied Science Publishers, pp. 279-300 (1980): London.

7 EPFM Testing

7.1 Introduction

In chapter 6 the two most widely known concepts of Elastic-Plastic Fracture Mechanics, the J integral and Crack Opening Displacement (COD) approaches, were discussed in general terms.

This chapter will deal with test methods for obtaining values of J and CTOD, including critical values J_{Ic} and $\delta_{t_{crit}}$. The chapter may be considered the EPFM counterpart of chapter 5, which discussed LEFM test methods.

The greater complexity of the J integral concept as compared to the COD concept is clearly demonstrated by the derivations in chapter 6. This difference in complexity is also found in the test methods. Therefore the discussion of J integral testing is subdivided into three sections:

1) The original J_{Ic} test method, section 7.2.
2) Alternative methods and expressions for J, section 7.3.
3) The standard J_{Ic} test, section 7.4.

The original J_{Ic} test method requires a large amount of data analysis. This problem led to the development of certain types of test specimen for which simple expressions for J could be derived, and ultimately to the standard J_{Ic} test.

Although it is not within the framework of EPFM testing, the K_{Ic} specimen size requirement (see section 5.2) is further discussed in section 7.5. The reason is that the J integral concept enables this criterion to be viewed from a different perspective.

The COD concept is much more straightforward than the J integral, at least from the experimental point of view. Thus only the standard $\delta_{t_{crit}}$ test itself will be described, namely in section 7.6.

With respect to standard test methods, it has already been remarked in sections 6.4 and 6.8 that the J integral concept was developed mainly in the USA and the COD concept in the UK. Consequently it is logical that the original standard for J_{Ic} was American (American Society for Testing and Materials) while the original COD test was the subject of an official British Standard (British Standards Institution, BSI), see references 1 and 2 of the bibliography to this chapter. At present both organizations have incorporated the J integral concept as well as the COD concept into their test standards.

7.2 The Original J_{Ic} Test Method

The first experimental method for determining J (more specifically J_{Ic}, the critical

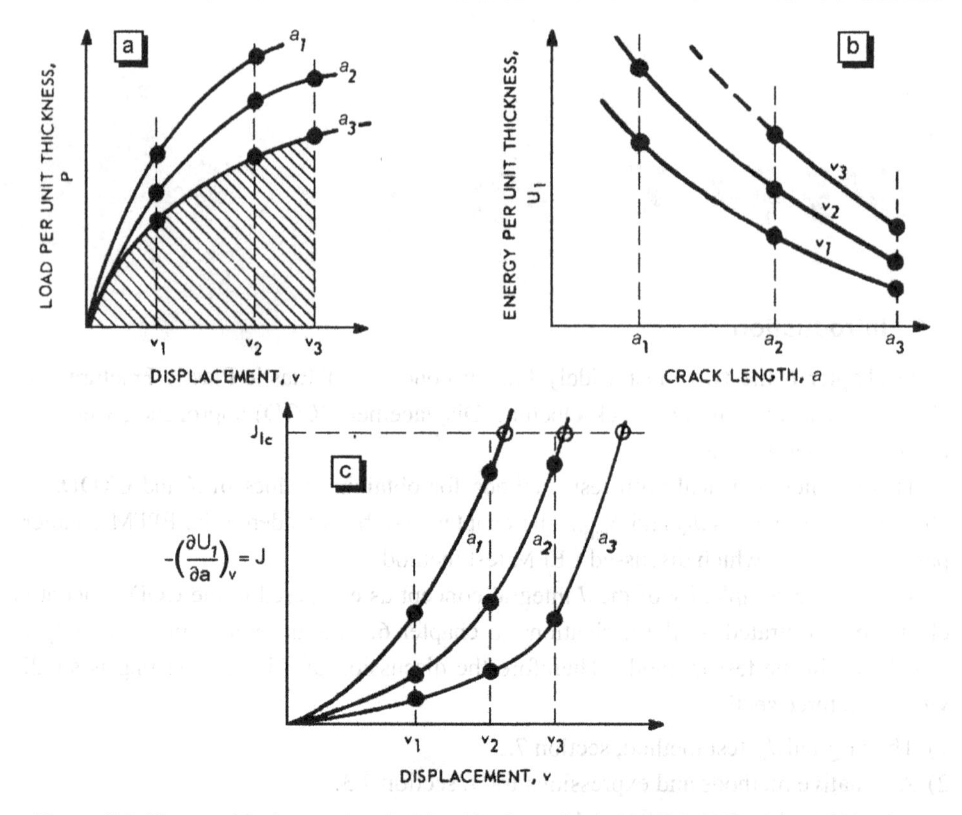

Figure 7.1. The graphical procedure involved in J_{Ic} testing according to Begley and Landes.

mode I value at the onset of crack extension) was published by Begley and Landes in 1971, reference 3 of the bibliography. The method is based on the definition of J as $-dU_p/da$, and requires graphical assessment of dU_p/da. The method will be illustrated with the help of figure 7.1, which schematically gives the graphical procedure for obtaining J_{Ic}.

The procedure is as follows:

1) Load-displacement diagrams are obtained for a number of specimens precracked to different crack lengths (a_1, a_2, a_3 in figure 7.1.a). Areas under the load-displacement curves represent the energy per unit thickness, U_1, delivered to the specimens. Thus the shaded area in figure 7.1.a is equal to the energy term U_1 for a specimen with crack length a_3 loaded to a displacement v_3.
2) U_1 is plotted as a function of crack length for several constant values of displacement, figure 7.1.b.
3) The negative slopes of the U_1–a curves, *i.e.* $-(\partial U_1/\partial a)_v$, are plotted against displacement for any desired crack length between the shortest and longest used in testing, figure 7.1.c. Since the elastic strain energy contents of a specimen is equal to the energy delivered to that specimen, it follows that $-(\partial U_1/\partial a)_v$ is equal to $-(\partial U_a/\partial a)_v$. In section 6.3 the energy definition of J was given as:

$$J = -\frac{dU_p}{da} = \frac{d}{da}(F - U_a) . \tag{6.1}$$

Since for crack extension under fixed grip conditions no work is performed by the loading system, it follows that:

$$J = -\left(\frac{\partial U_p}{\partial a}\right)_v = -\left(\frac{\partial U_a}{\partial a}\right)_v . \tag{7.1}$$

Hence figure 7.1.c in fact gives J-v curves for particular crack lengths.

4) Knowledge of the displacement v at the onset of crack extension enables J_{Ic} to be found from the J-v curve for each initial crack length. In figure 7.1.c the value of J_{Ic} is schematically shown to be constant as, ideally, it should be if J is an appropriate criterion for the onset of crack extension.

Knowledge of the critical displacement v is a weak step in the procedure. Begley and Landes used materials where the maximum in the load-displacement curve characterized the onset of crack growth. For other materials a crack extension measurement device (*e.g.* a potential drop measurement apparatus) is necessary. The method of Begley and Landes has the potential to find the applied J for an unknown geometry.

The graphical procedure described involves a large amount of data manipulation and replotting in order to obtain J-v calibration curves and hence J_{Ic}. There are thus many possibilities for errors, and so easier methods have been looked for, as will be discussed in section 7.3. However, the elegance of this original test method, in making direct use of the energy definition of J, remains and it is still used as a reference to check more recent developments.

7.3 Alternative Methods and Expressions for *J*

The main contribution to seeking alternatives for the Begley and Landes method was made by Rice *et al.*, reference 4 of the bibliography. Their analysis leads to simple expressions for J for certain types of specimen. However, before these expressions can be discussed it is necessary to consider alternative definitions of J.

Recall the expressions for U_p and J given in equations (4.2) and (6.1) respectively

$$U_p = U_o + U_a - F \tag{4.2}$$

$$J = -\frac{dU_p}{da} . \tag{6.1}$$

We will now consider the value of J for two extreme cases, namely for crack extension under constant displacement v and crack extension under constant load per unit thickness P. It follows that in both cases the change in potential energy due to a crack extension Δa is

$$\Delta U_p = U_p\big|_{a+\Delta a} - U_p\big|_a = \Delta U_a - \Delta F . \tag{7.2}$$

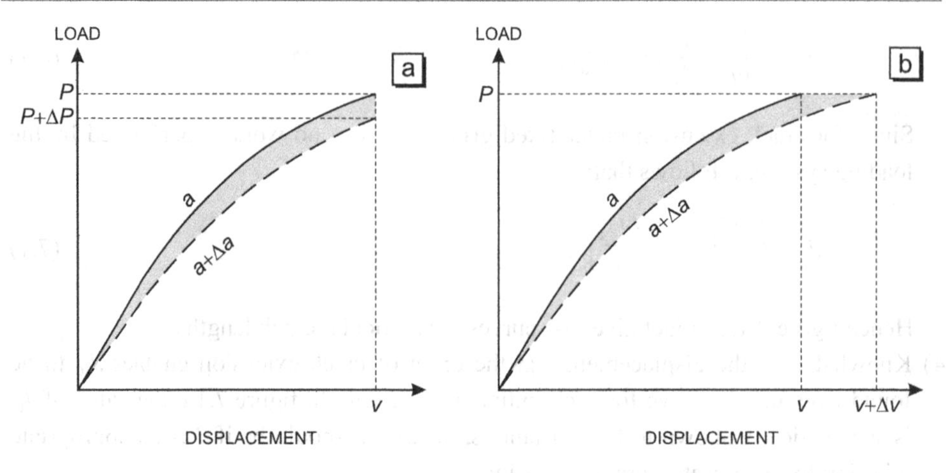

Figure 7.2. Crack extension in a nonlinear elastic body under (a) fixed grip and (b) constant load conditions.

For the case of a fixed grip condition, see figure 7.2.a, we may write

$$\Delta U_{\mathrm{a}} = \int_0^V P\Big|_{a+\Delta a} \mathrm{d}V - \int_0^V P\Big|_a \mathrm{d}V = \int_0^V \Delta P\,\mathrm{d}V \qquad \text{and} \qquad \Delta F = 0 \tag{7.3}$$

and thus, using equation (7.2), the change in potential energy is equal to

$$\Delta U_{\mathrm{p}} = \Delta U_{\mathrm{a}} - \Delta F = \int_0^V \Delta P\,\mathrm{d}V\,. \tag{7.4}$$

Note that ΔP is negative and that ΔU_{p} is equal to minus the shaded area between the curves for crack lengths a and $a+\Delta a$ in figure 7.2.a. From equation (6.1) it follows that

$$J = -\frac{\mathrm{d}U_{\mathrm{p}}}{\mathrm{d}a} = -\lim_{\Delta a \to 0} \frac{\Delta U_{\mathrm{p}}}{\Delta a} = -\int_0^V \left(\frac{\partial P}{\partial a}\right)_V \mathrm{d}V\,. \tag{7.5}$$

The case of a constant load condition, figure 7.2.b, is slightly more complicated, *i.e.*

$$\Delta U_{\mathrm{a}} = \int_0^{V+\Delta V} P\Big|_{a+\Delta a} \mathrm{d}V - \int_0^V P\Big|_a \mathrm{d}V \qquad \text{and} \qquad \Delta F = P\,\Delta V\,. \tag{7.6}$$

This leads to

$$\Delta U_{\rm p} = \Delta U_{\rm a} - \Delta F = \int_0^{v+\Delta v} P\Big|_{a+\Delta a} {\rm d}v - \left(\int_0^{v} P\Big|_{a} {\rm d}v + P\Delta v \right), \tag{7.7}$$

which, when regarded more closely, is equal to the shaded area in figure 7.2.b. Therefore we may rewrite $\Delta U_{\rm p}$ as

$$\Delta U_{\rm p} = -\int_0^{P} \Delta v \, {\rm d}P \,. \tag{7.8}$$

Note that Δv is positive now. Thus

$$J = -\frac{{\rm d}U_{\rm p}}{{\rm d}a} = -\lim_{\Delta a \to 0} \frac{\Delta U_{\rm p}}{\Delta a} = \int_0^{P} \left(\frac{\partial v}{\partial a} \right)_P {\rm d}P \,. \tag{7.9}$$

The same results can also be found purely algebraically. For crack extension under fixed grip conditions J is

$$J = -\left(\frac{\partial U_{\rm p}}{\partial a} \right)_v = \left(\frac{\partial F}{\partial a} - \frac{\partial U_{\rm a}}{\partial a} \right)_v = \left(0 - \frac{\partial U_{\rm a}}{\partial a} \right)_v = -\left(\frac{\partial U_{\rm a}}{\partial a} \right)_v = -\left(\frac{\partial}{\partial a} \int_0^{v} P \, {\rm d}v \right)_v = -\int_0^{v} \left(\frac{\partial P}{\partial a} \right)_v {\rm d}v \,,$$

while on the other hand, for the case of constant load conditions

$$J = -\left(\frac{\partial U_{\rm p}}{\partial a} \right)_P = \left(\frac{\partial F}{\partial a} - \frac{\partial U_{\rm a}}{\partial a} \right)_P = \left(P \frac{\partial v}{\partial a} - \frac{\partial}{\partial a} \int_0^{v} P \, {\rm d}v \right)_P = \left(P \frac{\partial v}{\partial a} \right)_P - \frac{\partial}{\partial a} \left(P v - \int_0^{P} v \, {\rm d}P \right)_P = \int_0^{P} \left(\frac{\partial v}{\partial a} \right)_P {\rm d}P \,.$$

Thus the alternative definitions of J, for crack extension under fixed grip or constant load conditions are

$$J = -\int_0^{v} \left(\frac{\partial P}{\partial a} \right)_v {\rm d}v = \int_0^{P} \left(\frac{\partial v}{\partial a} \right)_P {\rm d}P \tag{7.10}$$

Note the different sign for fixed grip and constant load conditions. This is analogous to the formulae for G, equations (4.23).

Using equation (7.10), Rice *et al.* showed in 1973 that it is possible to determine $J_{\rm Ic}$ from a single test of certain types of specimen. As an example, J for a deeply cracked bar in bending was derived as

$$J = \frac{2}{Bb} \int_0^{\theta_c} M \, d\theta_c \,, \tag{7.11}$$

where B is the thickness of the bar, b is the size of the uncracked ligament ahead of the crack, M is the bending moment and θ_c is the part of the total bending angle θ due to introduction of the crack. More recently, see reference 5 of the bibliography to this chapter, it was found that J is evaluated more accurately by simply using the total bending angle θ instead of θ_c, *i.e.*

$$J = \frac{2}{Bb} \int_0^{\theta} M \, d\theta \tag{7.12}$$

Equation (7.12) is important, since it applies to a basic cracked configuration. Therefore a derivation is given in some detail here with the help of figure 7.3. For this deeply cracked bar loaded in bending the ligament size, b, is chosen small compared to the width of the bar, W, so that it may be safely assumed that all plastic deformation is confined to this ligament.

M' is the bending moment per unit thickness, *i.e.* $M' = M/B$. We will use the definition of J for fixed grip conditions, *i.e.* the first form of equation (7.10). P and v are converted to M' and θ by assuming the moment is applied through three-point bending. The load per unit thickness, P, can be written as $4M'/L$, where L is the span of the bend specimen. Furthermore, since plasticity is confined to the ligament, the sides of the beam will remain straight and v is equal to $\theta L/4$. Finally, since $b = W - a$ it follows that $\partial/\partial a = -\partial/\partial b$. The first form of equation (7.10) can now be written as

$$J = -\int_0^{v} \left(\frac{\partial P}{\partial a}\right)_v dv = -\int_0^{\theta} \left(\frac{\partial M'}{\partial a}\right)_\theta d\theta = +\int_0^{\theta} \left(\frac{\partial M'}{\partial b}\right)_\theta d\theta \,. \tag{7.13}$$

Since this expression cannot be evaluated experimentally, an analytical relation must be found between θ, b and M'. Rice *et al.* argued that a dimensional analysis can be used to obtain this relation. However, here an alternative reasoning will be used.

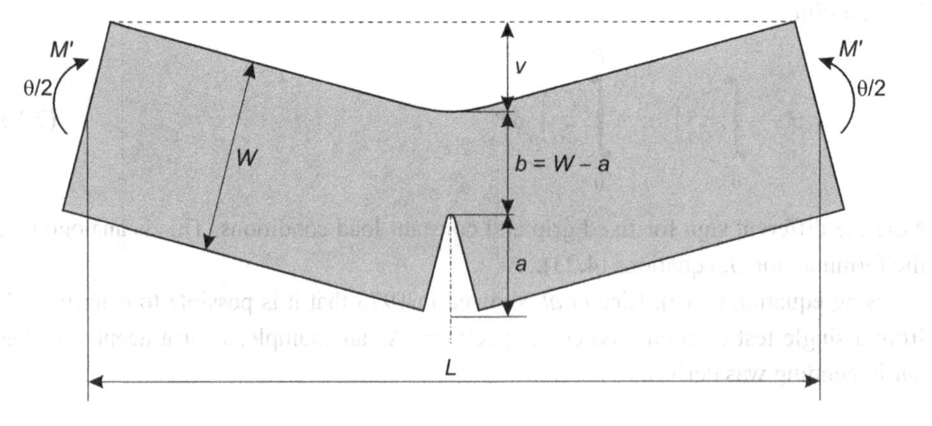

Figure 7.3. A deeply cracked bar loaded in bending.

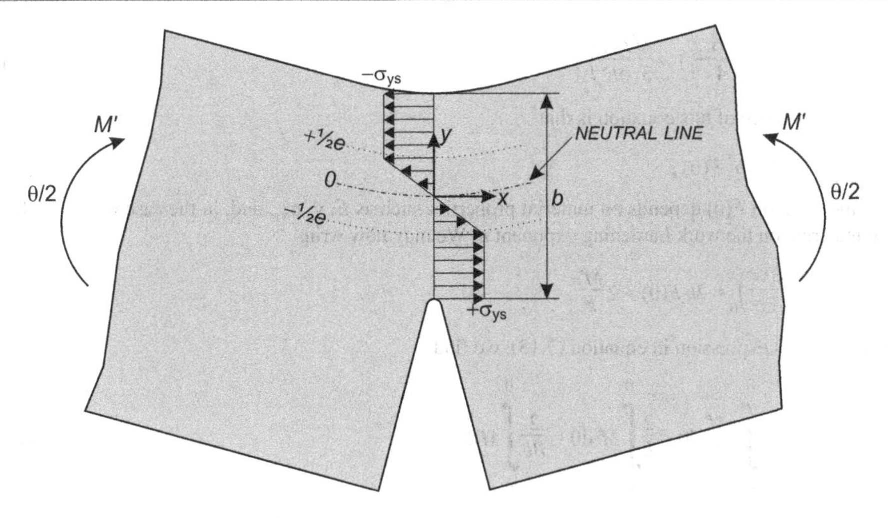

Figure 7.4. Stress distribution in the critical ligament

Elastic – perfectly plastic material behaviour is assumed, leading to a distribution in the ligament of the stress component parallel to the neutral line, σ_x, as follows (see figure 7.4):

$$\sigma_x(y) = \begin{cases} \dfrac{2y}{e}\sigma_{ys} & \text{for } |y| < \frac{1}{2}e \\ \sigma_{ys} & \text{for } \frac{1}{2}e < |y| < \frac{1}{2}b\,, \end{cases} \tag{7.14}$$

where y is the distance from the neutral line. The moment corresponding to this stress distribution can be straightforwardly calculated as

$$M' = \int_{-\frac{1}{2}b}^{+\frac{1}{2}b} y\,\sigma_x(y)\,\mathrm{d}y = \sigma_{ys}\left(\frac{b^2}{4} - \frac{e^2}{12}\right). \tag{7.15}$$

The width e of the elastic part of the ligament is now estimated by assuming that in the small region around the ligament the neutral line takes the shape of a circle segment with radius R and that all planes normal to the neutral line remain normal to that line. Under these assumptions the strain, ε_x, parallel to the neutral line can be written as a function of y:

$$\varepsilon_x(y) = \frac{2\pi(R-y) - 2\pi R}{2\pi R} = -\frac{y}{R}. \tag{7.16}$$

At the boundary between the elastic and plastic parts of the ligament, *i.e.* $y = \pm\frac{1}{2}e$, the absolute value of ε_x is approximately equal to the yield stress divided by E', *i.e.* E for plane stress and $E/(1-\nu^2)$ for plane strain. Thus using equation (7.16) it follows that

$$\frac{\sigma_{ys}}{E'} \approx \frac{\frac{1}{2}e}{R} \quad \text{or} \quad e \approx \frac{2R\sigma_{ys}}{E'}.$$

Now it is assumed that the length of the circular shaped segment of the neutral line in the region around the ligament is roughly of the order of the size of the ligament b, *i.e.* $R\theta \approx b$. This leads to

$$e \approx \frac{2b\sigma_{ys}}{\theta E'}. \tag{7.17}$$

Substitution of this expression in equation (7.15) leads to

$$M' = \frac{b^2 \sigma_{ys}}{4}\left\{1 - \frac{4}{3}\left(\frac{\sigma_{ys}}{\theta E'}\right)^2\right\}. \tag{7.18}$$

The significance of this equation is that

$$M' = b^2 F(\theta), \tag{7.19}$$

where the function $F(\theta)$ depends on material properties such as E, ν, σ_{ys}, and, in the case of a work hardening material, on the work hardening exponent n. We may now write

$$\left(\frac{\partial M'}{\partial b}\right)_\theta = 2b\,F(\theta) = 2\frac{M'}{b}. \tag{7.20}$$

Substituting this expression in equation (7.13), we find

$$J = \int_0^\theta 2\frac{M'}{b}\,d\theta = \frac{2}{b}\int_0^\theta M'\,d\theta = \frac{2}{Bb}\int_0^\theta M\,d\theta. \tag{7.21}$$

For a deeply cracked bar it is reasonable to assume that all plasticity is restricted to the ligament, and thus the two halves of the bar remain straight. This enables equation (7.12) to be rewritten as the more practical expression

$$J = \frac{2}{Bb}\int_0^v P\,dv, \tag{7.22}$$

where P is the load in terms of a force, *i.e.* no longer defined per unit thickness, and v is the displacement in the load line, termed the *load-line displacement*.

In a J_{Ic} test the load P acting on a cracked bar is measured as a function of the load-line displacement v. Using equation (7.22) J can then be determined for any displacement by calculating the area under the P-v curve up to that displacement, U. At the onset of crack extension, J is equal to J_{Ic}. Therefore

$$J = \frac{2U}{Bb} \quad \text{and} \quad J_{Ic} = \frac{2U_{cr}}{Bb}. \tag{7.23}$$

where U_{cr} is the area under the P-v curve at the onset of crack extension.

Hence in principle J_{Ic} can be determined by performing one test only in which the specimen is loaded until the onset of crack extension. However this is not normally done. The reason is that detection of the beginning of crack extension is difficult. It can only be done with costly apparatus as potential drop, acoustic emission, ultrasonic, eddy current etc., where each has its specific difficulties. An alternative is to make a number of tests whereby each specimen is loaded to give a small but different crack extension Δa. Then the values of J (which are, strictly speaking, invalid) are plotted versus Δa and extrapolated to $\Delta a = 0$ in order to obtain J_{Ic}. An example of this method is given in figure 7.5.

J-Δa lines like those in figure 7.5 are called J resistance curves, by analogy with the

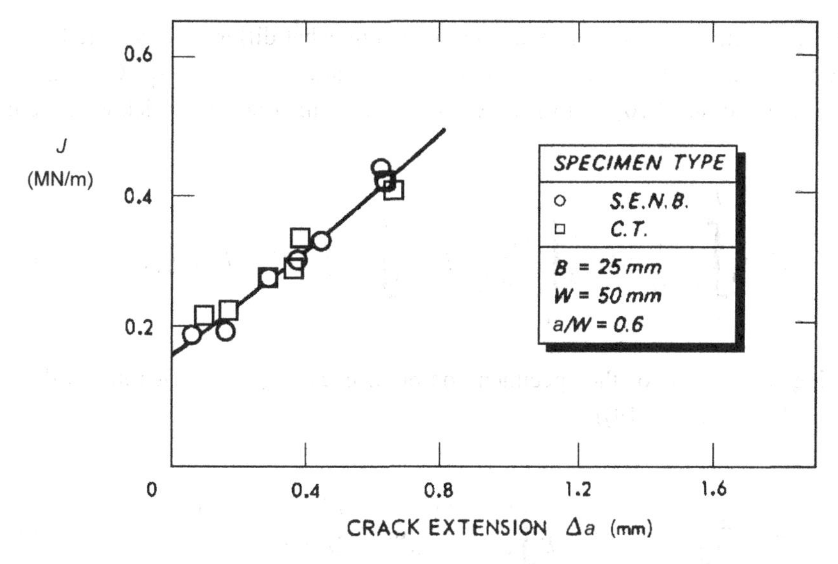

Figure 7.5. J-Δa plots for A-533B steel, after reference 6 of the bibliography.

LEFM R-curve. This seems slightly misleading, since J is strictly valid only up to the beginning of crack extension and not beyond it. However, it must be noted that under certain restrictions J resistance curves can be used to predict stable crack extension. This subject is discussed in chapter 8.

The J integral expression in equation (7.23) and the multiple specimen method just described, form the basis for the standard J_{Ic} test, which is discussed in the next section.

7.4 The Standard J_{Ic} Test

Before publication of the standard J_{Ic} test some ten different procedures had been used. Chipperfield (reference 7) reviewed these methods and showed that J_{Ic} values obtained in different ways varied by up to 20%. This clearly demonstrated the need for a standard test.

Original J_{Ic} Test Standard

A proposal for a standard J_{Ic} test was published in 1979. This proposal became an ASTM standard and was first published as such in 1981 under the designation ASTM E 813, reference 1 of the bibliography. This standard describes J_{Ic} determination using three-point notched bend (SENB) and compact tension (CT) specimens. Roughly these are the same specimen geometries as those for K_{Ic} testing (see figures 5.2 and 5.3), but there are a number of differences in detail. For both specimen configurations J is given simply by a form of equation (7.23), *i.e.* $J = (2U/Bb) \cdot f(a/W)$, where $f(a/W)$ depends on the specimen type.

Revised Test Standard

In 1989 a revised version of standard E 813 was published, which is referenced as number 8 of the bibliography. In this standard the same specimen geometries are described, but for experimental reasons J is evaluated in a somewhat different way. The load-line displacement is divided into an elastic and a plastic part, *i.e.* $v = v_{el} + v_{pl}$. Consequently, reverting to equation (7.10), with P now no longer defined per unit thickness, we may write

$$J = \frac{1}{B}\int_0^P \left(\frac{\partial v}{\partial a}\right)_P \mathrm{d}P = \frac{1}{B}\int_0^P \left(\frac{\partial v_{el}}{\partial a}\right)_P \mathrm{d}P + \frac{1}{B}\int_0^P \left(\frac{\partial v_{pl}}{\partial a}\right)_P \mathrm{d}P = J_{el} + J_{pl}\,. \tag{7.24}$$

Expressing v_{el} in terms of the specimen compliance, *i.e.* $v_{el} = C{\cdot}P$, it follows that (*cf.* equations (4.16.b) and (4.19))

$$J_{el} = \frac{1}{B}\int_0^P \left(\frac{\partial v_{el}}{\partial a}\right)_P \mathrm{d}P = \frac{1}{B}\int_0^P \left(\frac{\partial (C{\cdot}P)}{\partial a}\right)_P \mathrm{d}P = \frac{P^2}{2B}\frac{\partial C}{\partial a} = G = \frac{1-\nu^2}{E}K_I^2\,. \tag{7.25}$$

Since the SENB and CT specimens have the same geometry as the standard K_{Ic} specimens, K_I can be calculated using equations (5.1) and (5.2).

Using the same reasoning as given in the previous section, the plastic part of J, J_{pl}, can be related to the area under the P-v_{pl} curve up to the current value of v_{pl}, U_{pl}. The ASTM standard uses the relation

$$J_{pl} = \frac{\eta U_{pl}}{B_N b}\,, \tag{7.26}$$

where η = plastic work factor = $\begin{cases} 2 & \text{for SENB specimens} \\ 2 + 0.522\, b/W & \text{for CT specimens} \end{cases}$

B_N = net specimen thickness, which is equal to B if no side grooves are present.

Figure 7.6 illustrates how the plastic work U_{pl} is calculated. First the total work U is determined by integrating the P-v curve and then the elastic part of the work is subtracted. This elastic part is equal to $\frac{1}{2}v_{el}P$ or, using the elastic specimen compliance C, equal to $\frac{1}{2}CP^2$.

Clearly C has to be known to carry out this procedure. Note also that C depends on the current crack length. It can be determined either by calculating it using the formulae given in the ASTM standard that express C as a function of crack length, specimen dimensions and Young's modulus or by measuring it directly through partial unloading during the test (see also under the next subheading).

J_{Ic} Test Procedure

The steps involved in setting up and conducting a J_{Ic} test are:

1) Selection of specimen type (notch bend or compact tension) and preparation of shop drawings.

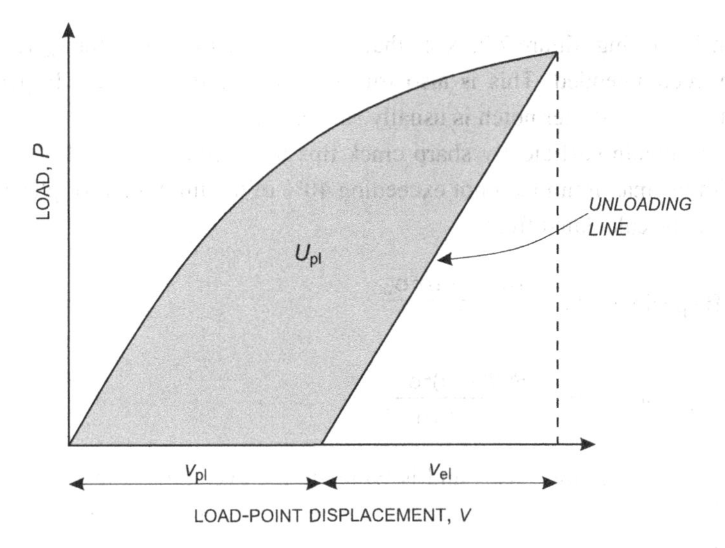

Figure 7.6 The part of the area under the P-v curve that represents the plastic work U_{pl}.

2) Specimen manufacture.
3) Fatigue precracking.
4) Obtain test fixtures and clip gauge for crack opening displacement measurement.
5) Testing.
6) Data analysis.
7) Determination of a provisional J_{Ic} (J_Q).
8) Final check for J_{Ic} validity.

Steps (1) – (5) will be concisely reviewed here insofar as they differ from similar steps for K_{Ic} testing in section 5.2. Steps (6) – (8) are considered under the next sub-heading in this section.

For both SENB and CT specimens the initial crack length (*i.e.* notch plus fatigue precrack) must be greater than 0.5 W to ensure validity of the formulae used to evaluate J. The maximum crack length is 0.75 W, while a value of 0.6 W is usually optimum from an experimental viewpoint.

A special feature of J_{Ic} testing is that the clip gauge has to be positioned in the load line. For the CT specimen this means that the shape of the starter notch is different to

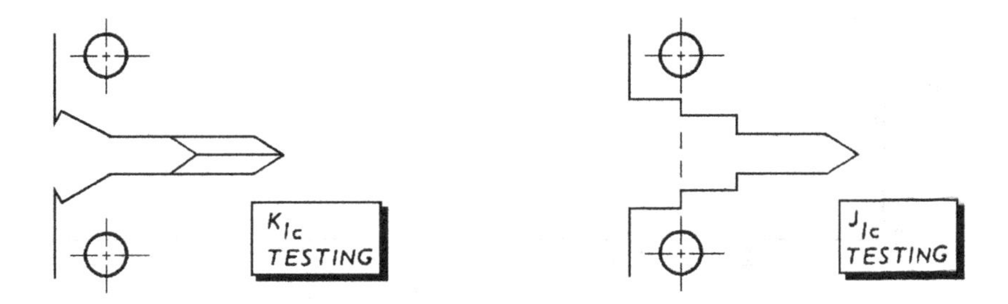

Figure 7.7. CT specimen starter notches.

that used in K_{Ic} testing, figure 7.7. Note that a chevron starter notch for J_{Ic} testing is not specifically recommended. This is also true for the SENB specimen. Experience has shown that a straight starter notch is usually sufficient.

In order to obtain sufficiently sharp crack tips the specimen should be fatigue pre-cracked with the maximum load not exceeding 40% of the limit load for plastic collapse P_L, which can be calculated from

$$\text{SENB specimen} \quad P_L = \frac{4B(W-a)^2\sigma_o}{3S} \tag{7.27}$$

$$\text{CT specimen} \quad P_L = \frac{B(W-a)^2\sigma_o}{(2W+a)}, \tag{7.28}$$

where σ_o is called the flow stress and is typically the average of the yield strength σ_{ys} and the ultimate tensile strength σ_{uts}, *i.e.* $\sigma_o = \frac{1}{2}(\sigma_{ys} + \sigma_{uts})$. The use of σ_o is to account for strain hardening.

The J_{Ic} tests must be carried out under controlled displacement conditions in order to obtain stable crack extension over the whole test range. This means that preferably an electro-mechanical testing machine must be used.

In section 7.3 it was stated that the basis for the standard J_{Ic} test is the multiple specimen method, *i.e.* a number of specimens are loaded to give small but different amounts of crack extension Δa. However, ASTM E 813 does allow a truly single specimen J_{Ic} determination. This involves the use of some technique for measuring the current crack extension during a test, enabling the determination of the J resistance curve defined in section 7.3.

A frequently used method for crack-length monitoring is the *unloading compliance technique*. After loading the specimen until a small amount of crack extension occurs,

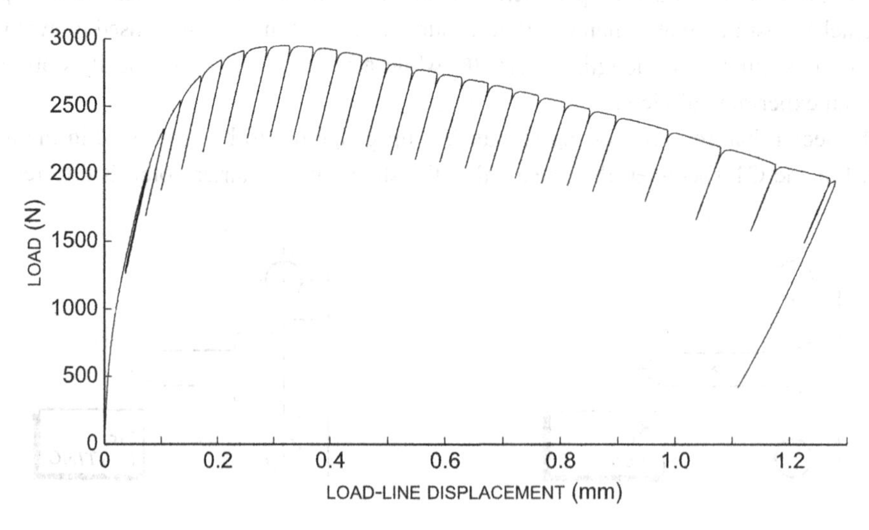

Figure 7.8. An example of the unloading compliance technique.

the load is partially removed and subsequently reapplied, see figure 7.8. To avoid reversed plasticity having any effect on the test results, the maximum unloading range is set to the smaller of 50% of the current load or 20% of P_L. In the load-displacement (P–v) diagram this loading procedure is reflected as the first part of the elastic compliance line for unloading. From the resulting elastic compliance, C, the instantaneous crack length, a, and thus also $\Delta a = a - a_0$ can be calculated. ASTM E 813 gives formulae equating the dimensionless crack length a/W to the dimensionless compliance for SENB and CT specimens. The current values for a, Δa, P and the P-v curve up to the current displacement lead to one point on the J–Δa curve. By repeating this process a number of times a J resistance curve can be obtained from a single specimen. A disadvantage of the method is that an accurate measurement of the unloading compliance line requires suitable equipment and sufficient experimental skill.

For both the multiple and the single specimen technique the specimen is broken afterwards to measure the crack extension visually from the crack surface. Note that for the single specimen technique this final crack extension is determined only to verify the accuracy of the unloading compliance technique. To be able to measure the crack extension a marking technique must be employed for distinguishing between Δa and the residual fracture due to breaking open the specimen after testing. One possibility is heat tinting, *i.e.* heating the specimen in air to cause oxide discoloration of existing crack surfaces. Another is to fatigue cycle after the J_{Ic} test. Details of these techniques are given in reference 8 of the bibliography.

The measurement of the crack extension gives specific problems. J integral test specimens are usually thick, such that 'crack front tunnelling' occurs during both precracking and J_{Ic} testing. This is illustrated schematically in figure 7.9. Experience has shown that to obtain consistent values of J and J_{Ic} it is necessary to take averages of at least nine measurements of a and Δa equally spaced across the specimen thickness, and to count the averages of side surface crack lengths as one measurement only.

Data Analysis and Determination of J_{Ic}

The data analysis consists of calculating J values for a number of crack extensions Δa.

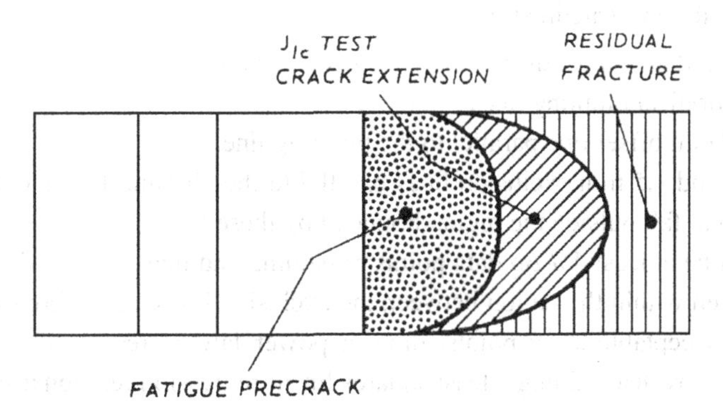

Figure 7.9. Schematic of a J_{Ic} test specimen broken open after testing.

The elastic part of each J value, J_{el}, is evaluated with equation (7.25) by substituting the K_I value corresponding to the load and the crack length at the moment the crack extension Δa was reached. For J_{pl} the load-displacement record is analysed to obtain the area U_{pl} under the curve up to the P-v point corresponding to crack extension Δa. Values of J_{pl} are then calculated by inserting U_{pl} and values of the crack length a into equation (7.26).

These J-Δa points are used in determining the provisional J_{Ic} (J_Q). However, depending on their value, some points may yet turn out to be unacceptable. To check for acceptability and at the same time determine J_Q a plot more or less similar to figure 7.5 must be constructed as shown schematically in figure 7.10.

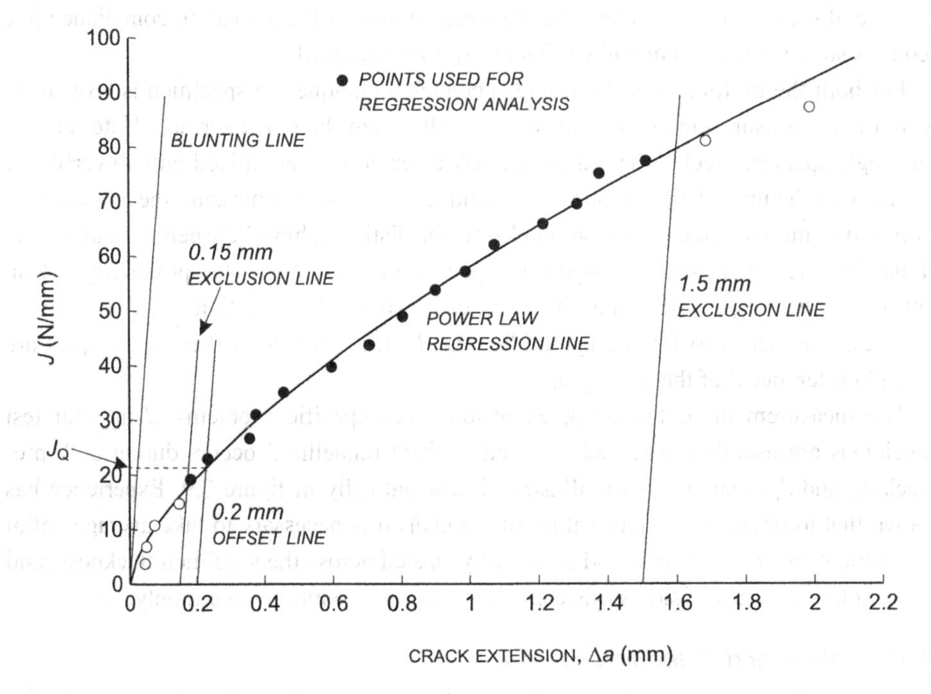

Figure 7.10. Schematic determination of acceptable J values and of J_Q.

The procedure for constructing this figure is:

1) Plot the J-Δa data points, discarding points with J values exceeding $b\sigma_o/15$.
2) Draw a theoretical blunting line $J = 2\sigma_o\Delta a$.
3) Draw a 0.2 mm offset line parallel to the blunting line.
4) Draw 0.15 and 1.5 mm exclusion lines parallel to the blunting line and discard all J-Δa points that fall outside the region bounded by these lines.
5) There must be at least 4 J-Δa data points remaining and they must be distributed sufficiently even within the region between the exclusion lines (see reference 8).
6) Using the acceptable J-Δa points, draw a power law regression line of the form $J = C_1(\Delta a)^{C_2}$ by determining a least squares linear regression relation according to:

$$\ln J = \ln C_1 + C_2 \ln (\Delta a) . \tag{7.29}$$

7) Determine the intersection of the power law regression line with the 0.2 mm offset line. The resulting J value is designated J_Q. The ASTM standard suggests an iterative procedure to determine the point of intersection with sufficient accuracy.
8) Draw two vertical lines through the intersections of the exclusion lines with the regression line. These vertical lines represent the minimum and maximum crack extensions. If data points fall outside this range they should be discarded and the procedure should be repeated starting at point 5.

Finally, for J_Q to qualify as a valid J_{Ic}, it is required that:

1) The specimen dimensions satisfy the equation

$$B \text{ and } W - a \text{ both} > \frac{25\, J_Q}{\sigma_o}. \tag{7.30}$$

2) The slope of the regression line at J_Q is smaller than σ_o.
3) None of the test specimens have experienced brittle fracture.
4) No excessive crack front tunnelling has occurred (see reference 8).
5) For the single specimen technique the predicted final crack extension does not deviate more than 15% from the crack extension measured directly from the crack surface.

Some Background to the J_{Ic} Determination

1) The minimum thickness requirement $B > 25\, J_Q/\sigma_o$ ensures that crack extension Δa occurs under plane strain. It is an empirical requirement based on tests with steels.
2) The minimum ligament length requirement $b = (W - a) > 25\, J_Q/\sigma_o$ is also empirical and is intended to prevent net section yield, see section 6.1. For this same reason all measured J values exceeding $b\sigma_o/15$ are discarded.
3) The blunting line procedure was adopted to account for the apparent increase in crack length owing to crack tip blunting. This apparent increase in crack length will be less than or equal to the blunted crack tip radius, which in turn is half the crack opening displacement δ_t. Thus the apparent $\Delta a \leq 0.5\delta_t$. Assuming $\delta_t = J/\sigma_o$, a relation discussed earlier in section 6.8, the apparent crack extension due to crack blunting can be accounted for by $\Delta a = 0.5\delta_t = J/2\sigma_o$, or

$$J = 2\sigma_o \Delta a\,. \tag{7.31}$$

   Although the concept of accounting for crack blunting is correct, the use of equation (7.31) can still be criticised for two reasons, which are discussed in points 5 and 6.
4) J_Q is not the J value at the initiation of crack extension, since it is determined as the intersection of the 0.2 mm offset line and the power law regression line. In the original ASTM standard (see reference 1) J_Q was determined as the intersection of a linear regression line and the blunting line and could thus be regarded as J at initiation. This procedure, however, was found to introduce much scatter in J_{Ic} values, because the transition between the blunting process and actual crack extension is not always distinct.

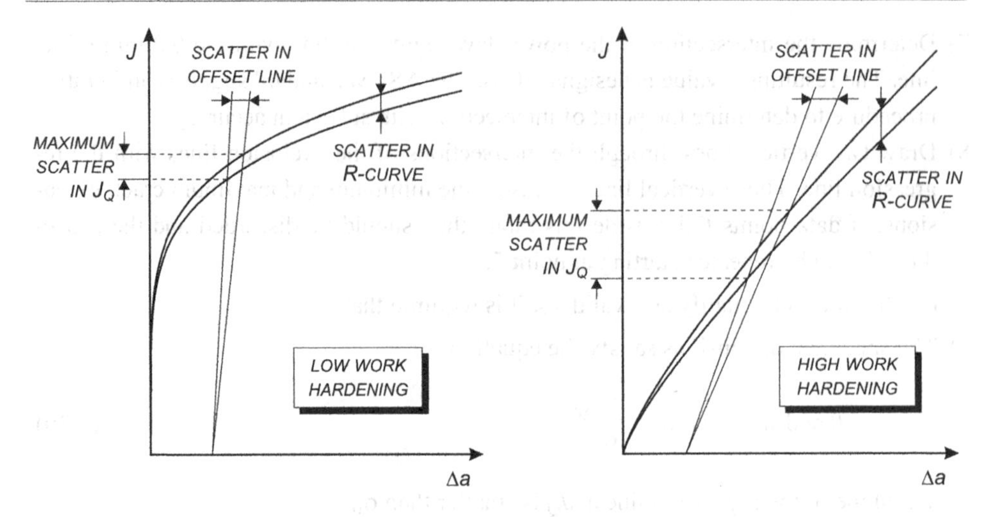

Figure 7.11. Influence of work hardening on J_Q estimation error.

Note that the current approach is analogous to that for quantities like the yield strength defined at 0.2% offset strain and K_{Ic} defined at 2% stable crack growth.

5) The blunting line and the J resistance curve are influenced by work hardening. With more work hardening the slope of the blunting line is less, while the J resistance curve is observed to be steeper. This leads to much more potential error in estimating J_Q, as is shown in figure 7.11. In the ASTM procedure this point is addressed by the requirement that the slope of the regression line at J_Q is smaller than σ_o.
6) The blunting line, equation (7.31), is based on $J = \delta_t\sigma_o$. As was discussed in more detail in chapter 6, relations of the form $J = M\delta_t\sigma_o$ are reasonable, but the factor M can vary between 1 and 3, and often has a value ~2. This means that the blunting line slope according to the ASTM standard may be too shallow, which results in an over-

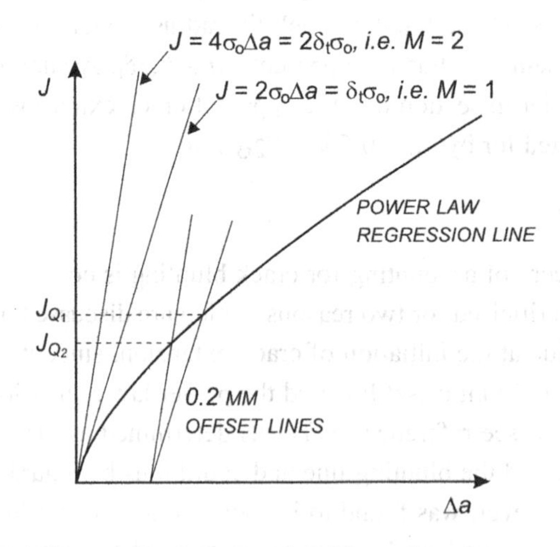

Figure 7.12. Influence of the relation between J and δ_t on J_Q.

estimation of J_Q, as figure 7.12 shows. Experiments have shown that the overestimation of J_Q may be as much as 10%, reference 7 of the bibliography.

It should be noted that in recent ASTM publications (*e.g.* reference 10) the use of a higher blunting line slope, obtained from experimental data, is suggested.

7) The 0.15 mm exclusion line ensures that Δa is at least 0.15 mm and so can be measured accurately enough. The 1.5 mm exclusion line ensures that Δa is generally less than 6% of the remaining ligament in the SENB and CT specimens proposed for J_{Ic} testing, and it has been shown that up to this amount of crack extension the J integral formula, equation (7.26), remains valid.
8) Steps 5 and 8 of the procedure to construct figure 7.10 and the final checking criteria 4 and 5 for J_{Ic} validity have been devised to minimise scatter and improve the reliability of the J resistance curve.

Concluding Remarks

It should be noted that the test procedure according to ASTM standard E 813 allows only J_{Ic} (or J_Q) to be determined. There are also standardized test procedures for determining the whole J resistance curve, involving larger amounts of stable crack extension than for the J_{Ic} determination. With the resulting curve the effect of stable crack growth on the material's crack resistance in the elastic-plastic regime is quantified. This type of test will not be discussed here, but the topics of J controlled crack growth and use of the J resistance curve will be elaborated on in chapter 8.

The J_{Ic} test procedure described in this section is restricted to cases of crack extension by means of a ductile failure mechanism (see chapter 12). However, J can also be used to characterize the onset of brittle fracture, before or during stable crack extension. The restrictions imposed on the amount of crack tip constraint are then much more severe (see reference 9).

It should be further noted that in 1997 the ASTM published a standard (see reference 10) that combines different types of fracture toughness measurements into a single set of test rules. It includes the determination of K_{Ic}, J_{Ic}, J resistance curve, $\delta_{t_{crit}}$ (see section 7.6) and also critical values for J and δ_t in the case of brittle fracture. The idea behind this new standard is to enable fracture toughness evaluation using a single experimental procedure, while minimising the risk of invalid test results because of unexpected material behaviour. If the evaluation of one critical fracture parameter fails it may be possible to evaluate another parameter using the same experimental data. However, the procedure for determining J_{Ic} described in this section 7.4 is more or less copied in this recent ASTM standard, and is therefore still relevant within the context of this course.

7.5 The K_{Ic} Specimen Size Requirement

Although it is not really part of EPFM testing, some attention will be paid to the evaluation of K_{Ic} for relatively tough materials. The reason is that J resistance curves enable a somewhat different view on this subject.

K_{Ic} is a workable fracture criterion for higher strength, lower toughness materials, see chapter 5, section 5.2. The specimen sizes required for a valid K_{Ic} are convenient to handle for these materials. For lower strength, higher toughness materials K_{Ic} cannot be measured so conveniently because the specimen size required for a valid test may be prohibitively large. However, Landes (see reference 11) argued that the assumption that a K_{Ic} always can be measured for any material provided that a large enough specimen is used is not true. He showed that for some materials it is impossible to measure a valid K_{Ic}.

For ductile materials, *i.e.* materials that exhibit stable crack extension prior to failure, the K_{Ic} is defined at the point where the stable crack extension Δa is 2% of the original specimen crack size a. The specimen size requirement in terms of crack length is given by

$$a \geq 2.5\left(\frac{K_{Ic}}{\sigma_{ys}}\right)^2 . \tag{7.32}$$

Combining this relation with $\Delta a = 0.02 \cdot a$ yields

$$\Delta a \geq 0.05\left(\frac{K_{Ic}}{\sigma_{ys}}\right)^2 , \tag{7.33}$$

a relation which should be fulfilled to obtain a valid K_{Ic}.

To further examine the size requirement it is convenient to write K in terms of J, using equation (6.30). For arbitrary values of J equation (7.33) can be rewritten as

$$\Delta a \geq 0.05 \frac{E}{1-\nu^2}\frac{J}{\sigma_{ys}^2} \tag{7.34}$$

or

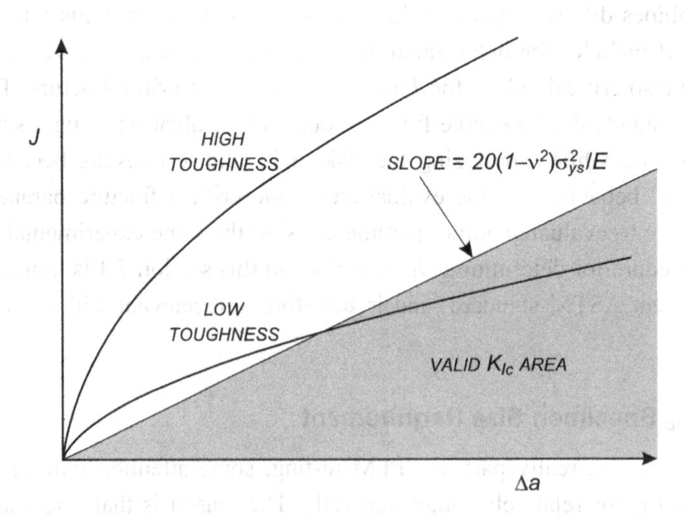

Figure 7.13. Schematic showing the K_{Ic} size requirement as an area in a J-Δa plot.

$$J \leq 20\,(1-\nu^2)\frac{\sigma_{ys}^2}{E}\Delta a\,. \tag{7.35}$$

As a function of the absolute amount of crack extension Δa, this relation gives the maximum J value for which the K_{Ic} size requirement with respect to the crack length, equation (7.32), would be fulfilled. This condition is represented by the shaded area in the J-Δa plot of figure 7.13. Also, in this figure J resistance curves are schematically plotted for materials with a high and a low fracture toughness.

Irrespective of specimen size, a valid K_{Ic} for a certain material can only be obtained if for some crack extension the J resistance curve enters the shaded area. The required specimen size then follows from equating the crack extension at which this occurs to 2% of the initial crack length. Clearly, for the tougher material K_{Ic} cannot be determined no matter how large a specimen is used. For the material with the lower toughness, if all other requirements are fulfilled also (see section 5.2), a valid K_{Ic} value can be determined, albeit that sometimes unrealistic specimen sizes would be required.

For aluminium alloys and for high strength steels the K_{Ic} size requirement will be fulfilled. However for lower strength, higher toughness steels this certainly will not be the case: no valid K_{Ic} can be determined, regardless the specimen size.

7.6 The Standard $\delta_{t_{crit}}$ Test

At the beginning of this chapter it was remarked that the original $\delta_{t_{crit}}$ test was the subject of an official British Standard. At present the most recent version, designated BS 7448, dates from 1991, see reference 12 of the bibliography.

The Standard COD Specimens

The standard COD test specimens conform to the three-point notched bend (SENB) and the compact tension (CT) configurations already described in section 5.2. For CT specimens a J_{Ic} type starter notch is allowed also (see figure 7.7). The preferred W/B ratio is 2, but deviation is allowed within certain limits. In principle the thickness B must be equal to that of the material as used in service, and the specimens are *not* side grooved. Exceptions are allowed if it can be shown that a lesser thickness does not affect fracture toughness or if a relation between thickness and fracture toughness can be established.

It is important to note that the $\delta_{t_{crit}}$ values resulting from this test method may be affected by the specimen geometry and size. Therefore caution is required when comparing results from different sources.

Expressions for Calculating δ_t

Direct measurement of δ_t at the crack tip is impossible. Instead a clip gauge is used to measure the crack opening displacement, v_g, at or near the specimen surface. It is then assumed that the ligament b ($= W - a$) acts as a plastic hinge. This implies a rotation point within the ligament at some distance $r \cdot b$.

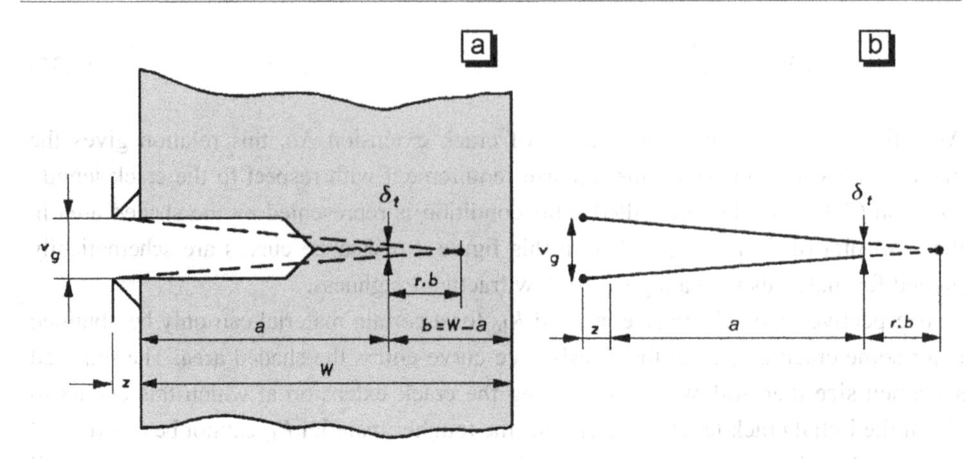

Figure 7.14. Relation between crack opening displacement v_g and crack tip opening displacement δ_t.

In figure 7.14.a an example is shown where the clip gauge is mounted on attachable knife edges on the specimen surface. Figure 7.14.b shows that δ_t can be expressed as

$$\delta_t = \frac{r \cdot b}{r \cdot b + a + z} v_g , \tag{7.36}$$

where the distance z corrects for the use of knife edges. In general $a + z$ should be interpreted as the distance between the position of the clip gauge and the crack tip. This possibly includes the size of attachable knife edges (see figure 5.5) and for CT specimens also depends on the type of notch used.

Although equation (7.36) is simple, there are two notable difficulties:

1) The value of the rotation factor r. Experiments show significant spread in the value to be used for r. This is because the determination requires complicated techniques, *e.g.* the double clip gauge method (reference 13) or infiltration of the crack with plastic or silicone rubber (reference 14). For the standard COD test the assigned r values are 0.4 for the SENB specimen and 0.46 for the CT specimen.
2) Interpretation of the clip gauge displacement v_g. The increase in v_g with loading from a null point setting is caused by two effects, namely elastic opening of the crack and rotation around $r \cdot b$. Thus to consider v_g as arising only from rotation, as in equation (7.36), would lead to erroneous results. Instead v_g must be separated into an elastic part v_{el} and a plastic part v_{pl} as shown schematically in figure 7.15.

Only the plastic part of the displacement is substituted into equation (7.36), *i.e.*

$$\delta_{pl} = \frac{v_{pl} \cdot r \cdot b}{r \cdot b + a + z} . \tag{7.37.a}$$

For reasons of accuracy the elastic part v_{el} is not used but the elastic contribution to δ_t is calculated according to the LEFM expression for CTOD, equation (3.20), modified for plane strain and a plastic constraint factor $C = 2$ (see also section 3.5), *i.e.*

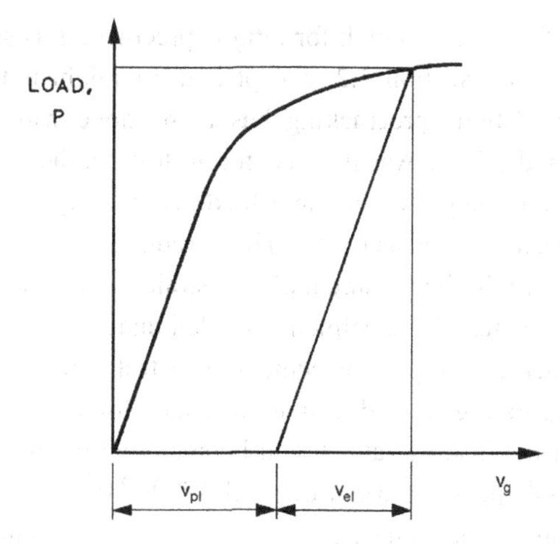

Figure 7.15. Separation of total crack opening displacement v_g into elastic (v_{el}) and plastic (v_{pl}) components.

$$\delta_{el} = \frac{K_I^2}{E\sigma_{ys}}\left(\frac{1-\nu^2}{2}\right). \tag{7.37.b}$$

and

$$\delta_t = \delta_{el} + \delta_{pl} = \frac{K_I^2(1-\nu^2)}{2E\sigma_{ys}} + \frac{r{\cdot}b}{r{\cdot}b + a + z}V_{pl}\,. \tag{7.38}$$

Note that the value of K_I in equation (7.38) is obtained from the standard formula for the SENB and CT specimens, equations (5.1) and (5.2), by substituting the initial crack length, a, and the load at which V_{pl} is measured.

As will be seen under the subheading "Analysis of Load-Displacement Records to Determine $\delta_{t_{crit}}$" several values of $\delta_{t_{crit}}$ can be defined. Note that the British Standard defines δ_t as the crack opening at the original crack tip, as shown in figure 6.7. This means that it is taken for granted that during loading the crack tip will displace and move forward owing to blunting, since at the very tip δ_t must always be zero.

COD Test Procedure

The steps involved in setting up and conducting a COD test are:

1) Prepare shop drawings of the specimen.
2) Specimen manufacture.
3) Fatigue precracking.
4) Obtain test fixtures and clip gauge for crack opening displacement measurement.
5) Testing.
6) Analysis of load-displacement records to determine $\delta_{t_{crit}}$.

Steps (1), (2) and (4) will not be considered further in view of previous discussions in section 5.2. Steps (3) and (5) will be reviewed here and step (6) will be dealt with under the next subheading.

The configuration of the starter notch for fatigue precracking is similar to that for the standard K_{Ic} specimens, see section 5.2, except that a straight notch is recommended rather than a chevron. Fatigue precracking has to be done with a stress ratio R (= $\sigma_{min}/\sigma_{max}$) between 0 and 0.1. As was the case for J_{Ic} testing, the maximum fatigue load should not exceed 40% of the plastic collapse load given in equations (7.27) and (7.28) for SENB and CT specimens respectively. These requirements are to ensure a sufficiently sharp precrack with limited residual plastic strain in the crack tip region.

During the actual COD test the specimen is loaded under displacement control while recording load and crack opening displacement. The test can be carried out with any testing machine incorporating a load cell to measure force electrically. The British Standard specifies that the loading rate should be such that the increase in stress intensity factor with time, dK_I/dt, is between 0.5 and 3.0 $MPa\sqrt{m}/s$. This is arbitrarily defined as 'static' loading, in the same way as for K_{Ic} testing. Again note that equations (5.1) or (5.2) may be used to calculate stress intensity factors.

Since the increase rate dK_I/dt is measured in the elastic region of the load-displacement curve this procedure can lead to large differences in loading rate for ductile specimens: if the loading rate of the testing machine is kept constant the rate of displacement will strongly increase in the plastic region of the load-displacement curve; if, on the other hand, the displacement rate of the testing machine is kept constant, the loading rate will decrease in the plastic region. It has been shown that low loading rates in the plastic region of the load-displacement diagram may lead to lower CTOD values, see reference 15 of the bibliography.

After the test the fracture surface must be examined. The procedure to determine the fatigue precrack length and the requirements that must be met to obtain a valid test result are the same as in J_{Ic} testing, see section 7.4. Furthermore, it is necessary to establish whether stable crack extension occurred during the test and to assess the amount of crack extension associated with possible pop-in behaviour, *i.e.* a small amount of unstable crack growth followed by crack arrest.

Analysis of Load-Displacement Records to Determine $\delta_{t_{crit}}$

The load-displacement records can assume six different forms. These are given schematically in figure 7.16. The assessment of $\delta_{t_{crit}}$ for each case will be briefly discussed.

Before classifying the measured load-displacement curve, it is necessary to decide whether possible pop-in behaviour must be considered significant. In all cases a pop-in is significant if post-test examination of the fracture surface reveals that the corresponding crack extension exceeded 4% of the uncracked ligament, b. Otherwise, a pop-in is only considered significant if at subsequent crack arrest the specimen compliance has dropped by more than 5%. A procedure for deciding this is suggested in the standard.

Cases 1, 2 and 3 are treated similarly. Cases 1 and 2 are monotonically rising load-displacement curves showing no or limited plasticity and no stable crack extension before fracture. Case 3 shows a (significant) pop-in owing to sudden crack extension and arrest. In all these cases $\delta_{t_{crit}}$ is taken to be δ_c, which is calculated according to equation (7.38) using P_c and V_c.

Cases 4 and 5 may also be treated similarly. Prior to instability, which again is either fracture or a (significant) pop-in, stable crack extension occurs. This should be revealed after the test by examination of the fracture surface. In these cases $\delta_{t_{crit}}$ is calculated as δ_u at (P_u,V_u).

Case 6 is relevant to extremely ductile materials for which stable crack extension proceeds beyond maximum load P_m: $\delta_{t_{crit}}$ is calculated as δ_m corresponding to (P_m,V_m).

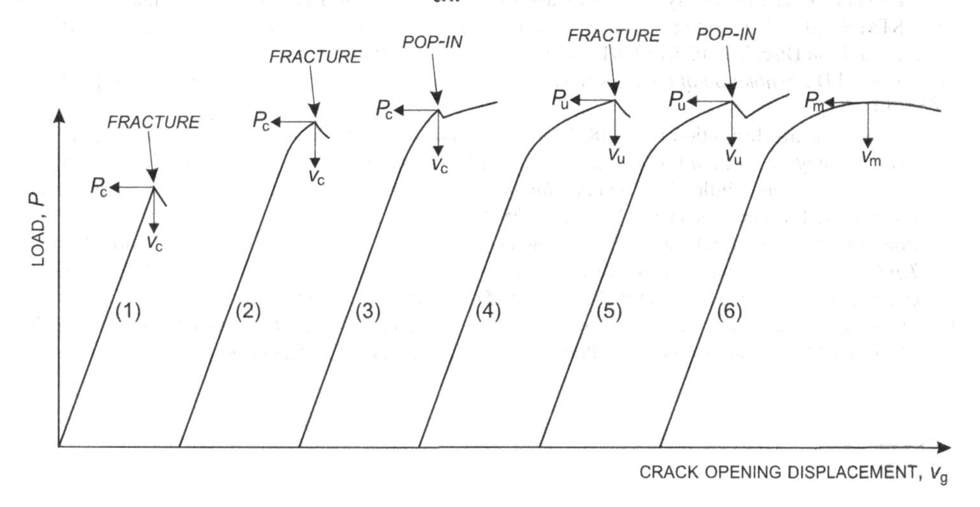

Figure 7.16. Types of load - crack opening displacement plots obtained during COD testing.

Concluding Remarks

The significance of $\delta_{t_{crit}}$ is somewhat limited in practice. Materials can be classified with it and to a certain extent $\delta_{t_{crit}}$ can be used in failure assessment procedures (see section 8.2). However, test results cannot be used to assess the effect of stable crack growth on crack resistance. For this purpose the British Standards Institution has published additional standards. These bear more resemblance to the J_{Ic} test procedure described in section 7.4.

7.7 Bibliography

1. ASTM Standard E 813-81, *Standard Test Method for J_{Ic}, A Measure of Fracture Toughness*, 1981 Annual Book of ASTM Standards. Part 10, pp. 822-840 (1982): West Conshohocken, Philadelphia.
2. British Standards Institution BS 5762, *Methods for Crack Opening Displacement (COD) Testing*, BSI (1979): London.
3. Landes, J.D. and Begley, J.A., *The Influence of Specimen Geometry on J_{Ic}*, Fracture Toughness, ASTM STP 514, American Society for Testing and Materials, pp. 24-39 (1972): Philadelphia.
4. Rice, J.R., Paris, P.C. and Merkle, J.G., *Some Further Results of J Integral Analysis and Estimates*, Progress in Flaw Growth and Fracture Toughness Testing, ASTM STP 536, American Society for Testing and Materials, pp. 231-245 (1973): Philadelphia.
5. Sumpter, J.D.G. and Turner, C.E., *Method for Laboratory Determination of J_c*, Cracks and Fracture, ASTM STP 601, American Society for Testing and Materials, pp. 3-18 (1976): Philadelphia.

6. Pickles, B.W., *Fracture Toughness Measurements on Reactor Steels*, Proceedings of the 2nd European Colloquium on Fracture, Darmstadt, reported in Vortschrittsberichte der VDI Zeitschriften, Series 18, Vol. 6, pp. 130-143 (1978).
7. Chipperfield, C.G., *A Summary and Comparison of J Estimation Procedures*, Journal of Testing and Evaluation, Vol. 6, pp. 253-259 (1978).
8. ASTM Standard E 813-89, *Standard Test Method for J_{Ic}, A Measure of Fracture Toughness*, 1996 Annual Book of ASTM Standards. Vol. 03.01, pp. 633-647 (1996): West Conshohocken, Philadelphia.
9. Anderson, T.L., Vanaparthy, N.M.R. and Dodds, R.H. Jr., *Predictions of Specimen Size Dependence on Fracture Toughness for Cleavage and Ductile Tearing*, Constraint Effects in Fracture, ASTM STP 1171, American Society for Testing and Materials, pp. 473-491 (1993): Philadelphia.
10. ASTM Standard E 1820-99a, *Standard Test Method for Measurement of Fracture Toughness*, ASTM Standards on Disc, Vol. 03.01 (2001): West Conshohocken, Philadelphia.
11. Landes, J.D., *Evaluation of the K_{Ic} size criterion*, International Journal of Fracture, Vol. 17, pp. R47-R51 (1981).
12. British Standards Institution BS 7448, *Fracture mechanics toughness tests — Part 1: Method for determination of K_{Ic}, critical CTOD and critical J values of metallic materials*, BSI (1991): London.
13. Veerman, C.C. and Muller, T., *The Location of the Apparent Rotation Axis in Notched Bend Testing*, Engineering Fracture Mechanics, Vol. 4, pp. 25-32 (1972).
14. Robinson, J.N. and Tetelman, A.S., *Measurement of K_{Ic} on Small Specimens Using Critical Crack Tip Opening Displacement*, Fracture Toughness and Slow-Stable Cracking, ASTM STP 559, American Society for Testing and Materials, pp. 139-158 (1974): Philadelphia.
15. Tsuru, S. and Garwood, S.J., *Some Aspects of Time Dependent Ductile Fracture of Line Pipe Steels*, Mechanical Behaviour of Materials, Pergamon Press, Vol. 3, pp. 519-528 (1980): New York.

8 Failure Assessment Using EPFM

8.1 Introduction

When dealing with structures containing postulated or actual flaws, there is a need for assessing the probability of failure. Obviously this is the case in the design phase and at in-service inspections. However, in the possible event of failure it is also important to reveal the cause and to determine how failure can be avoided in the future.

As was already mentioned in section 1.3, key questions that fracture mechanics deals with are

- what is the critical crack size for a given load or
- what is the maximum load for a given crack size?

To address these questions the concepts available in Elastic-Plastic Fracture Mechanics are the Crack Opening Displacement (COD) and the J integral (see chapters 6 and 7). In this chapter these parameters are used in the following three topics:

- The COD design curve, section 8.2.
- Stable crack growth and ductile instability, described by J, section 8.3.
- Failure assessment diagrams, section 8.4.

The very first method for assessing flawed structures under EPFM conditions was the COD design curve, developed in the 1960s. To outline the historic development and also because it is still in use, this approach will be briefly discussed in section 8.2.

It can be highly conservative to ignore the effect of stable crack growth in failure assessments under EPFM conditions. Therefore in section 8.3 the conditions for J-controlled crack growth are considered briefly. Furthermore the so-called tearing modulus is treated. This concept can be used to assess the onset of ductile instability that may follow after a certain amount of stable crack growth, which is also often referred to as stable tearing.

In section 8.4 an advanced method for failure assessment is discussed, based on the failure assessment diagram. This is a two-criteria approach in which failure is considered as a process resulting from both fracture and plastic collapse. Furthermore, it allows the effect of stable crack growth to be taken into account.

8.2 The COD Design Curve

In this section the development of the COD design curve is concisely reviewed.

More details are to be found in references 1, 2 and 3 of the bibliography. The basis of the original COD design curve is a relation between the CTOD and strains in the vicinity of the crack. Using this, critical CTOD values from test specimens could be related to maximum permissible strains near a crack with a certain size in an actual structure. In turn these maximum strains could be compared with the actual strain in order to determine whether the crack would be critical or not.

This approach has the disadvantage that nothing is said about how nearly critical a crack is, nor about the maximum permissible crack size (see figure 1.4). Later, however, critical COD values could be directly related to maximum permissible crack sizes. Nevertheless, to properly understand the COD design curve it is best to first briefly consider its historical development.

Analytical and Experimental Approach

For the COD design curve a dimensionless CTOD is introduced. This parameter, Φ, is obtained by dividing δ_t, by $2\pi\sigma_{ys}a/E$, which consists of known quantities. Thus

$$\Phi = \frac{\delta_t E}{2\pi\sigma_{ys}a} = \frac{\delta_t}{2\pi\varepsilon_{ys}a}, \tag{8.1}$$

where $\varepsilon_{ys} = \sigma_{ys}/E$, the elastic strain at the yield point. The numerical factor of 2π was added to the denominator for convenience in a later stage of analysis.

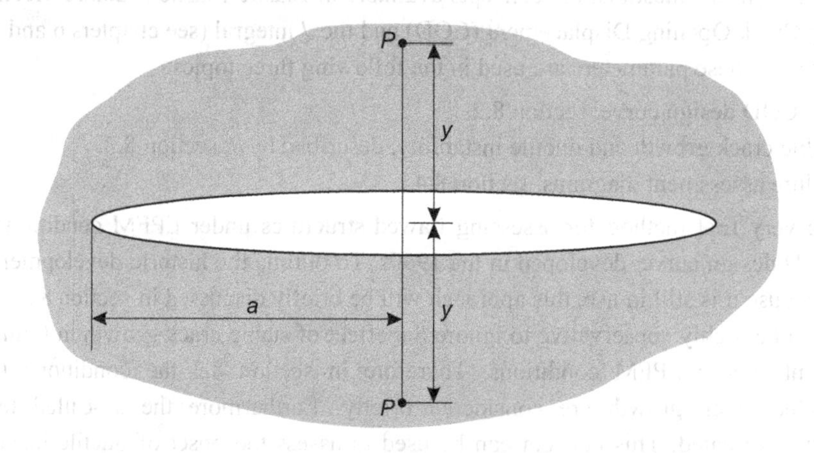

Figure 8.1. Points P at a distance *y* above and below the centre of a crack of length 2*a*.

In the original approach an analytical relation was established between the strain, ε_y, between two equidistant points P across a central crack, as shown in figure 8.1. The derivation of this relation is straightforward, although some complicated mathematics is involved. It was based on Dugdale's strip yield model using the expression given in section 3.3 for δ_t in an infinite centre cracked plate, *i.e.*

$$\delta_t = \frac{8\sigma_{ys}a}{\pi E}\ln\sec\frac{\pi\sigma}{2\sigma_{ys}}. \tag{3.19}$$

In figure 8.2 the final result is shown, taken from reference 1. For several values of the ratio of crack length to gauge length, a/y, the dimensionless CTOD, Φ, is plotted versus the relative strain $\varepsilon_y/\varepsilon_{ys}$.

This is the original COD design curve, an analytical one. The intention was to provide a design curve for each a/y value such that once the critical CTOD was known from specimen tests the maximum permissible strain in a cracked structure could be predicted. Computation of the actual strain in the structure should then indicate whether it were in danger of failing.

In the late 1960s tests on wide plates were done to check the predictive capability of the COD design curve. Measurements of critical CTOD and strain at fracture showed that the data fell into a single scatter band with no discernible dependence on a/y. Also

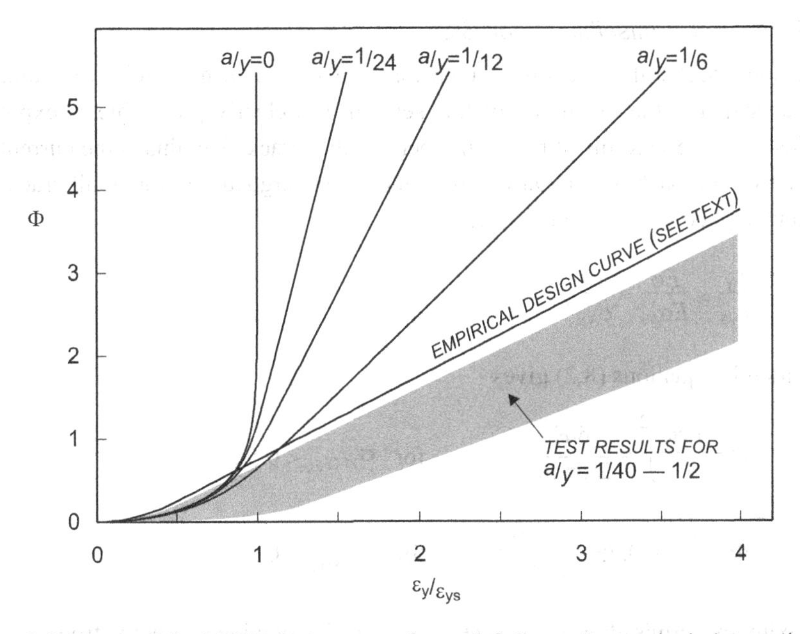

Figure 8.2. The analytical COD design curve for an infinite centre cracked plate and experimental data from tests on wide plates, from reference 1 of the bibliography.

the strain at fracture was much larger than would have been predicted. These test results are roughly indicated in figure 8.2.

Obviously, there is marked disagreement between theory and experiment when $\varepsilon_y/\varepsilon_{ys}$ exceeds 0.5, which may be explained as follows. For wide plates the relative crack length, a/W, is small, such that as $\varepsilon_y/\varepsilon_{ys}$ approaches unity the plates undergo net section yield and ultimately general yield. Net section yield causes the increase in CTOD to be equal to the increase in overall displacement, and a more or less linear relation between Φ and $\varepsilon_y/\varepsilon_{ys}$, independent of a/y, has to be expected. Ultimately, in the case of general yield, which is quite different from the assumption of a yielding strip ahead of the crack tip, the increase in $\varepsilon_y/\varepsilon_{ys}$ becomes much larger than the increase in Φ.

The problem that the analytical COD design curve is useless for $\varepsilon_y/\varepsilon_{ys}$ greater than about 0.5 was obviated simply by drawing a line just above the scatter band of experimental results, thereby obtaining the empirical design curve, which is also drawn in figure 8.2. In reference 2 the whole COD design curve is approximated as

$$\Phi = \left(\frac{\varepsilon_y}{\varepsilon_{ys}}\right)^2 \qquad \text{for} \quad \varepsilon_y/\varepsilon_{ys} < 0.5\,,$$
$$\Phi = \frac{\varepsilon_y}{\varepsilon_{ys}} - 0.25 \qquad \text{for} \quad \varepsilon_y/\varepsilon_{ys} \geq 0.5 \tag{8.2}$$

and so this concept has evolved to a purely empirical one, even though it has an analytical background.

The Maximum Permissible Crack Size

Equations (8.2) still express the dimensionless COD in terms of relative strain. From what was stated at the beginning of this section it is clearly preferable to express the COD design curve in terms of maximum permissible crack size: this is the current COD design curve approach due to Dawes, reference 3. He argued that for small cracks ($a/W < 0.1$) and applied stresses below yield

$$\frac{\varepsilon_y}{\varepsilon_{ys}} \approx \frac{E\sigma}{E\sigma_{ys}} = \frac{\sigma}{\sigma_{ys}}\,. \tag{8.3}$$

Substituting in equations (8.2) gives

$$\Phi = \left(\frac{\sigma_y}{\sigma_{ys}}\right)^2 = \frac{\delta_t E}{2\pi\sigma_{ys}a} \qquad \text{for} \quad \sigma_y/\sigma_{ys} < 0.5\,,$$
$$\Phi = \frac{\sigma_y}{\sigma_{ys}} - 0.25 = \frac{\delta_t E}{2\pi\sigma_{ys}a} \qquad \text{for} \quad \sigma_y/\sigma_{ys} \geq 0.5\,. \tag{8.4}$$

The maximum permissible crack size, a_{max}, can be obtained directly from equations (8.4) by substituting the critical COD value, $\delta_{t_{crit}}$:

$$a_{max} = \frac{\delta_{t_{crit}} E\sigma_{ys}}{2\pi\sigma_1^2} \qquad \text{for} \quad \sigma_1/\sigma_{ys} < 0.5\,, \tag{8.5.a}$$

$$a_{max} = \frac{\delta_{t_{crit}} E}{2\pi(\sigma_1 - 0.25\sigma_{ys})} \qquad \text{for} \quad 0.5 \leq \sigma_1/\sigma_{ys} < 2\,. \tag{8.5.b}$$

Note the designation σ_1. This will be explained in remark (2) below.

Remarks

1) The Dugdale approach implies (i) plane stress conditions and (ii) elastic – perfectly plastic material behaviour. The material is thus assumed to yield at σ_{ys}, while in reality most structural parts will yield at a somewhat higher stress level owing to work

hardening and plastic constraint. This means that the actual CTOD for a crack in a structural part will be smaller than predicted, and higher stresses will be needed to reach $\delta_{t_{crit}}$. Hence the COD design curve is conservative, *i.e.* its use will predict smaller maximum permissible strains and crack sizes than those in reality.

2) In equations (8.5) the designation σ_l was introduced instead of σ. This is a design-oriented convenience: using σ_l as the sum of all stress components (general and local) the effects of, for instance, residual stress in a weld or peak stress due to a geometrical discontinuity can be accounted for. σ_l may reach values as high as twice σ_{ys}. More information is given in reference 2 of the bibliography to this chapter.

3) Equations (8.5) are also used for predicting the maximum permissible defect size of elliptical and semi-elliptical defects. This is done by calculating the LEFM stress

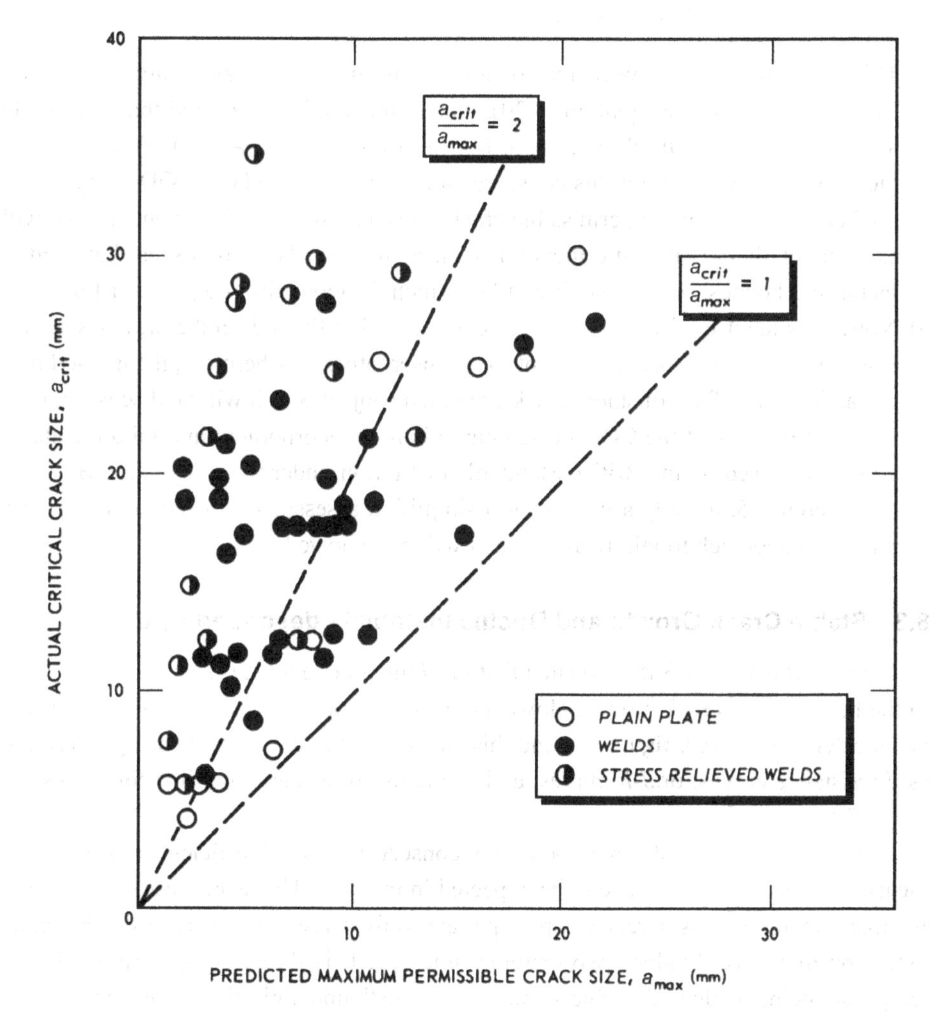

Figure 8.3. Correlation of COD design curve predictions of maximum permissible crack size with actual critical crack size for a structural steel, after reference 3 of the bibliography.

intensity factor for such defects, compare section 2.5. The result is set equal to $K = \sigma_I\sqrt{\pi a}$ for a through-thickness defect. From this equation an equivalent through-thickness crack length, a, follows and this is compared to a_{max} in equations (8.5).

4) The British Standards Institution has published guidelines to assess the significance of weld defects (which are considered as elliptical flaws) based on the COD design curve. From service loads and measured CTOD values the tolerable defect sizes can be predicted. For more information the reader is referred to the official documents, *e.g.* reference 4 of the bibliography.

5) The COD design curve is generally considered to be conservative. For example from equation (8.5.a):

$$\delta_{t_{crit}} = \frac{2\pi\sigma_I^2 a_{max}}{E\sigma_{ys}} = \frac{2K_I^2}{E\sigma_{ys}}. \tag{8.6}$$

This value of $\delta_{t_{crit}}$, is twice that obtainable from the Dugdale analysis assuming LEFM conditions, see equation (3.20), and so it should be expected that at least the lower part of the COD design curve (*i.e.* up to $\varepsilon_y/\varepsilon_{ys} = \sigma_I/\sigma_{ys} = 0.5$) has a safety factor of 2. As a check on this conservatism figure 8.3 correlates COD design curve predictions of maximum permissible crack sizes, based on small specimen tests, with experimentally determined critical crack sizes in wide plates. It is seen that a safety factor of 2 bounds most of the data, all of which lie above the $a_{crit}/a_{max} = 1$ line.

6) Nowadays the use of the COD design curve is rather limited. In the previous remark it was already noted that assessments are conservative. Furthermore, the method does not address the effect of stable crack growth, a subject which will be discussed in the next section. In fact the COD design curve is now incorporated in a failure code recently published by the British Standards Institution under the designation BS 7910, see reference 5, as only a means for a simplified assessment. In section 8.4 a more advanced approach to failure assessment will be reviewed.

8.3 Stable Crack Growth and Ductile Instability described by *J*

In sections 6.4 and 7.3 it was stated that the J integral concept is strictly valid only up to the beginning of crack growth. However, J shows a well-defined rise with increasing crack extension Δa, *e.g.* figure 7.5, and this has resulted in $J - \Delta a$ plots being referred to as J resistance curves, or *J-R* curves, and to the use of a regression line for J_{Ic} testing, figure 7.10.

A related aspect is that it may be highly conservative (*i.e.* inefficient) to use J_{Ic} as a measure of the crack resistance to be expected in practice. This is because the *J-R* curve for many materials has a very steep slope, and only a few millimetres of stable crack extension may give J values two or three times J_{Ic}. It is therefore no surprise that attempts are being made to describe stable crack growth under elastic-plastic conditions.

Depending on material behaviour, geometry and loading conditions ductile instability may occur after a certain amount of crack growth. Obviously it is important to be able to predict this phenomenon.

It should be noted that, especially in a material such as low-strength steel at relatively low temperatures, the process of blunting and stable crack growth can also be interrupted by unstable crack growth due to cleavage (see section 12.5). In the present context only the more ductile behaviour, *i.e.* stable tearing without cleavage, will be considered.

J and Stable Crack Growth

An outline will be given of the conditions for which J can be adequately used beyond crack initiation, *i.e.* J-controlled crack growth. A more thorough discussion of this subject is given in reference 6 of the bibliography.

In section 6.5 the stresses and strains near the crack tip in a material that exhibits power law hardening were expressed in terms of J, the so-called HRR solution. The conditions for J-controlled crack growth are now interpreted as the ability of J to describe the crack tip fields in the presence of crack growth, at least in some annular region around the tip.

In the J integral concept nonlinear elastic material behaviour is used to model actual (plastic) behaviour. This is referred to as the deformation theory of plasticity. In section 6.3 it was stated that to obtain an adequate description of the material behaviour no unloading may occur. However, this is not the only restriction that should be imposed. The reason is that strain hardening in a material not only depends on the amount of plastic strain, but also on the path followed in 'strain space' to arrive at this strain. Deformation theory can describe strain hardening appropriately as long as plastic deformation is 'proportional'. This is the case if during deformation all strain increments $d\varepsilon_{ij}$ are proportional to the increment in the same parameter, *e.g.* an infinitesimal increase in J, dJ.

In figure 8.4 a schematic is shown of the different regions around a crack tip that has extended by Δa in elastic-plastic material. The distance from the crack tip beyond which

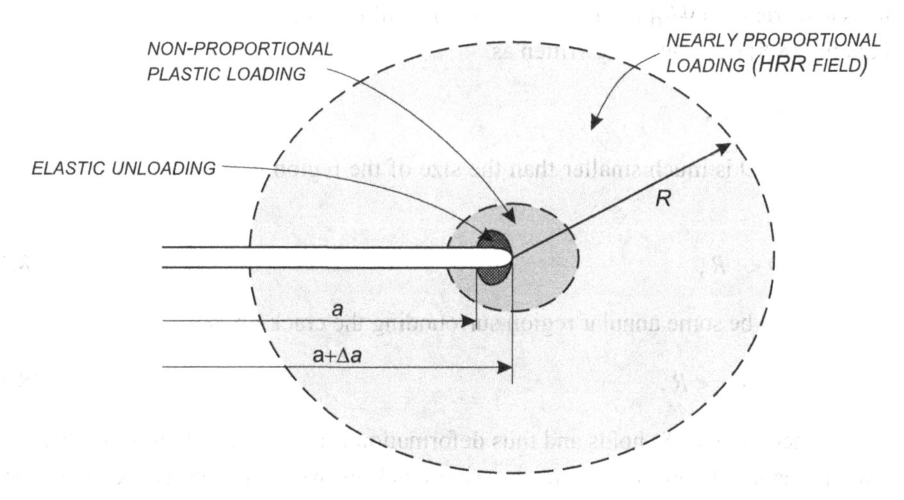

Figure 8.4. Different regions around a growing crack tip in elastic-plastic material.

the HRR solution is no longer valid, even for a stationary crack, is denoted by R. Along the newly formed crack flanks there is a region in which elastic unloading occurs. Finally there is a so-called process zone immediately surrounding the tip in which non-proportional plastic loading occurs due to large geometry changes, *e.g.* blunting and void formation.

Obviously, J will be invalid in the process zone as well as in the region where unloading occurs. Both regions have a size of the order of Δa, and therefore a first requirement for valid application of J under stable crack growth conditions must be:

$$\Delta a << R\,. \tag{8.7}$$

Consider some point at a distance r from the crack tip, where $r < R$. When a crack extends as the result of an increasing load, and thus an increasing J value, the strain at that point will change. This change can be attributed to the crack extension as well as to the increase in J. It can be derived (see reference 6) that the deformation is predominantly proportional (see above) if it is mainly caused by the increase in J and only slightly by crack extension. This condition is expressed by

$$\frac{\mathrm{d}a}{r} << \frac{\mathrm{d}J}{J}\,. \tag{8.8}$$

The question whether this inequality holds for a certain r value depends entirely on material behaviour: J must increase sufficiently during crack growth. At this stage it is convenient to define a length quantity

$$D = \frac{J}{\mathrm{d}J/\mathrm{d}a}\,. \tag{8.9}$$

At crack initiation, *i.e.* $J \approx J_{\mathrm{Ic}}$, D is roughly equal to the crack extension corresponding to a doubling in J. As the crack extends as the result of an increase in J, which also involves a decrease in $\mathrm{d}J/\mathrm{d}a$, it is obvious that D will increase.
Inequality (8.8) can now be rewritten as

$$D << r\,. \tag{8.10}$$

As long as D is much smaller than the size of the region in which the HRR solution is valid, *i.e.*

$$D << R\,, \tag{8.11}$$

then there will be some annular region surrounding the crack tip, defined by

$$D << r < R\,, \tag{8.12}$$

in which inequality (8.8) holds and thus deformation is approximately proportional.

Summarising, J can be used to describe crack growth if (i) the crack extension is limited, *cf.* equation (8.7), and (ii) if a region exists in which strain is mainly determined by the increase in J rather than by crack extension, *cf.* equation (8.11). These conditions for J-controlled crack growth can also be expressed somewhat more quantitatively. For this it is convenient to consider R as being some fraction of the uncracked ligament b (= $W - a$). Equations (8.7) and (8.11) may then be rewritten as

$$\Delta a << b \quad \text{or} \quad \frac{\Delta a}{b} = \alpha << 1 , \tag{8.13}$$

and

$$D << b \quad \text{or} \quad \frac{b}{D} >> 1 \quad \text{and thus} \quad \frac{\mathrm{d}J}{\mathrm{d}a} \cdot \frac{b}{J} = \omega >> 1 . \tag{8.14}$$

Based on both experimental and analytical studies using bend type geometries, *i.e.* SENB and CT specimens, a maximum α value of 0.06 - 0.10 and a minimum ω value of 10 is suggested by Shih *et al.* (reference 7). However, they also found that for tensile type geometries, *e.g.* CCT specimens, these limit values are 0.01 and 80 for α and ω respectively. Thus it seems that the range over which J controls crack growth is geometry-dependent.

More recently this geometry dependence has been attributed to the fact that stresses and strains near the crack tip are not in all cases accurately described by the HRR solution, see reference 8. Differences arise from the development of so-called *in-plane constraint* around a growing crack (see also under the subheading "Applicability of the Tearing Modulus" in this section).

The general tendency is that in the case of bending loads, large specimens and/or a high degree of strain hardening, crack growth will be mainly J-controlled.

Ductile Instability

The first notable attempt to describe ductile instability under elastic-plastic conditions is the tearing modulus concept developed by Paris *et al.*[1], reference 9. This concept, which will be treated in some detail here, is based on the elastic-plastic analogue of the R-curve concept in LEFM. In this case of J-controlled crack growth we must therefore distinguish between the applied J value on the one hand and the material's resistance to crack growth on the other.

If a structural component containing a crack is loaded, a certain value of J is applied which can be considered as the 'crack driving force'. For a certain geometry this J value, which will be denoted as J_{app}, depends on the load level and the crack length. For a limited number of geometries J_{app} values can be calculated analytically. Otherwise numerical methods, such as the finite element method, must be used.

On the other hand, as is schematically shown in figure 8.5, the J-R curve represents the material's crack resistance as a function of crack extension. This crack resistance in terms of J, denoted as J_{mat}, is assumed to be independent of the initial crack length.

---

[1] P.C. Paris is generally credited with developing the tearing modulus concept. However, J.R. Rice and J.W. Hutchinson have made important contributions, as Paris acknowledges in his publications on the subject.

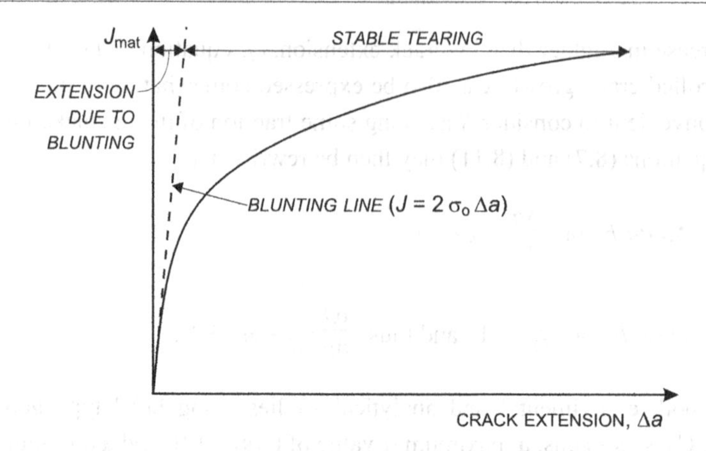

Figure 8.5. A schematic *J–R* curve, *i.e.* J_{mat} versus crack extension.

In figure 8.6 calculated values for J_{app} as a function of crack length at discrete load levels P_1 to P_4 are schematically represented by solid lines. Since J_{mat} is independent of the initial crack length, the crack initiation and stable crack growth behaviour of a component containing a crack of length a_o can be studied by inserting the material's *J-R* curve in figure 8.6, starting at this initial crack length a_o.

Assume that in figure 8.6 initiation of crack growth occurs at load level P_3. By increasing the load, stable tearing takes place as long as there is equilibrium in the sense that

$$J_{app}(P,a) = J_{mat}(\Delta a)\ . \tag{8.15}$$

Analogous to the LEFM *R*-curve, tearing will become unstable as soon as

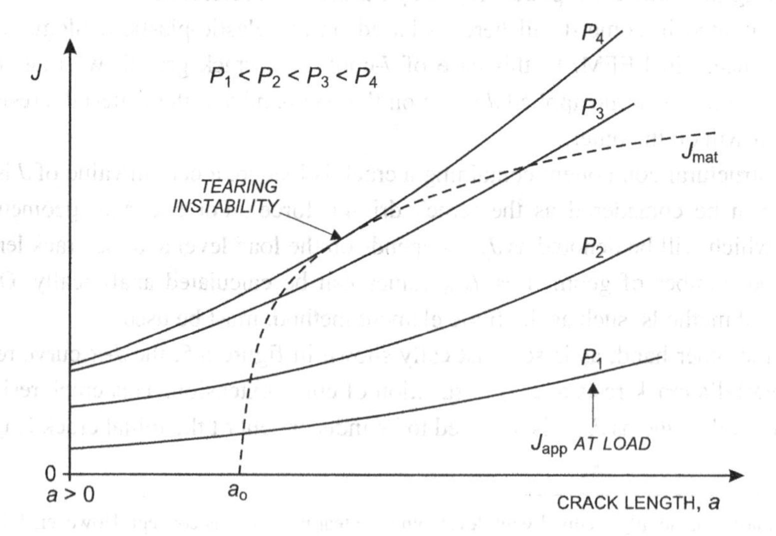

Figure 8.6. The crack driving force diagram.

$$\frac{\partial J_{\rm app}}{\partial a} > \frac{{\rm d}J_{\rm mat}}{{\rm d}a} . \tag{8.16}$$

From figure 8.6 it can be seen that at this point the load has increased to P_4, while the amount of stable ductile tearing can be assessed also. This type of graph is referred to as a 'crack driving force diagram', see reference 7.

The Tearing Modulus Concept

A serious drawback of the approach adopted in figure 8.6 is its lack of accuracy: the J value at instability and the amount of crack extension have to be obtained from a graph. Paris formulated the criterion for ductile instability, equation (8.16), as

$$T_{\rm app} > T_{\rm mat} , \tag{8.17}$$

using the non-dimensional quantities

$$T_{\rm app} = \frac{E}{\sigma_0^2}\frac{\partial J_{\rm app}}{\partial a} \quad \text{and} \quad T_{\rm mat} = \frac{E}{\sigma_0^2}\frac{{\rm d}J_{\rm mat}}{{\rm d}a} , \tag{8.18}$$

which he termed the applied and the material tearing modulus respectively. As before, σ_0 is the flow stress equal to the average of the yield strength σ_{ys} and the ultimate tensile strength σ_{uts}.

The material tearing modulus, T_{mat}, is completely determined by material properties, *i.e.* the slope of the *J-R* curve, the Young's modulus and the flow stress, and is therefore unique for a given material and fully describes its (J-controlled) tearing behaviour.

Paris also chose the non-dimensional formulation of equation (8.18), because he found that the material tearing modulus T_{mat} depends less on temperature than the slope of the *J-R* curve, dJ_{mat}/da, does. Furthermore, for a limited number of idealized configurations ductile instability can be analytically assessed, leading to expressions that contain the definition of T_{mat}.

An approximate analysis of ductile instability for a centre cracked plate under plastic limit load is given here as an illustration (taken from reference 9). Consider figure 8.7 in which a plate-shaped specimen is shown with width, W, length, L and thickness, B, containing a central crack with length, $2a$. It is assumed that material behaviour is elastic – perfectly plastic and that the ligaments adjacent to the crack have become fully plastic. For this case the specimen load is equal to the plastic limit load, P_L, and is given by

$$P_{\rm L} = \sigma_0(W - 2a)B . \tag{8.19}$$

Owing to the load the length of the specimen, L, increases by ΔL. This specimen lengthening can be divided into an elastic part

$$\Delta L_{\rm el} = \frac{P_{\rm L}L}{BWE} \tag{8.20}$$

and a plastic part, ΔL_{pl}. The plastic part is caused solely by the presence of the crack and is therefore approximated by the crack tip opening displacement, δ_t. Using equation (6.34), δ_t can be related to J, leading to

$$\Delta L_{\rm pl} = \delta_{\rm t} = \frac{J}{\sigma_0} . \tag{8.21}$$

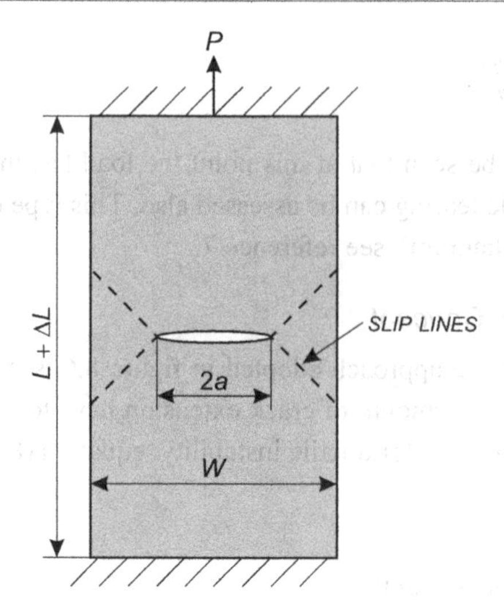

Figure 8.7. Centre cracked plate under limit load.

If J is sufficiently high, an increase dJ will involve the extension of each crack tip over a distance da. Note that the material's tearing behaviour determines the relation between dJ and da.

Increasing J by dJ implies an increase in plastic lengthening

$$\mathrm{d}(\Delta L_{\mathrm{pl}}) = \frac{\mathrm{d}J}{\sigma_{\mathrm{o}}}, \tag{8.22}$$

while at the same time a crack extension, da, causes a change in the (limit) load

$$\mathrm{d}P_{\mathrm{L}} = -2\sigma_{\mathrm{o}}B\mathrm{d}a, \tag{8.23}$$

leading to a decrease in elastic lengthening, *i.e.*

$$\mathrm{d}(\Delta L_{\mathrm{el}}) = \frac{\mathrm{d}P_{\mathrm{L}}L}{BWE} = -\frac{2\sigma_{\mathrm{o}}L}{WE}\mathrm{d}a. \tag{8.24}$$

Assume that the specimen is loaded under fixed grip conditions, *i.e.* the crack extends while the total specimen lengthening, ΔL, remains constant. Instability will now occur as soon as the elastic shortening, *i.e.* $-\mathrm{d}(\Delta L_{\mathrm{el}})$, is larger than the plastic lengthening. Thus, using equations (8.22) and (8.24), the instability criterion reads

$$\frac{2\sigma_{\mathrm{o}}L}{WE}\mathrm{d}a > \frac{\mathrm{d}J}{\sigma_{\mathrm{o}}}. \tag{8.25}$$

After rearranging this criterion to

$$\frac{2L}{W} > \frac{E}{\sigma_{\mathrm{o}}^2}\frac{\mathrm{d}J}{\mathrm{d}a}, \tag{8.26}$$

it can directly be compared with equation (8.17), *i.e.* $T_{\mathrm{app}} > T_{\mathrm{mat}}$.

The applied tearing modulus for the specific case considered here turns out to depend on geometrical quantities only. The tendency is that tearing instability will occur sooner, *i.e.* at a lower amount of crack growth, if the specimen has a high length to width ratio. This seems reasonable in view of the higher amount of elastic energy stored in the specimen.

Applicability of the Tearing Modulus

Within the validity range of J-controlled crack growth, the tearing modulus concept should be capable of predicting ductile instability. It requires the calculation of T_{app} for a certain structural component and experimental data on T_{mat} for the component material. However, its usefulness in practice must still be considered.

First it must be noted that criteria for *initiation* of crack extension, based on the K, G, J or COD concept, are based only on stress, crack length and a single material parameter. In general, no further information is needed for predicting initiation of fracture in a component or structure.

In contrast with this, predicting ductile instability is very complex. Firstly, the applied tearing modulus, T_{app}, not only depends on crack length and the load level, but also on quantities such as specimen geometry, flow stress and strain hardening characteristics of the material. In reference 10 of the bibliography J_{app} – crack length a solutions are presented for a number of standard geometries. Material properties such as E, σ_0 and strain hardening exponent n (*cf.* equation (6.32)) have to be substituted to obtain the $J_{app} - a$ curve for a cracked component of a specific material under a given load. For other geometries calculations are required which relate to that specific case and thus have no general validity.

Secondly, for most materials, the slope of the *J-R* curve becomes less steep with increasing crack extension. This means that T_{mat} is not a constant material parameter. Furthermore the *J-R* curve is found to depend strongly on the state of stress near the crack tip or, more specifically, on the constraint in the thickness direction (*i.e.* plane strain, plane stress), but also on the in-plane constraint. Numerical analyses show that the in-plane constraint around a growing crack depends strongly on specimen geometry and type of loading: for example, a deeply cracked SENB specimen shows a significantly larger amount of in-plane constraint than a CCT specimen.

To understand why the *J-R* curve and thus the process of ductile crack growth is affected by the amount of constraint, one needs to consider the hydrostatic stress component, σ_h, defined as the average of the principle stresses, *i.e.*

$$\sigma_h = \tfrac{1}{3}(\sigma_1 + \sigma_2 + \sigma_3)\,. \tag{8.27}$$

This hydrostatic stress component is responsible for the nucleation and growth of voids, which is an essential step in the ductile tearing mechanism (see chapter 12). Since constraint determines the magnitude of σ_h, $\mathrm{d}J/\mathrm{d}a$ for a given material can be expected to be affected also.

Evidently, the practical use of the tearing modulus concept is subject to many restrictions.

8.4 The Failure Assessment Diagram: CEGB R6 Procedure

Accurate failure assessments in an elastic-plastic context cannot be solely based on fracture mechanics concepts but should also consider effects of plastic deformation. In fact the Feddersen approach discussed in section 5.3 is an example in which the effect of (contained) yield on failure is already considered.

During the last decades considerable effort has been put into the development of a procedure that is able to deal with more widespread plasticity and the interaction with fracture. This procedure is focused around a *two-criteria approach* which was originally introduced in 1975 by Dowling and Townley, see reference 11. In 1976 the Central Electricity Generating Board in the UK (CEGB) published the first procedure for failure assessment using the two-criteria approach. It has gained widespread attention and is used in various countries.

Since the first publication numerous changes have been implemented. Here revision 3 of what is commonly referred to as the *R6 procedure* will be treated, see reference 12. Note that roughly the same procedure is adopted by the British Standards Institution and issued as a standard, see reference 5.

Principle

The load carrying capacity of a flawed structure is limited by two criteria. First of all the linear elastic stress intensity factor K_I must not be greater than the fracture toughness. In the notation of the R6 procedure this is expressed in terms of the dimensionless parameter K_r, *i.e.*

$$K_r = \frac{K_I}{K_{mat}} \leq 1 . \tag{8.28}$$

Here K_{mat} is a fracture toughness value whose precise definition depends on the type of analysis, as will be discussed later.

The second criterion is that the applied load that actually contributes to plastic collapse must not be larger than the plastic collapse load of the flawed structure. To express this criterion the parameter L_r is introduced, which describes the proximity to plastic yielding of the structure, *i.e.*

$$L_r = \frac{\text{applied load that contributes to plastic collapse}}{\text{plastic yield load of the flawed structure}} \leq L_r^{max} . \tag{8.29}$$

Note that the load needed for plastic yield is used and not that for plastic collapse. These loads will be discussed further under the subheading "Evaluation of K_r and L_r".

The value of L_r^{max} depends on the plastic behaviour of the material and more specifically on the amount of strain hardening, *i.e.*

$$L_r^{max} = \frac{\sigma_o}{\sigma_{ys}} , \tag{8.30}$$

where σ_o is again the flow stress, equal to the average of the yield strength, σ_{ys}, and the ultimate tensile strength, σ_{uts}. For ideal plastic (*i.e.* non-hardening) material behaviour L_r^{max} would be 1, but in general L_r^{max} is somewhat larger than 1.

The criteria (8.28) and (8.29) represent two distinct failure mechanisms. However, some interaction between the mechanisms is to be expected and is included in the R6 procedure by replacing the inequality in equation (8.28) by

$$K_r \leq f(L_r) . \tag{8.31}$$

The criteria represented by inequalities (8.29) and (8.31) may be represented by a

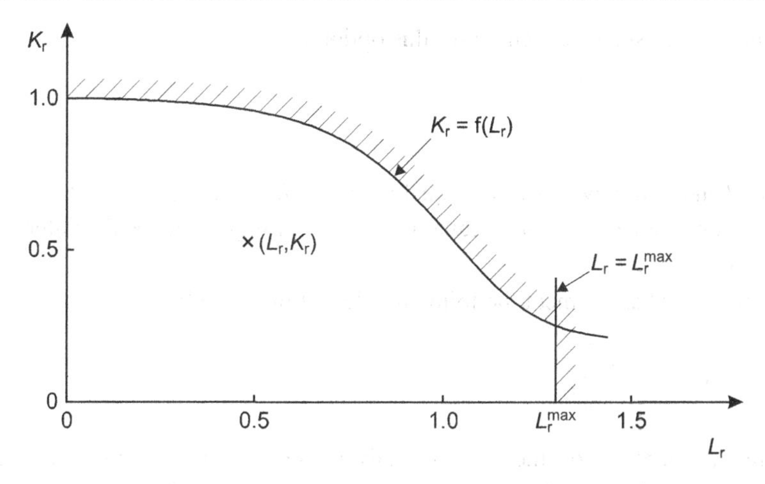

Figure 8.8. A Failure Assessment Diagram.

Failure Assessment Diagram (FAD) as depicted in figure 8.8. If the point (L_r, K_r) calculated for the cracked body at extreme loading conditions lies within the region bounded by equations (8.29) and (8.31), it is considered safe.

In the first version of the R6 procedure the limiting curve of the FAD was based on Dugdale's strip yield model (see chapter 3). Work hardening was incorporated implicitly by using an alternative definition of L_r (then still called S_r). Instead of the plastic yield load being used, as in equation (8.29), the plastic limit load was chosen, which is a quantity based on the flow stress acting on the net section. At the same time L_r was limited to 1. Although this is satisfactory for most structural steels, this is not the case for materials like austenitic steels, which show a low initial rate of, and a high capacity for work hardening. The need arose to explicitly incorporate the details of the material's stress-strain behaviour.

Revision 3 of the R6 procedure uses the J integral to calculate K_r, an idea originally introduced by Bloom, reference 13, and Shih *et al.*, reference 14. The method developed is based on the explicit formulation of J given in reference 10.

Failure Assessment Curve

In the R6 procedure three options are given for establishing a failure assessment curve. These options require an increasing amount of material data and effort, but the results are increasingly accurate (less conservative). For convenience, these options will be treated in reversed order.

- Option 3: J integral analysis

  In this option an assessment curve is obtained that is specific for the material and the geometry considered. Stress analyses must be performed for the cracked structure, for example using the finite element method, and values for the J integral should be evaluated. For an appropriate range of L_r values (and thus load values) both elastic and elastic-plastic analyses should be performed, resulting in J_e and J values respectively.

The failure assessment curve for this option is

$$f_3(L_r) = \sqrt{\frac{J_e}{J}}, \tag{8.32}$$

where J and J_e are evaluated for equal values of L_r. Clearly, for low L_r values the assessment curve will be almost equal to 1, while at higher values the value can be expected to drop.

In the actual assessment the following definition is used:

$$K_r = \sqrt{\frac{J_e}{J_{mat}}}, \tag{8.33}$$

where J_{mat} depends on the type of analysis (see below). By combining equations (8.32) and (8.33), it can be seen that in fact the assessment is based on the criterion $J \leq J_{mat}$. The advantage in using the assessment curve lies in the fact that once this curve is established for the geometry and material under consideration, the actual assessment is based on J_e, for which only an elastic calculation needs to be performed.

Although the Option 3 assessment curve requires a J analysis of the cracked structure, it has the potential for greater accuracy than the approximate curves of Options 2 and 1.

- Option 2: Approximate J integral analysis
  As stated before, reference 10 gives J integral solutions for standard geometries. These require the material stress-strain behaviour to be described in terms of the Ramberg-Osgood relation, equation (6.32). However, not all materials are well described by this relation. For this reason Ainsworth (reference 15) reformulated the J solutions in such a way that the actual stress-strain relation of the material could be used. In combination with equation (8.32) he obtained the assessment curve

$$f(L_r) = \left(\frac{E\,\varepsilon_{ref}}{\sigma_{ref}} + \frac{L_r^2}{2(1+L_r^2)}\right)^{-1/2}. \tag{8.34}$$

In this expression a so-called *reference stress*, σ_{ref}, is introduced. This stress is defined such that the ratio of σ_{ref} to σ_{ys} is equal to L_r, *i.e.* $\sigma_{ref} = L_r\sigma_{ys}$. From the reference stress a *reference strain*, ε_{ref}, is obtained using the material's true stress-strain curve.

In the case where the structure behaves elastically the first term in equation (8.34) is equal to 1, and since $L_r << 1$ the second term is negligible. On the other hand, for fully plastic behaviour the first term is much larger than 1, while the second term still remains smaller than 1. Therefore in these two extreme cases $f(L_r)$ is determined only by the first term in equation (8.34). For the intermediate case where only small-scale yielding occurs, the bulk behaviour of the structure is still elastic, *i.e.* $E\varepsilon_{ref} \approx \sigma_{ref}$. The fact that J now already exceeds its elastic value must be reflected in a value for $f(L_r)$ somewhat lower than 1. The second term in equation (8.34) accounts for this correction.

To diminish the correction of the second term in the case of fully plastic behaviour the denominator includes the term $(1 + L_r^2)$. However, it was found that this diminishing effect is not in all cases satisfactory, since it is based on L_r. It should be based on the extent to which the structure behaves fully plastically, which is described by $E\varepsilon_{ref}/\sigma_{ref}$. Substituting this term for $(1 + L_r^2)$ in equation (8.34) leads to the failure assessment curve for Option 2:

$$f_2(L_r) = \left(\frac{E\,\varepsilon_{ref}}{L_r\,\sigma_{ys}} + \frac{L_r^3\,\sigma_{ys}}{2E\,\varepsilon_{ref}} \right)^{-1/2} . \tag{8.35}$$

Using the Option 2 assessment curve avoids the necessity of a full J analysis. However, it still requires full knowledge of the (true) stress-strain curve of the material.

- Option 1: General failure assessment curve
The Option 1 curve is derived as an empirical fit to the Option 2 curves for several commonly used materials, but biased towards the lower bound. The result of this fit is:

$$f_1(L_r) = (1 - 0.14\,L_r^2)\left\{0.3 + 0.7\,\exp(-0.65\,L_r^6)\right\} . \tag{8.36}$$

This curve is plotted in Figure 8.9. Note that this curve needs no material data apart from σ_{ys} and the flow stress, σ_0. These are used to calculate L_r^{max} according to equation (8.29). Typical values for some steel categories are indicated in figure 8.9.

For materials that show a discontinuous yield point, as is the case for a number of structural steels, the use of the Option 1 curve should be restricted to $L_r \leq 1$. The Op-

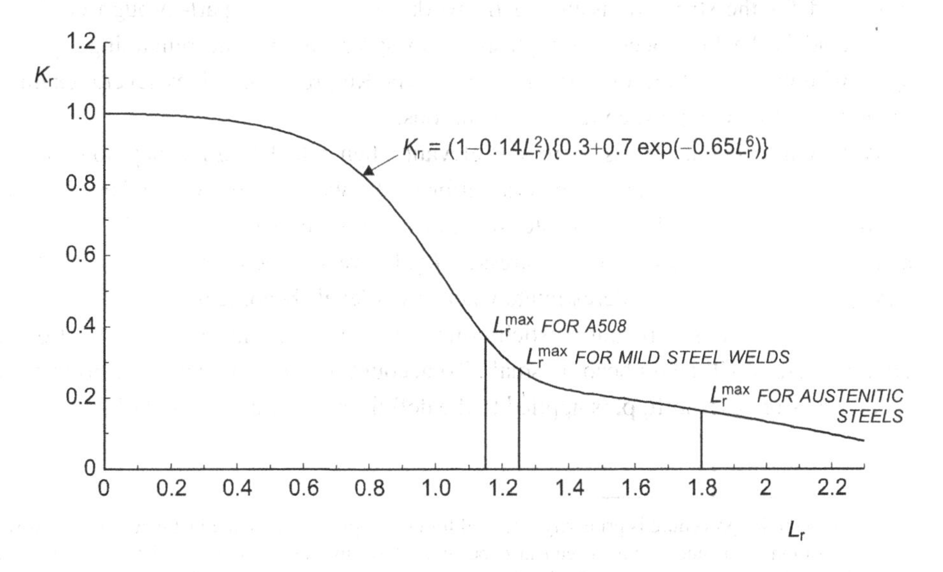

Figure 8.9. The Option 1 failure assessment curve of the R6 procedure (reference 12).

tion 2 curve for these materials predicts a sharp drop near $L_r = 1$, which is not described by equation (8.36).

Obviously, the Option 1 assessment curve requires the least amount of material data and effort, but also leads to the most conservative failure assessment.

Evaluation of K_r and L_r

Before the assessment parameters K_r and L_r can be evaluated, stress analyses must be performed. For Options 1 or 2 an elastic analysis of the uncracked structure suffices, using any suitable method that yields the stresses in the region of the defect. As stated before, Option 3 requires both an elastic and elastic-plastic stress analysis of the cracked structure.

It is important to categorize the loads or resulting stresses with respect to their nature. A distinction must be made between

- *primary stresses*, σ^p, defined as those stresses arising from loads which actually contribute to plastic collapse, such as pressure, dead-weight or interaction with other components,
- *secondary stresses*, σ^s, which are self-equilibrating stresses not contributing to plastic collapse and caused by, for example, local thermal gradients or welding.

In view of the definition of L_r, equation (8.29), it is obvious that secondary stresses need not be considered to evaluate L_r, *i.e.*

$$L_r = \frac{\text{applied load giving rise to } \sigma^p \text{ stresses}}{\text{plastic yield load of the flawed structure}}. \tag{8.37}$$

The definition of the plastic yield load also depends on the nature of the defect. For through-thickness cracks this is the global yield load, which can be determined as the limit load for the structure assuming no work hardening. For part-through cracks the yield load is the load needed for plasticity to spread across the remaining ligament, again without accounting for work hardening. The R6 procedure gives several examples of, and references to, plastic yield load solutions.

Although secondary stresses are not relevant when calculating L_r, they do contribute to K_r. Since this contribution is not straightforward it will be concisely reviewed here.

In the elastic range the stress intensity caused by secondary stresses, K_I^s, can simply be added to that caused by primary stresses, K_I^p. However, due to crack tip plasticity this superposition will be an underestimate when stress levels become higher.[2] Ultimately, if the structure shows significant plasticity due to high primary stresses, the effect of secondary stresses will again become small. To account for this interaction of primary and secondary stresses a shift, ρ, is applied to the definition of K_r in equation (8.28), *i.e.*:

---

[2] Although the R6 procedure is primarily intended for elastic-plastic cases, the fact that a straightforward superposition is an underestimate can also be understood by considering the LEFM correction for crack tip plasticity suggested by Irwin. From equation (3.7) it is easily verified that $K_I(\sigma_1 + \sigma_2) > K_I(\sigma_1) + K_I(\sigma_2)$.

$$K_r = \frac{K_I}{K_{mat}} + \rho\,, \tag{8.38}$$

where $K_I = K_I^p + K_I^s$.

The value of the shift depends on the magnitude of both primary and secondary stresses. Based on finite element analyses the R6 procedure gives values for ρ depending on the magnitude of the primary stresses relative to the yield strength, expressed by the value for L_r:

- $L_r \le 0.8$
  The shift ρ is equal to ρ_1, a value which is independent of L_r, but is a function of a parameter χ according to

$$\rho_1 = 0.1\chi^{0.714} - 0.007\chi^2 + 0.00003\chi^5\,, \tag{8.39}$$

  where

$$\chi = \frac{K_I^s}{K_I^p} L_r\,. \tag{8.40}$$

  The parameter χ can be considered a measure for the level of the secondary stresses relative to the yield strength.[3]
- $0.8 < L_r \le 1.05$
  For increasing primary loads the shift is decreased linearly to zero according to

$$\rho = 4\rho_1 (1.05 - L_r)\,. \tag{8.41}$$

- $1.05 < L_r$
  The shift ρ is set to zero. This is a conservative estimate since finite element analyses suggest negative values for ρ in this L_r range.

Note that for high levels of secondary stresses, *i.e.* for $\chi > 4$, the approach described here may lead to a significant overestimation of the effect of secondary stresses, because of plastic relaxation of peak elastic stresses that exceed the yield stress. However, to account for this an elastic-plastic analysis is required. Such an analysis is also required when only secondary stresses are present.

The evaluation of K_r also requires a value for K_{mat} to be set. This, however, depends on the type of analysis that is chosen and will be treated under the subheading "Types of Analyses".

Flaw Characterisation

Generally, the geometry of flaws will not allow a straightforward analysis and some simplifications must be made. This process, termed flaw characterisation, should be performed with care since it must lead to conservative results. The following aspects need to be considered:

---

[3] Equation (8.40) suggests a dependency of χ (and thus of ρ_1) on L_r. However, this is only seemingly so since K_I^p increases proportionally with L_r.

- Orientation

  Arbitrarily shaped flaws are represented by equivalent planar crack-like defects. A flaw is projected on a plane either:

  1) through its principal plane,
  2) normal to the direction of the maximum principal stress or
  3) normal to the surface and parallel to the principal axis of the flaw.

  Note that only in the second case a pure mode I situation is obtained.
- Shape

  A distinction is made between through-thickness defects, semi-elliptical surface defects and elliptical embedded defects. If the assessment indicates ligament failure for either an embedded or a surface defect it may be re-characterised as a surface defect or a through-thickness defect, respectively.
- Interaction

  The interaction between a defect and a neighbouring defect or a free surface can be accounted for by assuming the ligaments to be part of the defect. If this does not yield satisfactory results, one may use the more extensive interaction criteria given in the Boiler and Pressure Vessel Code of the American Society of Mechanical Engineers (ASME), Section XI (see reference 16) or the British Standards Institution (BSI) Published Document 6493 (see reference 4). As a last resort, the calculation of K_r can be based on appropriate K_I solutions, if available, while L_r follows from the procedure already briefly described under the subheading "Evaluation of K_r and L_r".

Types of Analyses

In the R6 procedure three analysis categories are defined:

- Analysis against Crack Initiation Criteria (Category 1)

  This type of analysis is appropriate when failure is either brittle or is preceded by only a limited amount of ductile tearing. It can also be used when the material shows significant ductile tearing prior to failure, but then the increase in toughness involved in crack growth cannot be taken into account. This can, however, be advantageous in view of the relative simplicity of this type of analysis.

  In LEFM cracks are assumed to initiate when $K_I > K_{Ic}$. If a valid K_{Ic} value can be determined for the material, for example using the test method given in ASTM standard E 399, K_{mat} in equations (8.28) or (8.38) is equal to K_{Ic}. If no valid K_{Ic} can be obtained but the slope of the load-displacement test record does not deviate more than 5% from the initial value, the conditional K_{Ic} value, K_Q, may be used for K_{mat}.

  If the nonlinearity of the load-displacement test record is larger, indicating significant plastic deformation, K_{mat} must be determined using the J integral. If the total crack extension prior to failure, Δa, which is defined as the sum of crack tip blunting and stable tearing, is less than 0.2 mm, then

$$K_{mat} = \sqrt{\frac{EJ}{1-\nu^2}}, \tag{8.42}$$

where J is evaluated at failure. When the total crack extension is larger, J in equation (8.42) is evaluated at $\Delta a = 0.2$ mm. This is done by constructing a J-Δa regression line more or less similar to that used for the J_{Ic} determination according to ASTM standard E 813. Note, however, that the R6 procedure is different, in the sense that no distinction is made between blunting and actual crack growth.

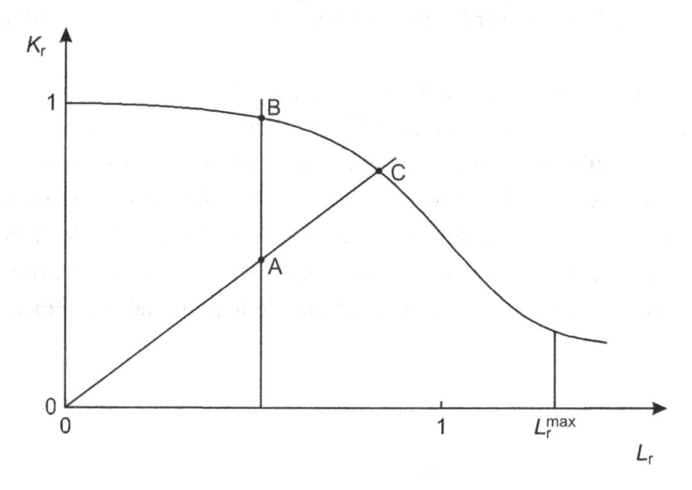

Figure 8.10. FAD for a category 1 analysis against crack initiation (reference 12).

The calculated K_r and L_r values represent a point in the FAD. For the condition represented by point A in figure 8.10 crack initiation will not occur, since this point lies on the safe side of the assessment line. Crack initiation can occur as a result of changes in one or more parameters affecting the assessment. Such changes can be represented on the FAD by the loci of points. For example, the line AB represents a decrease of the initiation toughness for the material, and the line AC shows what happens if the load applied to the structure is increased. Since points B and C lie on the assessment curve, they represent different limiting conditions for the avoidance of crack initiation.

- Ductile Tearing Analysis (Categories 2 and 3)

If the assessment point of a Category 1 analysis lies outside the safe region of the FAD, this does not always indicate a failure condition. For materials that exhibit stable crack growth by ductile tearing, the fracture toughness increases with crack growth. As already expressed in equations (8.15) and (8.16), the crack will remain stable as long as

$$J_{app} \leq J_{mat} \tag{8.43.a}$$

and

$$\frac{\partial J_{app}}{\partial a} \leq \frac{dJ_{mat}}{da} . \tag{8.43.b}$$

A Category 3 ductile tearing analysis is performed by calculating L_r and K_r for a range of postulated crack extensions, Δa, starting from the initial crack length a_o. In the calculation of K_r, according to equation (8.38), K_{mat} is derived from the material's J resistance (J-R) curve for a crack growth increment Δa. As a result of crack extension both K_I (or $\sqrt{J_{app}}$) and K_{mat} (or $\sqrt{J_{mat}}$) will increase. However, as long as inequality (8.43.b) holds, K_r will decrease. At the same time L_r will increase somewhat as a result of crack growth, so in the FAD the locus of assessment points will be directed roughly downwards from the point corresponding to the initial crack length a_o.

Figure 8.11 gives an example of a Category 3 assessment. For equal initial crack lengths assessment curves AB and CD are calculated at a load P_1 and a much higher load P_2, respectively. Curve AB lies entirely below the assessment curve and so no cracking will occur at load level P_1. Curve CD is first above and then tangent to the assessment curve. This means that load level P_2 is the limiting load for this analysis, since for any load smaller than P_2 the assessment points would eventually drop below the assessment curve as a result of ductile tearing and thus crack growth would stop.

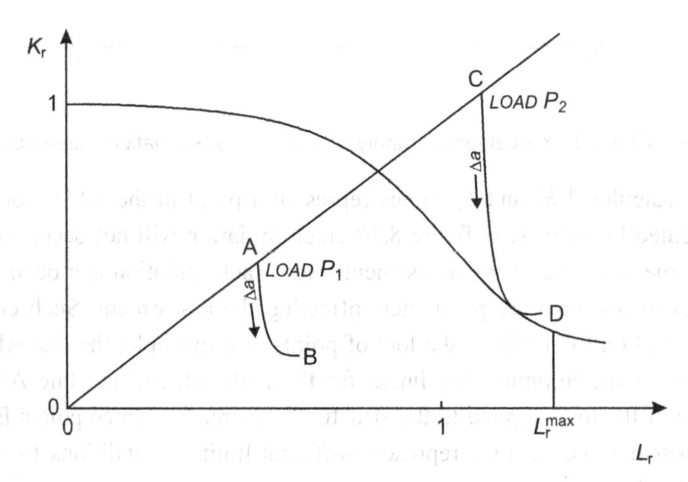

Figure 8.11. FAD for a category 3 ductile tearing analysis (reference 12).

To be able to actually assess the tangency condition, it is imperative that the J resistance data for the material, *i.e.* the J-R curve, extends over a sufficiently large crack extension range. Very often such data are not available due to the limited dimensions of the test specimens used for determining the J-R curve. Specimen dimensions restrict the maximum allowable J value to the smaller of

$$J_{max} = \frac{b\sigma_o}{25} \tag{8.44.a}$$

and

$$J_{max} = \frac{B\sigma_o}{25}\,, \tag{8.44.b}$$

cf. equation (7.30). Furthermore, crack growth must be J-controlled in order for an experimentally determined J-R curve to be applicable to an actual structure. For this reason the maximum amount of crack growth, Δa_{max}, is related to the ligament size, *cf.* equation (8.13). In the R6 procedure

$$\Delta a_{max} = 0.06(W - a_o) + 0.2 \, . \tag{8.45}$$

At the same time it is required that

$$\omega = \frac{b}{J} \cdot \frac{dJ}{da} > 10 \, , \tag{8.46}$$

cf. equation (8.14).[4] Note that if the onset of instability cannot be assessed, the limiting load resulting from a Category 3 analysis will be lower.

A Category 2 analysis requires less effort, while still providing a safeguard against instability. In a Category 2 analysis L_r and K_r are evaluated only for two crack lengths, *i.e.* at the initial size a_o and at the size after a certain amount of crack growth, $a_o+\Delta a_g$. The crack growth Δa_g is the validity limit for the J-R test imposed by either the maximum allowable J value or the maximum crack growth.

Both the Category 2 and 3 analyses consider ductile tearing due to a certain load. It is assumed that no form of *subcritical crack growth* is involved during this tearing, *e.g.* fatigue crack growth or sustained load fracture (see chapters 9 and 10 respectively). If these crack growth mechanisms cannot be excluded, Category 2 or 3 analyses should only be applied to overload conditions. (Obviously subcritical crack growth under normal loading should be taken into account to estimate the crack size after a certain service time.)

Significance of Analysis Results

In practice it is not sufficient to define a limiting condition, *e.g.* a critical crack size or a maximum load. It is important to obtain insight into the sensitivity of the result to variations in input parameters, such as material data, loads and/or crack sizes. To do this it is convenient to define reserve factors with respect to these parameters.

An important reserve factor is that with respect to the load, F^L, which is defined as

$$F^L = \frac{\text{Load producing a limiting condition}}{\text{Load actually applied}} \, . \tag{8.47}$$

The load producing a limiting condition corresponds to an assessment point (L_r, K_r) lying on the failure assessment curve. In the absence of secondary stresses F^L simply follows from scaling the actual assessment point along a line from the origin in the failure assessment diagram. Referring to figure 8.10, the reserve factor $F^L = \text{OC}/\text{OA}$. In the case where both primary and secondary stresses are present the R6 procedure gives a graphical procedure to account for the interaction between these two types of stress.

---

[4] Under certain conditions the limits for J-controlled growth can be relaxed. This is the case if tests show that specimen size is not relevant for the J-R curve or if the cross-section of the structure is thin and the test specimens have the same thickness.

For example, the reserve factors on crack size, F^{a}, and on fracture toughness, F^{K}, are

$$F^{\mathrm{a}} = \frac{\text{Crack size producing limiting condition}}{\text{Actual crack size}}, \tag{8.48}$$

$$F^{\mathrm{K}} = \frac{\text{Material fracture toughness}}{\text{Fracture toughness producing limiting condition}}. \tag{8.49}$$

The minimum values for the relevant reserve factors needed to establish whether a loading condition is acceptable follow from sensitivity analyses. In such analyses the sensitivity of reserve factors to variations in load, secondary stresses, crack size, material properties, etc. are evaluated. As an example, figure 8.12 schematically shows a preferred and a non-preferred variation of F^{L} with crack length. In the non-preferred situation F^{L} would most probably be required to be higher than 1.4 in order to reliably avoid the limiting condition. Obviously the range of uncertainty in the parameter under consideration should also be taken into account.

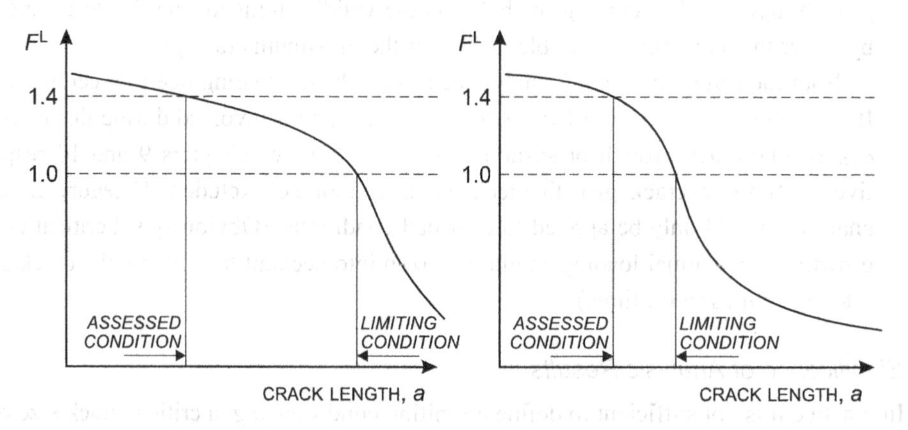

Figure 8.12. Preferred (left) and non-preferred (right) variation of load reserve factor with crack length (reference 12).

For a Category 2 analysis a specific sensitivity analysis is suggested in the R6 procedure. Two load reserve factors, $F^{\mathrm{L}}_{\mathrm{o}}$ and $F^{\mathrm{L}}_{\mathrm{g}}$, are evaluated:

- $F^{\mathrm{L}}_{\mathrm{o}}$ at the initiation of crack growth, *i.e.* at crack length a_{o} and the initial fracture toughness,
- $F^{\mathrm{L}}_{\mathrm{g}}$ after Δa_{g} of ductile crack growth, *i.e.* at crack length $a_{\mathrm{o}}+\Delta a_{\mathrm{g}}$ and an increased fracture toughness.

Since for a Category 2 analysis the acceptance of the loading condition generally relies on the increase in fracture toughness caused by ductile tearing, $F^{\mathrm{L}}_{\mathrm{o}}$ will be relatively small and probably smaller than 1. To ensure that F^{L} increases sufficiently during ductile tearing, it is required that

$$\frac{F^{\mathrm{L}}_{\mathrm{g}}}{F^{\mathrm{L}}_{\mathrm{o}}} \geq 1.2\,. \tag{8.50.a}$$

Furthermore, at the validity limit of crack extension, Δa_{g}, the load reserve factor should have a certain minimum value, *i.e.*

$$F^{\mathrm{L}}_{\mathrm{g}} \geq 1.1\,. \tag{8.50.b}$$

8.5 Bibliography

1. Burdekin, F.M. and Stone, D.E.W., *The Crack Opening Displacement Approach to Fracture Mechanics in Yielding Materials*, Journal of Strain Analysis, Vol. 1, pp. 145–153 (1966).
2. Dawes, M.G., *Fracture Control in High Yield Strength Weldments*, The Welding Journal, Research Supplement, Vol. 53, pp. 369s–379s (1974).
3. Dawes, M.G., *The COD Design Curve*, Advances in Elasto-Plastic Fracture Mechanics, ed. L.H. Larsson, Applied Science Publishers, pp. 279–300 (1980): London.
4. British Standards Institution PD6493: 1980, *Guidance on Some Methods for the Derivation of Acceptance Levels for Defects in Fusion Welded Joints*, BSl (1980): London.
5. British Standards Institution BS 7910: *Guide on Methods for Assessing the Acceptability of Flaws in Fusion Welded Structures*, BSl (1999): London.
6. Hutchinson, J.W. and Paris, P.C., *Stability Analysis of J Controlled Crack Growth*, Elastic-Plastic Fracture, ASTM STP 668, American Society for Testing and Materials, pp. 37–64 (1979): Philadelphia.
7. Shih, C.F., German, M.D. and Kumar, V., *An Engineering Approach for Examining Crack Growth and Stability in Flawed Structures*, International Journal of Pressure Vessel and Piping, 9, pp. 159–196 (1981).
8. Hackett, E.M., Schwalbe, K.-H. and Dodds, R.H. (Eds.), *Constraint Effects in Fracture*, ASTM STP 1171, American Society for Testing and Materials (1993): Philadelphia.
9. Paris, P.C., Tada, H., Zahoor, A. and Ernst, H., *The Theory of instability of the Tearing Mode of Elastic Plastic Crack Growth*, Elastic-Plastic Fracture, ASTM STP 668, American Society for Testing and Materials, pp. 5–36 (1979): Philadelphia.
10. Kumar, V., German, M.D. and Shih, C.F., *An Engineering Approach for Elastic Plastic Fracture Analysis*, Report Prepared for Electric Power Research Institute, EPRI NP 1931 (1981): Palo Alto, California.
11. Dowling, A.R. and Townley, C.H.A., *The Effect of Defects on Structural Failure: A Two-Criteria Approach*, International Journal of Pressure Vessel and Piping, 3, pp. 77–107 (1975).
12. Milne, I., Ainsworth, R.A., Dowling, A.R., and Stewart, A.T., *Assessment of the Integrity of Structures Containing Defects*, CEGB Report R/H/R6 — Revision 3 (1986).
13. Bloom, J.M., *Validation of a Deformation Plasticity Failure Assessment Diagram Approach to Flaw Evaluation*, Elastic-Plastic Fracture, Second Symposium, ASTM STP 803, American Society for Testing and Materials, pp. II.206–II.238 (1983): Philadelphia.
14. Shih, C.F., Kumar, V., and German, M.D., *Studies on the Failure Assessment Diagram Using the Estimation Method and J-Controlled Crack Growth Approach*, Elastic-Plastic Fracture, Second Symposium, ASTM STP 803, American Society for Testing and Materials, pp. II.239–II.261 (1983): Philadelphia.
15. Ainsworth, R.A., *The Assessment of Defects in Structures of Strain Hardening Material*, Engineering Fracture Mechanics, 19, pp. 633–642 (1984).
16. ASME Boiler and Pressure Vessel Code Section XI: *Rules for in-service inspection of nuclear power plant components*, ASME (1983).

Part IV
Fracture Mechanics Concepts for Crack Growth

9 Fatigue Crack Growth

9.1 Introduction

In the middle of the nineteenth century failures were observed in bridges and railway components that were subjected to repeated loading. Because the loading was such that statically it would pose no problem, it was accepted very soon that the failures were a consequence of the cyclic nature of the loading. A complicating factor was that most failures occurred without any obvious warning. The problem was defined as metal fatigue, which was considered as a fracture phenomenon caused by repeated or cyclic loading. A rigorous definition of metal fatigue is difficult. In reference 1 it is defined as: "Failure of a metal under a repeated or otherwise varying load which never reaches a level sufficient to cause failure in a single application".

In 1860 Wöhler, a German railway engineer, proposed a method by which failure of

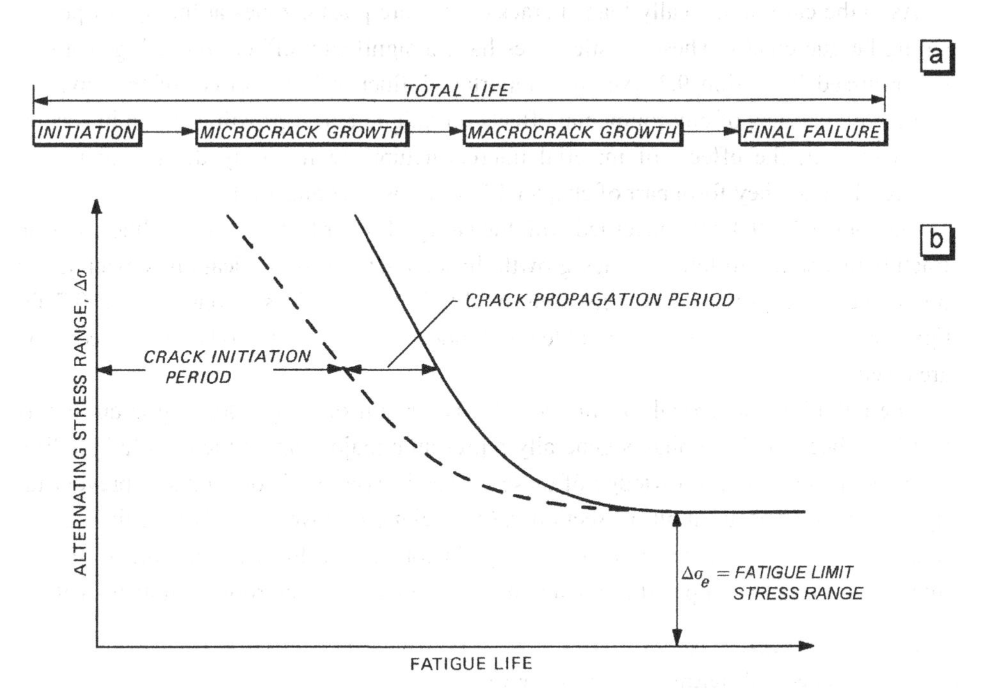

Figure 9.1. Schematic representation of the fatigue life and its dependence on stress levels.

cyclically loaded components could be avoided. Figure 9.1 shows a so-called Wöhler curve, which gives the fatigue lifetime as a function of the applied alternating stress range.[1] Wöhler found that by limiting the alternating stress range to a certain level the life of the load bearing components would become virtually infinite. This safe level was called the *fatigue limit stress range*[2] and was considered a material property provided some special precautions were taken.

At the beginning of the 20th century it became increasingly clear that fatigue failure is a progressive and localized process, involving both initiation of a crack and its growth until instability. However, this was not generally accepted. For a long time fatigue was considered as a gradual deterioration of a material subjected to repeated loads. A considerable confusion about the nature of fatigue resulted. This changed around 1950 when a considerable interest in the initiation and growth of fatigue cracks arose. This interest was stimulated by the understanding that the fatigue lifetime of a cyclically loaded structure comprises stages of both crack initiation and propagation, as is also indicated in figure 9.1.

In the context of this fracture mechanics course the division into microcrack and macrocrack (or long crack) growth periods is of basic importance. This division can be defined in various ways. An apparently very reasonable definition is that a macrocrack has dimensions sufficient for its growth to depend only on bulk properties and conditions rather than on local ones. In LEFM terms this means that macrocrack growth can be described by the stress intensity factor concept. The merits of this approach will be discussed in section 9.2.

As in the case of statically loaded cracks, there are plastic zones at the tips of propagating fatigue cracks. These plastic zones have a significant effect on crack growth, to be discussed in section 9.3. Another important influence is the effect of the environment. An overview of environmental effects on fatigue crack growth is given in section 9.4. However, the effects of material microstructure are not fully dealt with in this chapter. Instead they form part of chapter 13, sections 13.3 and 13.4.

Sections 9.2 – 9.4 are concerned with the straightforward application of linear elastic fracture mechanics to fatigue crack growth. In section 9.5 this application is extended to predicting crack growth under constant amplitude loading. In sections 9.6 and 9.7 fatigue crack propagation under variable amplitude loading and methods for its prediction are discussed.

The initiation and growth of microcracks as a result of fatigue loading is an important issue because these phases generally represent a major part of the total fatigue lifetime and also because knowledge of these phases is essential if one aims to prevent fatigue damage to occur at all. In section 9.8 the relation between the fatigue limit stress range and the threshold stress intensity range is considered. Furthermore, the growth of microcracks (small compared to the notch dimensions) from the roots of notches will be

---

[1] This is also frequently referred to as an S/N curve.

[2] In the literature the term 'fatigue limit' is often used, which is half the stress range at the fatigue limit, *i.e.* the amplitude of the alternating stress at the fatigue limit.

treated briefly. Finally, an engineering approach to the effect of defects on the fatigue limit is discussed.

9.2 Description of Fatigue Crack Growth Using the Stress Intensity Factor

A very important advance in metal fatigue during the past decades is the general understanding that structures can contain crack-like defects that are either introduced during manufacturing, especially in case of welding, or form early during service. Virtually the whole life of some structures can be occupied by fatigue crack growth from flaws. Despite the fact that cyclic loading can change the deformation response of a metal and its microstructure, leading to fatigue crack initiation, it is now generally accepted that in engineering terms fatigue damage is best dealt with by a combination of 'traditional' lifing approaches (fatigue initiation) and lifing based on fatigue crack growth. Knowledge about fatigue crack growth is essential for the understanding and prediction of fatigue behaviour of many structures.

The main question concerning fatigue crack growth is: how long does it take for a crack to grow from a certain initial size to the maximum permissible size, *i.e.* the crack size at which failure of a component or structure is just avoidable. This is one of the five basic questions posed in section 1.3. There are three aspects to this question:

- the initial crack size, a_d,
- the maximum permissible or critical crack size, a_{cr},
- the period of crack growth between a_d and a_{cr}.

The initial crack size, a_d, corresponds to the minimum size that can be reliably detected using non-destructive inspection (NDI) techniques, or it corresponds to a crack size that cannot be detected but is assumed to be present. Secondly, the maximum permissible crack size, a_{cr}, can be determined, at least in principle, using LEFM or EPFM analysis to predict the onset of unstable crack extension. The third aspect requires knowledge of a fatigue crack growth curve, schematically shown in figure 9.2. Note that

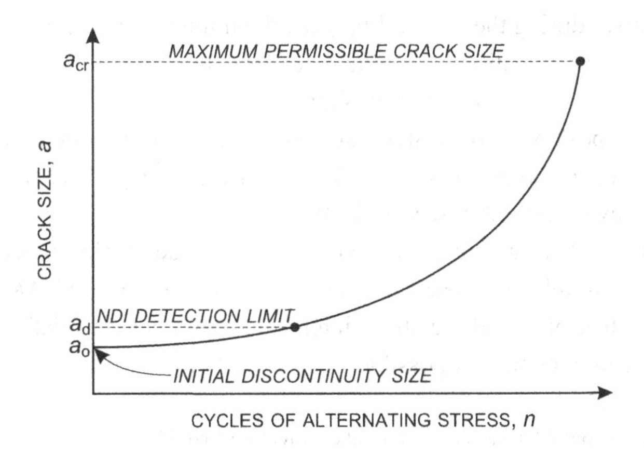

Figure 9.2. Schematic fatigue crack growth curve.

there is also an initial discontinuity size, a_o, which is not the same as a_d and is non-zero. This is because real components and structures often contain initially non-detectable discontinuities (voids, flaws, damage, inhomogeneities).

At this point it is convenient to define some parameters that are used to describe fatigue crack growth. In figure 9.3 a varying load is shown with a constant stress amplitude and mean stress.[3] The stress ratio R is defined as $\sigma_{min}/\sigma_{max}$, the stress range $\Delta\sigma = \sigma_{max} - \sigma_{min}$ and the mean stress $\sigma_{mean} = ½(\sigma_{max} + \sigma_{min})$. Note that any combination of two of the parameters R, $\Delta\sigma$, σ_{min}, σ_{max} and σ_{mean} completely defines the load.

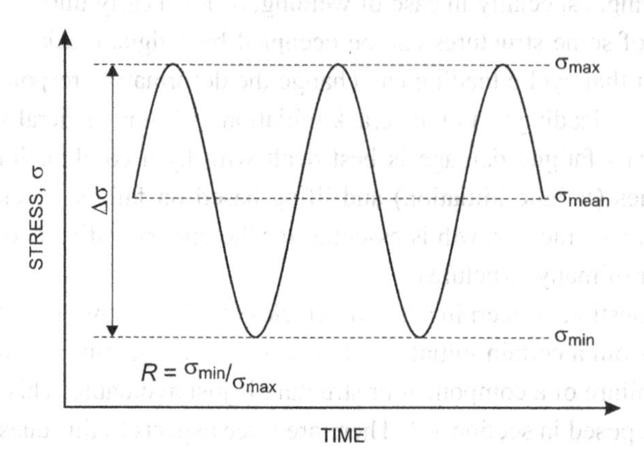

Figure 9.3. Variables describing fatigue loading.

If the maximum load level in fatigue is well below the critical value for the onset of unstable crack extension, as is generally the case, LEFM can be applied. For a given crack length all stress levels can then be converted to corresponding K_I values using the appropriate relation, *e.g.* $K_I = C\sigma\sqrt{\pi a}\cdot f(a/W)$.[4]

The fatigue crack propagation rate is defined as the crack extension, Δa, during a small number of cycles, Δn, *i.e.* the propagation rate is $\Delta a/\Delta n$, which is usually written as the differential da/dn. Because crack growth during one cycle is discontinuous, *i.e.* there is an enhanced crack length increase during the rising part of the load and far less or no crack growth during the descending part, the minimum value of dn is one cycle.

Experimental determination of fatigue crack growth curves for every type of component, loading condition, and crack size, shape and orientation in a structure is impractical, not to say impossible. Fortunately, at least for constant amplitude loading, one can use the correlation between fatigue crack growth rate, da/dn, and the stress intensity range, ΔK, already discussed in section 1.10.

The correlation of ΔK and da/dn for constant amplitude loading was a very important discovery. For example, suppose the relation between da/dn and ΔK is known from standard tests. Then provided the stress intensity – crack length relationship can be determined for a component, it is possible to specify da/dn for each crack length, and the

---

[3] In the figure a sinusoidal load is shown, but other waveforms could also be assumed.

[4] In this chapter, as is usual in fatigue, the mode I stress intensity factor will be conveniently denoted as K, *i.e.* without the subscript I.

required $a - n$ curve can be constructed by integrating the standard $da/dn - \Delta K$ data over the appropriate range of crack length (a_d to a_{cr} in figure 9.2).

The notion that da/dn is fully determined by ΔK is known as the similitude approach, with ΔK as a similitude parameter. The approach can be defined as: "similar conditions applied to the same system will have similar consequences", see reference 2. More specifically, a similar K cycle applied to a crack in a standard specimen will induce the same crack length increment as when applied to a crack in a structure with an arbitrary geometry consisting of the same material.

This rule seems logical and physically sound, but careful examination is needed to assure that similar conditions do indeed apply. In the case of fatigue crack growth, as will be discussed below, it is often found that besides ΔK the crack growth rate also depends on stress ratio, load frequency, environment, shape of the load cycle, temperature, and load history. Moreover, in view of the stress state, material thickness and crack geometry can also be significant.

The Fatigue Crack Growth Rate Curve $da/dn - \Delta K$

The characteristic sigmoidal shape of a $da/dn - \Delta K$ fatigue crack growth rate curve is shown in figure 9.4, which divides the curve into three regions according to the curve shape, the mechanisms of crack extension and various influences on the curve. In region I there is a threshold stress intensity range, ΔK_{th}, below which cracks either propagate at an extremely low rate or do not propagate at all. Knowledge of ΔK_{th} permits the calculation of permissible crack lengths and/or applied stresses in order to avoid fatigue crack growth. Above the threshold value the crack growth rate increases relatively rapidly

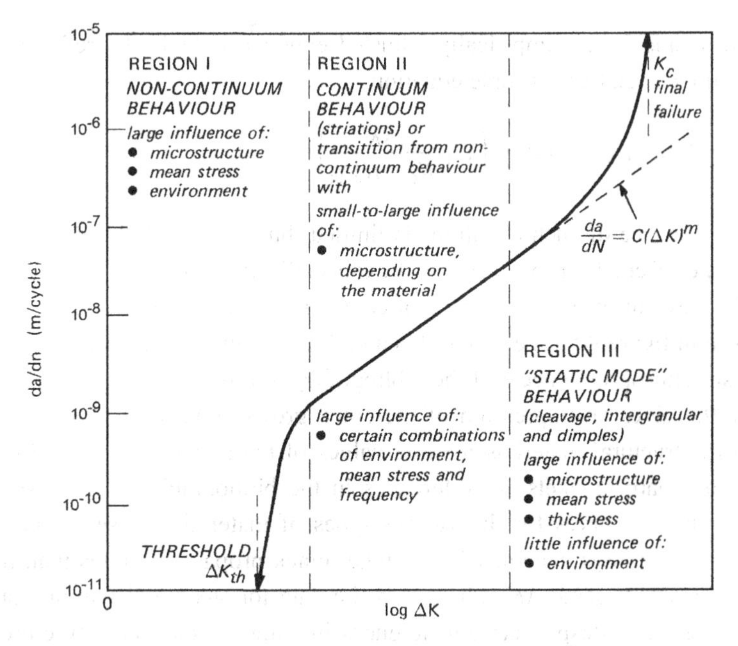

Figure 9.4. Characteristics of the fatigue crack growth rate curve $da/dn - \Delta K$.

with increasing ΔK.

In region II the crack growth rate, da/dn, is often some power function of ΔK, leading to a linear relation between $\log(da/dn)$ and $\log(\Delta K)$. Finally, in region III the crack growth rate curve rises to an asymptote where the maximum stress intensity factor, K_{max}, in the fatigue stress cycle becomes equal to the critical stress intensity factor, K_c.

There have been many attempts to describe the crack growth rate curve by 'crack growth laws', which usually are semi or wholly empirical formulae fitted to a set of data. The most widely known is the Paris equation

$$\frac{da}{dn} = C(\Delta K)^m. \tag{9.1}$$

Forman proposed the following well-known 'improved' relation:

$$\frac{da}{dn} = \frac{C(\Delta K)^m}{(1-R)K_c - \Delta K}. \tag{9.2}$$

Paris' equation describes only the linear log-log (region II) part of the crack growth curve, as indicated in figure 9.4, while Forman's equation also describes region III.

It is also possible to describe the complete $da/dn - \Delta K$ curve by an expression like

$$\frac{da}{dn} = C(\Delta K)^m \left\{ \frac{1 - \left(\frac{\Delta K_{th}}{\Delta K}\right)^{n_1}}{1 - \left(\frac{K_{max}}{K_c}\right)^{n_2}} \right\}^{n_3}, \tag{9.3}$$

where n_1, n_2 and n_3 are empirically adjusted exponents. McEvily, reference 3, developed yet another relatively simple equation:

$$\frac{da}{dn} = C(\Delta K - \Delta K_{th})^2 \left(1 + \frac{\Delta K}{K_c - K_{max}}\right). \tag{9.4}$$

The significance of such equations is limited, but they can be useful in providing a first estimate of crack growth behaviour, especially if the material concerned exhibits region II behaviour over a wide range of crack growth rates (see section 9.5). An example is given in figure 9.5. The material is a higher strength structural steel often used for offshore structures, reference 4 of the bibliography to this chapter.

Figure 9.6 gives an impression of the crack growth rate behaviour of a number of well-known structural materials for low values of the stress ratio, $R = \sigma_{min}/\sigma_{max}$ (the numbers in square brackets are references in the bibliography). The positions of the crack growth rate curves for the various types of material represent a general trend. Thus aluminium alloys generally have higher crack propagation rates than titanium alloys or steels at the same ΔK values, and the data for steels fall within a surprisingly narrow scatter band despite large differences in composition, microstructure and yield strength.

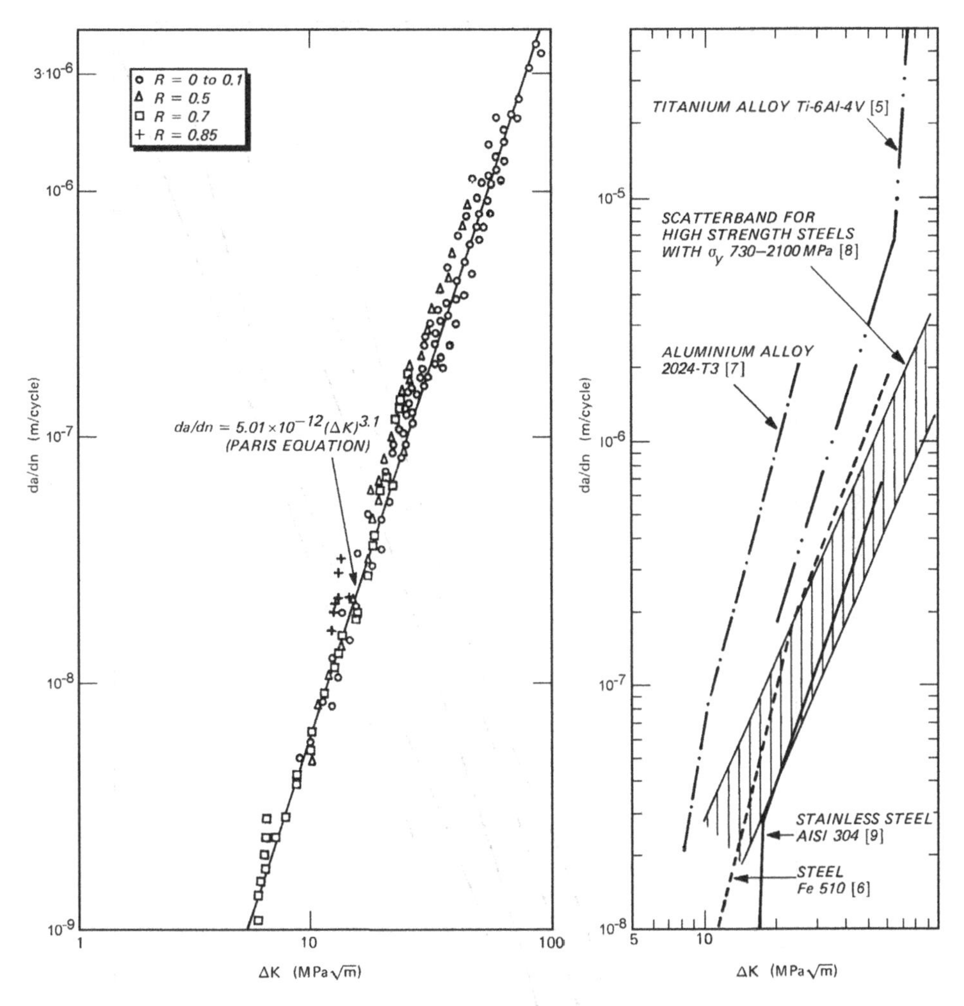

Figure 9.5. Fatigue crack growth for structural steel BS4360 at room temperature and with cycle frequencies 1 – 10 Hz.

Figure 9.6. Fatigue crack growth rates as functions of ΔK for various structural materials at low R values.

The ability of ΔK to correlate fatigue crack growth rate data depends to a large extent on the fact that the alternating stresses causing crack growth are small compared to the yield strength. Therefore crack tip plastic zones are small compared to crack length or other relevant dimensions, *e.g.* ligament size, even in very ductile materials like stainless steels.

However, even though they are small, fatigue crack plastic zones can significantly affect the crack growth behaviour. This will be shown in the next section.

9.3 The Effects of Stress Ratio and Crack Tip Plasticity: Crack Closure

The Paris equation states that da/dn is solely determined by ΔK. Other influences were thought to be secondary to that of ΔK and as a consequence they were neglected.

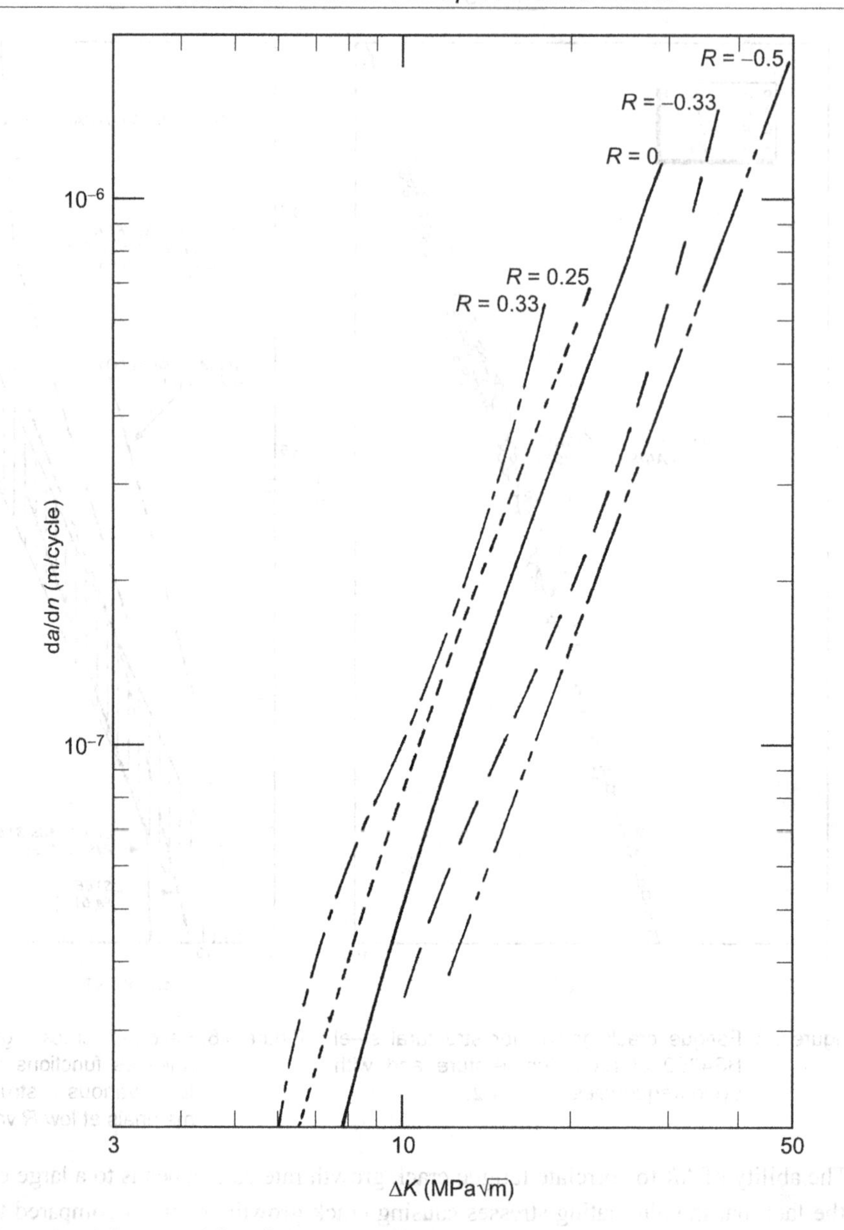

Figure 9.7. Influence of R on fatigue crack growth in aluminium alloy 2024-T3 Alclad sheet.

The effect of the stress ratio, R, was also assumed to be a secondary effect, but later experiments showed that for many materials R can significantly affect fatigue crack growth behaviour. This is expressed by the equation already given in section 1.10, *i.e.*

$$\frac{da}{dn} = f(\Delta K, R) . \tag{9.5}$$

In other words, besides the stress intensity range, ΔK, there is an influence of the relative values of K_{max} and K_{min}, since $R = \sigma_{min}/\sigma_{max} = K_{min}/K_{max}$. This is illustrated in figure 9.7, which shows that crack growth rates at the same ΔK value are generally higher when R is more positive.

There is no immediately obvious explanation for the effect of R. A proper explanation requires the understanding of crack closure, to be discussed next. However, it is worth noting here that the effect of R has proved to be strongly material dependent, a fact readily observed by comparing figures 9.5 and 9.7.

Crack Tip Plasticity and Crack Closure

In the early 1970s Elber (reference 10 of the bibliography) discovered the phenomenon of crack closure, which can help in explaining the effect of R on crack growth rates. He found that fatigue cracks are closed for a significant portion of a tensile load cycle, probably owing to residual plastic deformation left in the wake of a growing crack. This phenomenon is normally designated as plasticity-induced crack closure.

At the tip of a growing fatigue crack each loading cycle generates a monotonic plastic zone during increased loading and a much smaller reversed (cyclic) plastic zone during unloading. The reversed plastic zone is approximately one-quarter of the size of the monotonic plastic zone. This means there is residual plastic deformation consisting of monotonically stretched material.

The relative sizes of the monotonic and cyclic plastic zones can be understood from a superposition model. Figure 9.8.a shows Irwin's solution for the (forward) monotonic plastic zone with a size given by (*cf.* section 3.2)

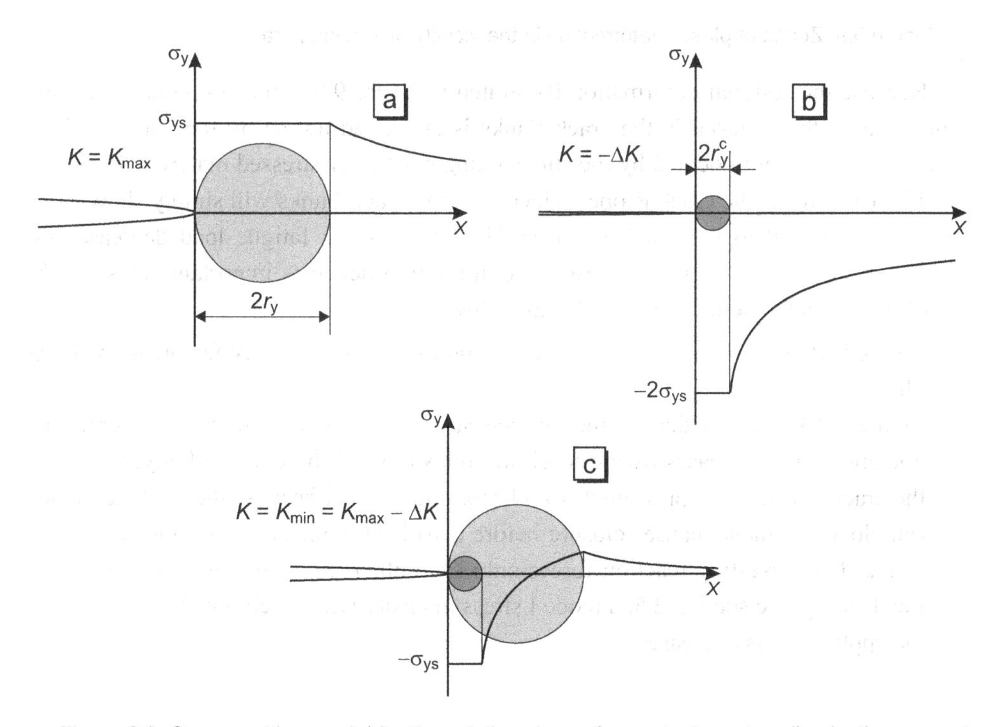

Figure 9.8. Superposition model for the relative sizes of monotonic and cyclic plastic zones at the tip of a fatigue crack.

$$2r_{\mathrm{y}} = \frac{1}{\pi}\left(\frac{K_{\max}}{\sigma_{\mathrm{ys}}}\right)^2 . \tag{9.6}$$

To obtain the solution (figure 9.8.c) for a zone of reversed plastic flow, *i.e.* σ_y changing from $+\sigma_{ys}$ to $-\sigma_{ys}$ on unloading, it is necessary to subtract the solution in figure 9.8.b from that in figure 9.8.a. In the solution shown in figure 9.8.b the yield stress is 'doubled'. Therefore it can be argued that the size of the reversed (cyclic) plastic zone, $2r_y^c$, depends on the magnitude of ΔK and *twice* the yield strength, *i.e.*

$$2r_{\mathrm{y}}^{\mathrm{c}} = \frac{1}{\pi}\left(\frac{\Delta K}{2\sigma_{\mathrm{ys}}}\right)^2 . \tag{9.7}$$

Consequently, for $R = 0$ the ratio of the monotonic and cyclic plastic zone sizes is 4, because now $\Delta K = K_{max}$. For positive R the ratio is smaller than 4 and for negative R it is larger.

As the crack grows the residual plastic deformation forms a wake of monotonically stretched material along and perpendicular to the crack flanks. This is depicted in figure 9.9 for the case of gradually increasing K levels and hence gradually increasing plastic zone sizes.

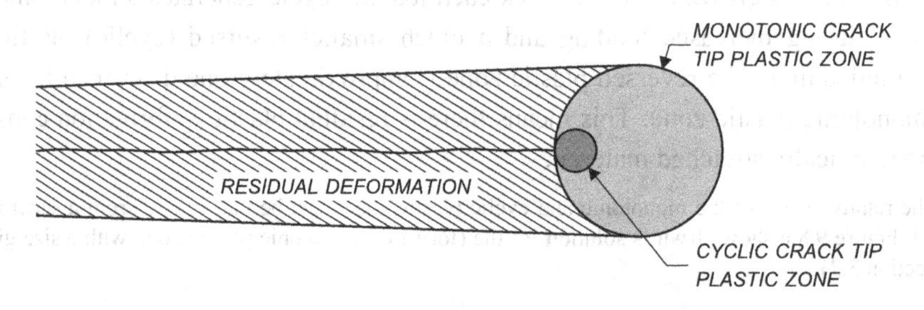

Figure 9.9. Zones of plastic deformation in the vicinity of a fatigue crack.

Because the residual deformation illustrated in figure 9.9 is the consequence of tensile loading, the material in the crack flanks is elongated normal to the crack surfaces and has to be accommodated by the surrounding elastically stressed material. This is no problem as long as the crack is open, since then the crack flanks will simply show a displacement normal to the crack surfaces. However, as the fatigue load decreases the crack will tend to close and the residual deformation becomes important. This will be illustrated with the help of figure 9.10, as follows:

1) Figure 9.10.a shows the variation in the nominal stress intensity factor, K, with applied stress, σ.
2) Figure 9.10.b shows that as the applied stress decreases from σ_{max} the crack tip opening angle decreases owing to elastic relaxation of the cracked body. However, the crack surfaces are prevented from becoming parallel because the stretched material along the flanks causes closure before zero load is reached. This closure results in the flanks exerting reaction forces onto each other. Since this is a case of crack-line loading, see section 2.5, a mode I stress intensity will develop which increases as the applied stress decreases.

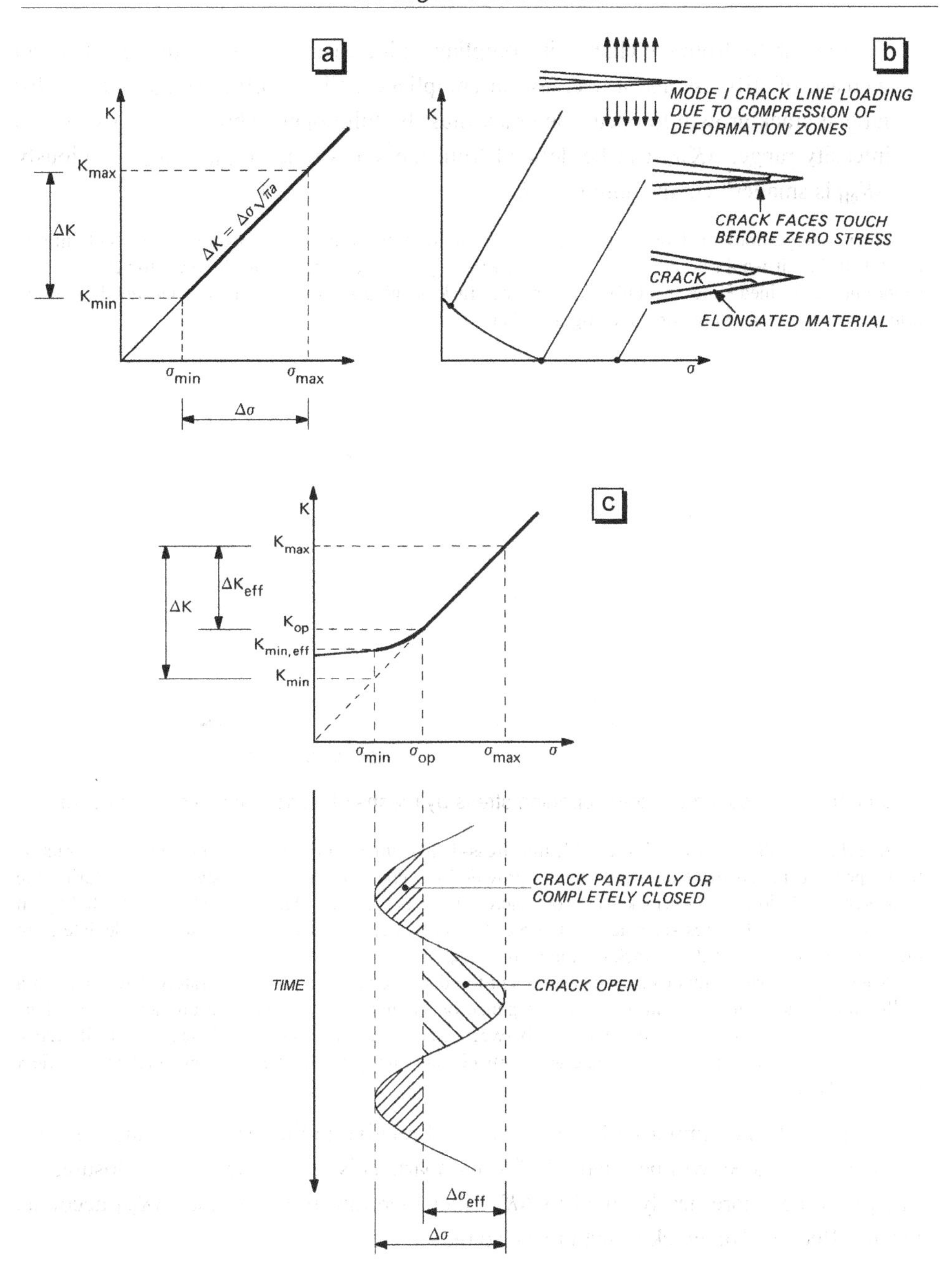

Figure 9.10. Principal of crack closure: (a) nominal K-σ plot, (b) residual deformation due to crack tip plasticity results in mode I crack-line loading K values, compare section 2.5, (c) superposition of K values shows the effect of crack closure.

3) Figure 9.10.c shows superposition of the stress intensities due to the applied stress and due to reaction forces (the latter defined as crack closure). Also the crack opening stress, σ_{op}, is indicated. This important concept was defined by Elber as that value of applied stress for which the crack is just fully open. It can be determined

experimentally from the change in compliance: increasing crack closure results in an increase of stiffness and a decrease in compliance. Elber further suggested that for fatigue crack growth to occur the crack must be fully open. Thus an effective stress intensity range, ΔK_{eff}, can be defined from the stress range $\sigma_{max} - \sigma_{op}$. Obviously ΔK_{eff} is smaller than the nominal ΔK.

Elber's suggestion that fatigue crack growth occurs only when the crack is fully open is not entirely correct. In fact it would be better to define ΔK_{eff} as $K_{max} - K_{min,eff}$, see figure 9.10.c. However, $K_{min,eff}$ cannot be determined experimentally, whereas the crack opening stress, σ_{op}, can be determined by measuring a stress-displacement curve (see figure 9.11).

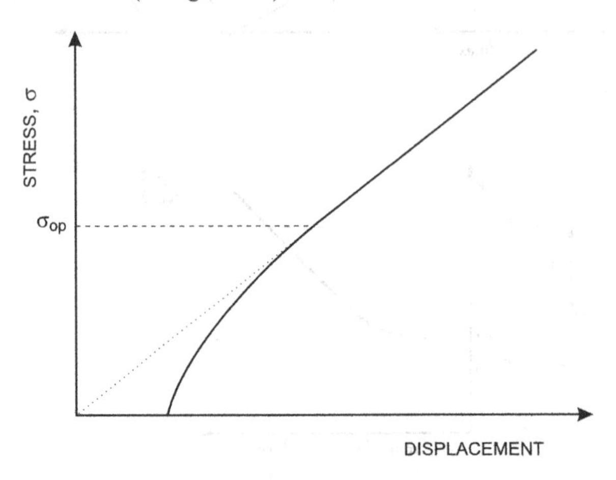

Figure 9.11. Measuring the crack opening stress by means of a stress-displacement curve.

As schematically shown in figure 9.11, the stress-displacement curve is linear above σ_{op}: the crack is fully open and the stiffness does not change provided the crack does not grow significantly during the measurement. Below σ_{op} the crack will close increasingly as the stress decreases. This is reflected by an increasing slope of the stress-displacement curve. The stress below which the curve starts to deviate from linearity can be considered the crack opening stress, σ_{op}.

Note that to perform this closure measurement the displacement has to be accurately determined at a well-chosen place using a strain gauge, a clip gauge or another displacement measurement device. Furthermore, a slight hysteresis is often visible between loading and unloading. This leads to a difference between σ_{op} and the stress at which the crack starts closing during unloading. For convenience this effect is ignored here.

A higher R value implies a higher mean load relative to the load amplitude. As a result the crack flanks will be relatively further apart. This leads to less crack closure, *i.e.* ΔK_{eff} becomes more nearly equal to ΔK. Elber therefore proposed that ΔK_{eff} accounts for the effect of R on crack growth rates, so that

$$\frac{da}{dn} = f(\Delta K_{eff}) \,, \tag{9.8}$$

cf. equation (9.5). He also obtained the empirical relationship

$$\frac{\Delta K_{eff}}{\Delta K} = U = 0.5 + 0.4R \quad \text{for} -0.1 \leq R \leq +0.7 \,, \tag{9.9}$$

which enabled crack growth rates for the indicated range of R values to be correlated by

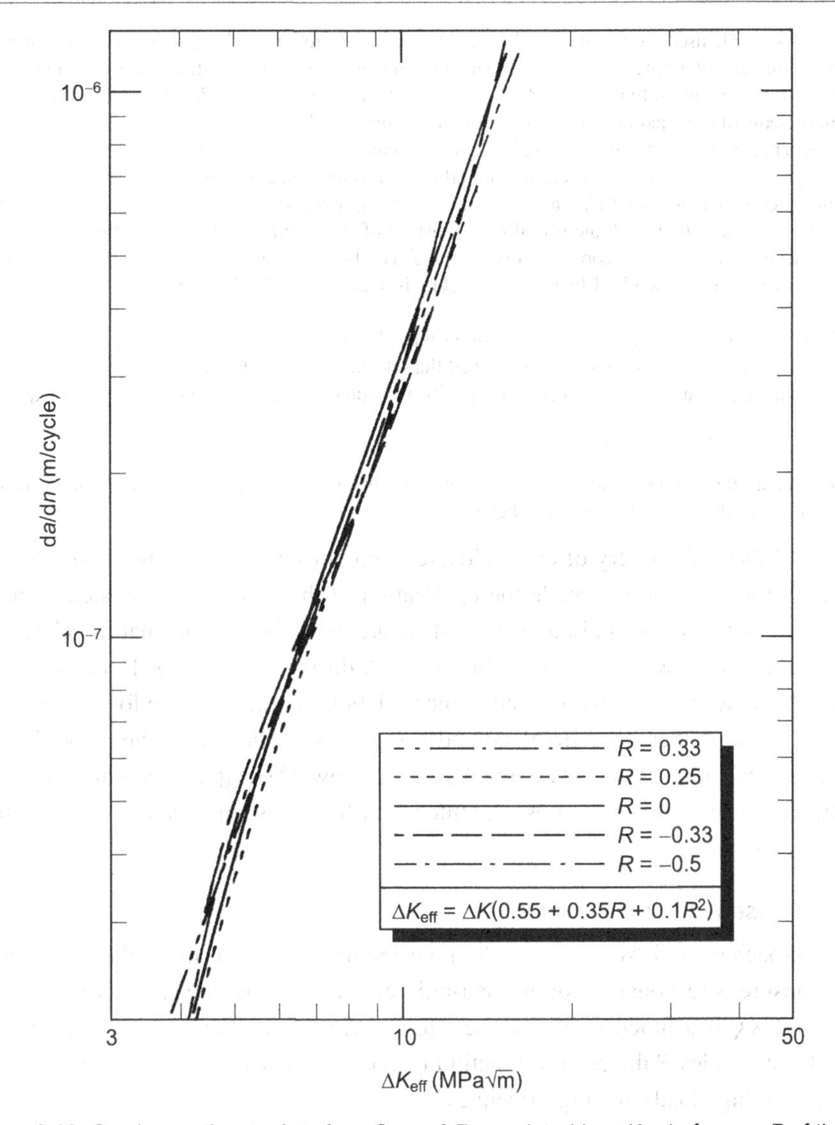

Figure 9.12. Crack growth rate data from figure 9.7 correlated by ΔK_{eff} (reference 7 of the bibliography).

ΔK_{eff}. This *crack closure function*, which was obtained for the aluminium alloy 2024-T3, has been modified by Schijve (reference 11 of the bibliography) as follows:

$$U = 0.55 + 0.35R + 0.1R^2 \quad \text{for } -1 \leq R \leq +1 \,. \tag{9.10}$$

The usefulness of equation (9.10) is demonstrated by figure 9.12, in which the crack growth rate data from figure 9.7 are plotted against ΔK_{eff}. Although equations (9.9) and (9.10) were found for aluminium alloy 2024-T3, they can often be used for other metal alloys too.

Elber measured σ_{op} values from stress-displacement curves (the principle is shown in figure 9.11). In contrast with this, Schijve used a similitude approach by using the assumption that an equal ΔK_{eff} leads to an equal da/dn. He used a quadratic crack closure function, U, to let the $da/dn - \Delta K_{eff}$ curves for different R values coincide, *cf.* figures 9.7 and 9.12. An additional and necessary constraint was added to obtain the result, *i.e.* the assumption that $\Delta K_{eff} = \Delta K$ for $R = 1$. This implies $U = 1$ for $R = 1$ and thus that the sum of the coefficients of the quadratic crack closure function must be 1.

In general, exactly the same correlation will be found between da/dn and either $\Delta K_{eff} = U(R)\,\Delta K$ or $\Delta K^*_{eff} = q\,U(R)\,\Delta K = U(R)^*\,\Delta K$, where q is an arbitrary constant (see reference 2). The resulting ratios of the coefficients of $U(R)$ and $U(R)^*$ are the same, *i.e.* for a quadratic U: $a/b/c = qa/qb/qc$. In the $da/dn - \Delta K_{eff}$ plot given in figure 9.12, the use of $U(R)^*$ instead of $U(R)$ only results in a shift to the left or to the right, depending on q, but the correlation is the same. To obtain a definite solution either a physical measurement, such as performed by Elber (see fig. 9.11), is needed or a U value must be assumed for a certain R value.

Note that it is not always correct to assume that closure occurs for all $R < 1$. For example, recent measurements on aluminium alloy 5083 showed that closure is absent for $R \geq 0.5$ (reference 12). In general to obtain a quadratic closure function of the form of equation (9.10) the coefficients must satisfy

$$a + bR_c + cR_c^2 = 1\ ,$$

where R_c is the R value above which closure is absent. The closure function U now becomes 1 for $R = R_c$, while for $R > R_c$ the U value should be taken as 1.

Since Elber's discovery of crack closure, numerous papers have been written on this subject. Crack closure has made the application of the similitude approach more difficult. A prediction of da/dn in a cracked structure not only requires that the K applied to the structure is known, but also the history of K during previous load cycles. Although models have been developed to predict the variation of crack closure for any arbitrary K history, many problems are involved and several assumptions must be made. For loading situations with a slowly changing K_{max}, *i.e.* a low dK_{max}/da, the situation is less difficult, since standard relations as presented by Elber or Schijve can be adopted to calculate ΔK_{eff}.

Other Causes for Crack Closure

The approaches used by Elber and Schijve implicitly assume that (plasticity-induced) crack closure is responsible for all load-ratio effects, and that the crack closure function $U = \Delta K_{eff}/\Delta K$ is a function of R alone. However, some authors, *e.g.* reference 13, observed that besides R the closure function U may also depend on K_{max}. This is specially the case for high loads and high R values.

Another factor that may play a role in the amount of closure is the crack front geometry. The experimental conditions under which both Elber and Schijve measured crack growth and closure for the aluminium alloy 2024-T3 were such that *shear lips* (see section 3.6) are likely to have been present on part of the fracture surfaces. This may cause additional closure as is illustrated in figure 9.13.

More recently it was shown that the roughness of shear lips can also affect the amount of crack closure, see reference 15. Figure 9.14 shows fracture surfaces of specimens of aluminium alloy 2024 subjected to identical fatigue loads but at different frequencies. At the higher frequency a lower crack growth rate was found, which was attributed to roughness-induced crack closure. Note that this type of closure can also result from the microstructure. For example, a large grain size may lead to crack path tortuosity and thus to more roughness-induced closure.

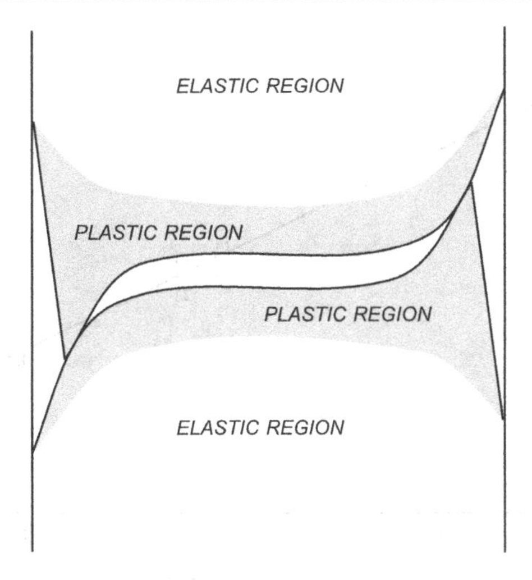

Figure 9.13. Cross-section of a through-thickness crack showing additional closure due to shear lips (from reference 14).

Other closure mechanisms have also been reported, such as (microstructural) transformation of material near the crack, formation of oxides or corrosion products on the crack surfaces or a (viscous) fluid resisting the closing and opening of the crack.

Stress Ratio Effect on Fatigue Threshold Stress Intensity Range

Besides the influence on crack growth rate in the ΔK range where the Paris equation is valid, crack closure plays an even more important role in the threshold region. In this region of very low crack growth rates the loads are low leading to small average crack opening values. This means that different closure mechanisms can be operative. For materials with only limited closure effects ΔK_{th} is more or less independent of R. On the other hand, for example, ΔK_{th} for the aluminium alloy 2024-T3 decreases with increas-

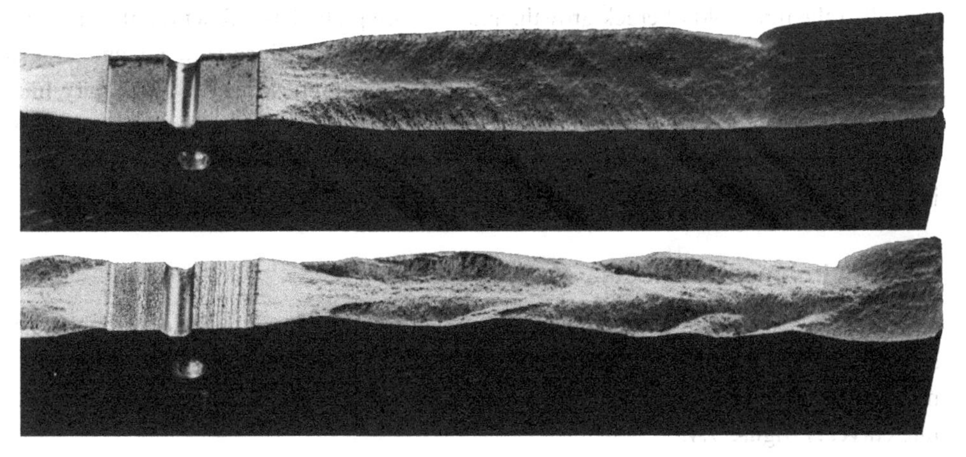

Figure 9.14. Smooth (top) and rough (bottom) shear lips on aluminium alloy 2024 fatigue fracture surfaces formed at frequencies of 0.2 and 20 Hz, respectively, both at a constant $\Delta K = 20\ \mathrm{MPa}\sqrt{\mathrm{m}}$, $R = 0.11$ and 42-55% relative humidity.

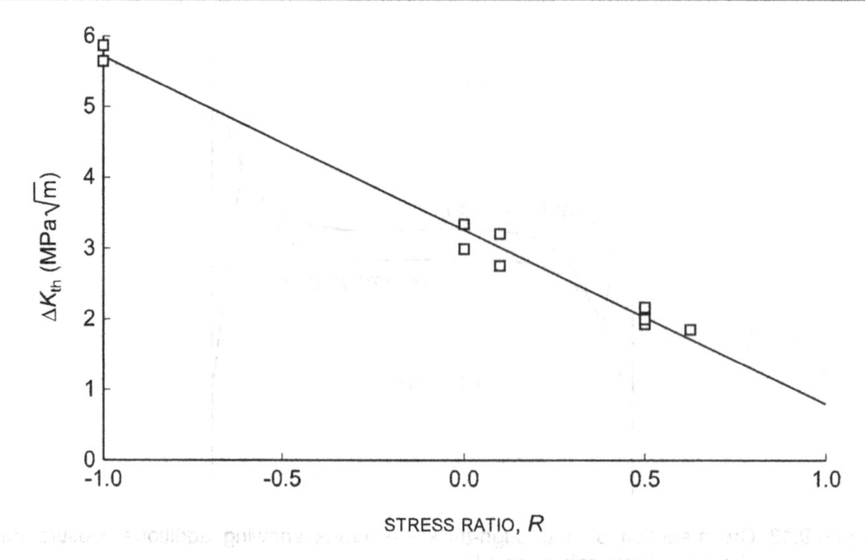

Figure 9.15. Threshold stress intensity range for fatigue crack growth as a function of R for aluminium alloy 2024-T3 (reference 16).

ing stress ratio R, as is clearly shown in figure 9.15. In this alloy closure plays a significant role (see also figures 9.7 and 9.12).

A simple description of the effect of R on the fatigue crack growth threshold is given by Schmidt and Paris (reference 17). They assumed that

- the threshold condition for fatigue crack growth corresponds to a constant effective stress intensity range, $\Delta K_{th,eff}$, that is independent of R,
- no crack closure will occur at R values above a certain critical value, R_c,
- the opening stress intensity level, K_{op}, is independent of R for $R < R_c$, *i.e.* when crack closure does occur.

The consequences of these assumptions are depicted in figure 9.16, where K_{max}, K_{min}, K_{op}, all at the threshold of crack growth, and ΔK_{th} are plotted as a function of R. For $R < R_c$, *i.e.* in the presence of closure, $K_{max} = K_{op} + \Delta K_{th,eff}$, and since both K_{op} and $\Delta K_{th,eff}$ are assumed constant K_{max} must be constant too. This implies that ΔK_{th} will vary linearly with R according to $\Delta K_{th} = (1 - R)K_{max}$. For $R > R_c$, *i.e.* in the absence of crack closure, ΔK_{th} will be constant and equal to $\Delta K_{th,eff}$. Clearly both K_{max} and K_{min} will now increase with R.

9.4 Environmental Effects

Fatigue crack growth is a complex process influenced by a number of variables besides the effective crack tip stress intensity range. In particular the environment and material microstructure can have large influences in various regions of the crack growth rate curve, *cf.* figure 9.4.

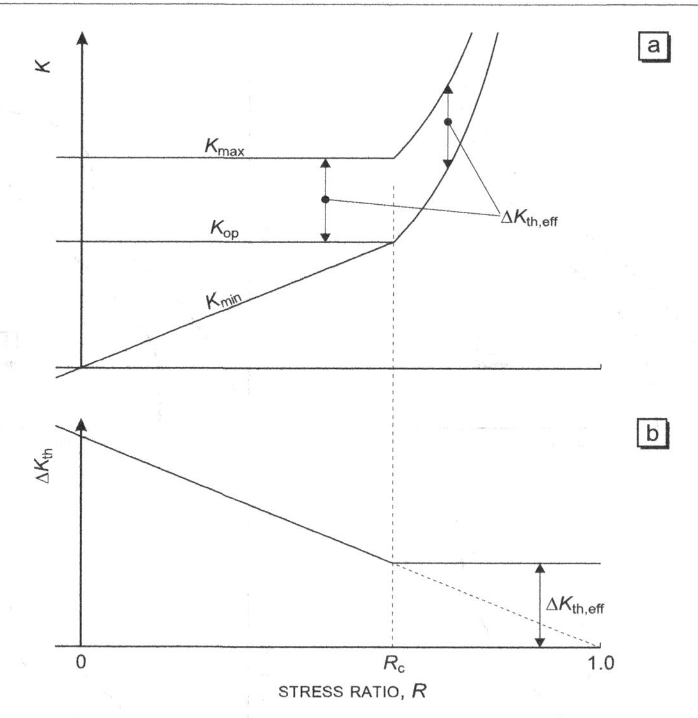

Figure 9.16. K_{max}, K_{min} and K_{op} at the fatigue crack growth threshold (a) and ΔK_{th} (b) as a function of R.

Environmental effects are very important in regions I and II of the crack growth rate curve. Several types of environmental effects are schematically illustrated in figure 9.17. The first type, figure 9.17.a, is a reduction or elimination of ΔK_{th} and an overall enhancement of crack growth rates until high ΔK levels, at which the purely mechanical contribution to crack growth predominates.

A second type of environmental effect, figure 9.17.b, differs from the first in that above an intermediate ΔK level there is an additional enhancement of crack growth rates owing to monotonic, sustained load fracture during the tensile part of each stress cycle. This behaviour is typical for fatigue in liquid environments that cause stress corrosion. On the other hand a gaseous environment, usually hydrogen, that causes sustained load fracture can give the type of behaviour shown in figure 9.17.c. (Sustained load fracture is discussed as a separate topic in chapter 10.)

Finally, figure 9.17.d shows a type of environmental effect in which there is an overall enhancement of crack growth rates except near the threshold. In such cases ΔK_{th} in the aggressive environment can even be higher! This can be explained in some instances by local corrosion on the crack surfaces. The corrosion products increase the volume of material contributing to crack closure, thereby raising σ_{op} and decreasing ΔK_{eff}. This will be discussed in more detail in chapter 13, section 13.4.

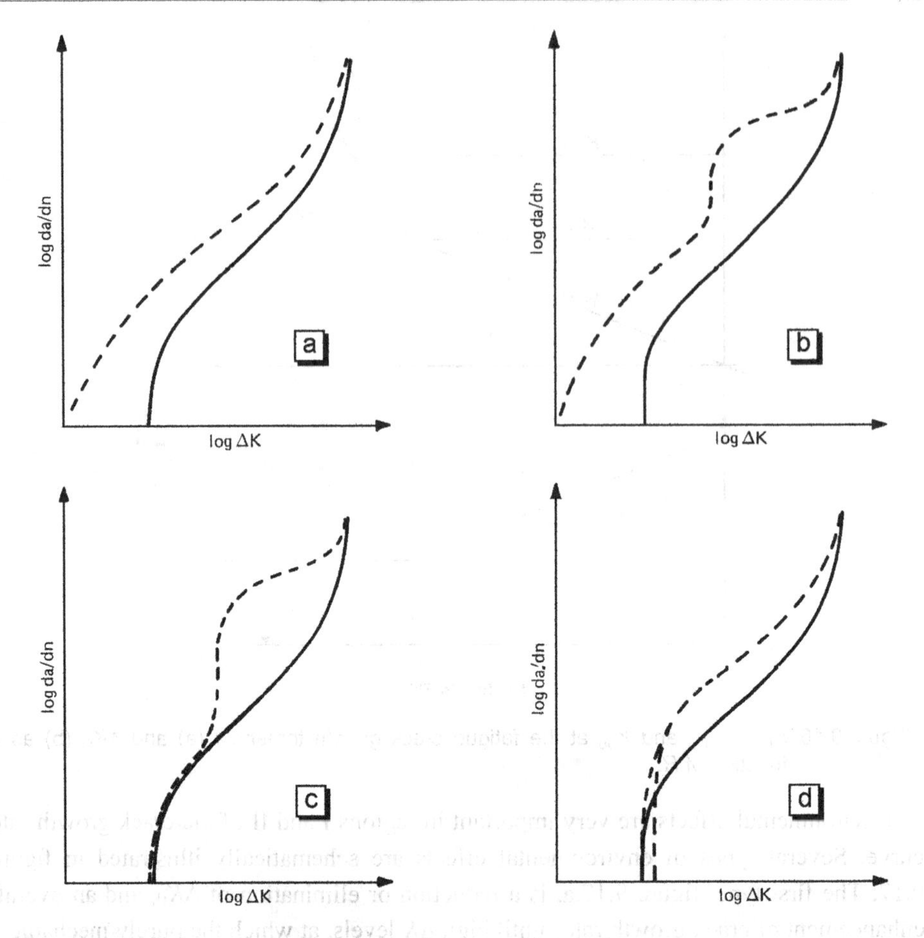

Figure 9.17. Environmental effects on the fatigue crack growth rate curve: normal environment (*e.g.* air) ———— ; aggressive environment – – – – –.

Remarks

1) Despite the last statement about corrosion products and crack closure it is not intended here to discuss possible mechanisms of environmental fatigue crack growth. This is a controversial subject, since material – environment interactions at crack tips are difficult to observe and apparently highly complex.
2) Environmental effects on fatigue crack growth strongly depend on the specific material – environment combination and also on several other factors. The main ones are:
   - Frequency of the fatigue stress cycle: low frequencies generally result in greater environmental effects.
   - Waveform of the fatigue stress cycle: in general, a crack grows more per cycle if the increasing part of the load cycle occurs more slowly. For example, a positive sawtooth waveform /\/\/\/\/\/ results in higher environmental fatigue crack growth rates than a negative sawtooth waveform \/\/\/\/\/\ .
   - Temperature: environmental effects are usually greater at higher temperatures.

- A very high growth rate (high ΔK) diminishes the importance of the environmental effect (K_{max} during fatigue nears K_c, see figure 9.4).

3) The unusual shapes of the environmental fatigue crack growth rate curves in figures 9.17.b and 9.17.c deserve comment. At intermediate ΔK levels the overall crack growth rate suddenly increases greatly, followed by crack growth rates with a relatively weak dependence on ΔK. This is a consequence of a major contribution to the overall crack growth by sustained load fracture. As will be discussed in chapter 10, sustained load fracture crack growth rates can be almost constant over a wide range of stress intensity.

9.5 Prediction of Fatigue Crack Growth Under Constant Amplitude Loading

The main purpose of crack growth prediction is to construct a crack growth curve, a versus n, like that shown schematically in figure 9.2. For constant amplitude loading this is a fairly straightforward procedure of integrating $^{da}/_{dn}$ – ΔK curves.

It is often possible to use 'crack growth laws' like equations (9.1) – (9.4). These are useful because they are closed form expressions that can be integrated analytically. For example, the crack growth rate data for BS4360 structural steel shown in figure 9.5 can be expressed as

$$\frac{da}{dn} = 5.01 \times 10^{-12} (\Delta K)^{3.1} \ ^{m}/_{cycle} \tag{9.11}$$

if ΔK is in $\mathrm{MPa}\sqrt{\mathrm{m}}$. For a wide plate one can substitute $\Delta K = \Delta\sigma\sqrt{\pi a}$ in equation (9.11), giving

$$da = 5.01 \times 10^{-12} \left(\Delta\sigma\sqrt{\pi a}\right)^{3.1} dn$$

or

$$dn = \frac{da}{5.01 \times 10^{-12}\left(\Delta\sigma\sqrt{\pi a}\right)^{3.1}}$$

and

$$n = \frac{10^{12}}{5.01\left(\Delta\sigma\sqrt{\pi}\right)^{3.1}} \int_{a_d}^{a_{cr}} a^{-1.55}\, da\,. \tag{9.12}$$

Note that the crack length a has to be in metres (m) because of the units for ΔK.

However, if finite width correction factors or more complicated crack growth laws are necessary it becomes difficult to integrate crack growth rate data analytically. Instead a numerical approach must be used. This can be done in the following way:

- choose a suitable increment of crack growth, $\Delta a_i = a_{i+1} - a_i$, *i.e.* small enough to obtain sufficient accuracy but not too small to cause an excessive calculation effort,

- calculate ΔK for the crack length corresponding to the mean of the crack growth increment, *i.e.* $(a_{i+1} + a_i)/2$,
- determine da/dn for this value of ΔK,
- calculate Δn_i from $\Delta a_i/(da/dn)_i$,
- repeat the previous steps over the required range of crack growth and sum the values of Δn_i.

Thus it is possible to predict crack growth from any type of $da/dn - \Delta K$ curve as long as the relation between crack length, a, and ΔK is known for the structure or component under consideration. Herein lies the difficulty: in practice it may be difficult to estimate ΔK owing to the complex geometry of the crack (*e.g.* semi-elliptical surface cracks or quarter-elliptical corner cracks) and to the occurrence of load shedding. In built-up structures cracked elements will shed load to uncracked elements because cracking causes a decrease in stiffness and the displacements in each element are mutually constrained in a kind of 'fixed grip' condition (see also section 4.2).

9.6 Fatigue Crack Growth Under Variable Amplitude Loading

Although constant amplitude loading does occur in practice (*e.g.* pressurisation cycles in transport aircraft cabins, rotating bending stresses in generators, thermal stress cycles in pressure vessels) the vast majority of dynamically loaded structures actually experience variable amplitude loading that is often fairly random. An example is given in figure 9.18.

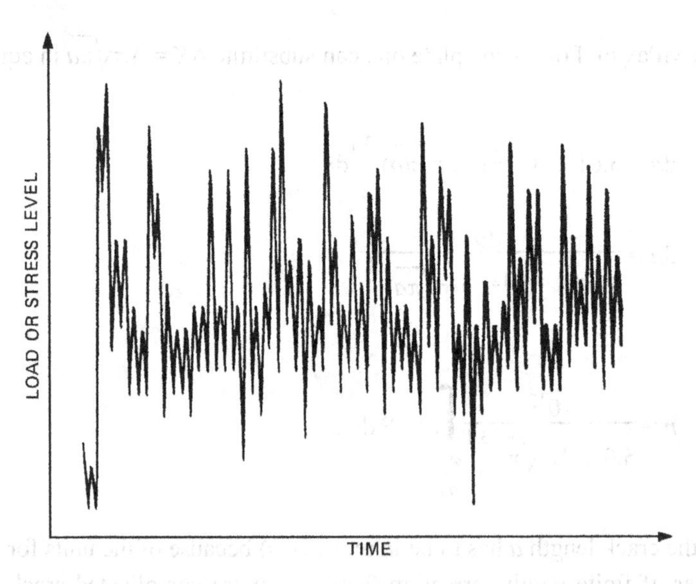

Figure 9.18. Example of a fairly random cyclic load history (part of a flight for a tactical aircraft).

An important consequence of variable amplitude loading is that ΔK need not increase gradually with increasing crack length. For example, large and small stress ranges follow each other directly in figure 9.18 and so the instantaneous ΔK will vary greatly. This variation results in load interaction effects which may strongly influence fatigue crack growth rates. The nature of such interaction effects is best demonstrated by some simple examples.

Simple Examples of Variable Amplitude Loading

The simplest type of variable amplitude loading is the occurrence of occasional peak loads in an otherwise constant amplitude loading history. Figure 9.19 shows two such simple variable amplitude loading histories with their effects on fatigue crack growth as compared to constant amplitude loading. There is a profound effect on crack growth owing to the occurrence of positive peak stresses only (curve C). However, the effect is much less when positive peak stresses are immediately followed by negative peak stresses (curve B). The explanation for these effects is as follows:

1) Positive peak stresses (curve C). Each peak stress opens up the crack tip much more than the normal maximum stress, but also creates a larger plastic zone ahead of the crack. There are three consequences when normal load cycling is resumed. Initially the amount of crack closure is reduced (due to the large plastic zone before the tip the crack may be fully open at minimum stress with $\Delta K_{\mathrm{eff}} = \Delta K$) and the crack will grow somewhat faster than before the peak stress occurred. However, the crack soon grows into the large plastic zone and encounters high residual compressive stresses because this plastic zone must be accommodated by the surrounding elastically stressed material. Also a wake of enhanced residual deformation forms behind the crack tip and causes an increase in crack closure.

The overall result is a significant retardation of crack growth. This is often called

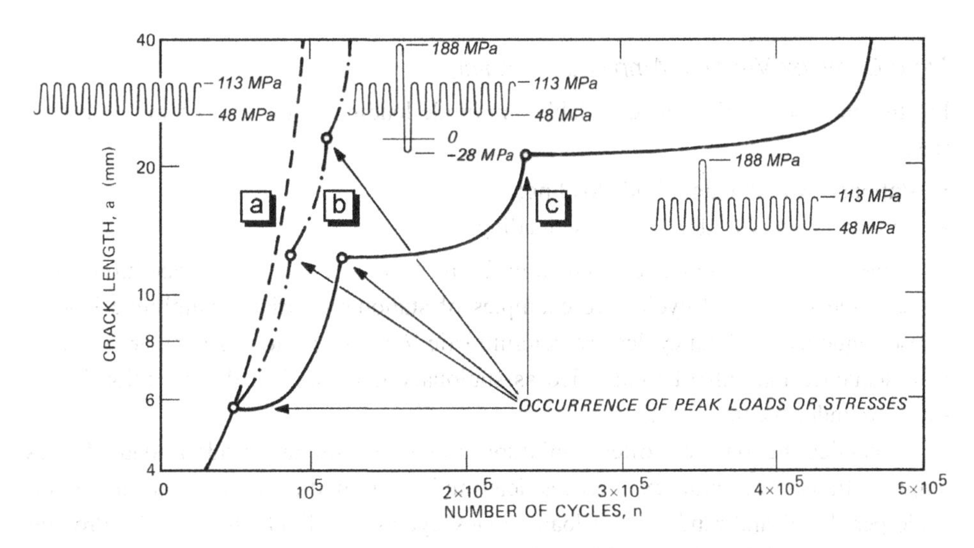

Figure 9.19. Crack growth curves under constant amplitude loading and constant amplitude loading + occasional peak loads for a centre cracked panel of aluminium alloy 2024-T3 (reference 18 of the bibliography).

delayed retardation, owing to the initial acceleration of crack growth following the peak stress.

2) Positive + negative peak stresses (curve B). The negative peak stress reverses most of the tensile plastic deformation due to the positive peak stress and the amount of crack growth retardation is greatly diminished.

Since residual stresses and residual plastic deformation in the vicinities of crack tips are responsible for interaction effects, the magnitude of these effects will depend on the ratios of the peak stresses to the normal maximum stresses, the material yield strength and strain hardening characteristics, and the stress state (plane strain or plane stress). The larger plastic zones in plane stress result in much greater interaction effects.

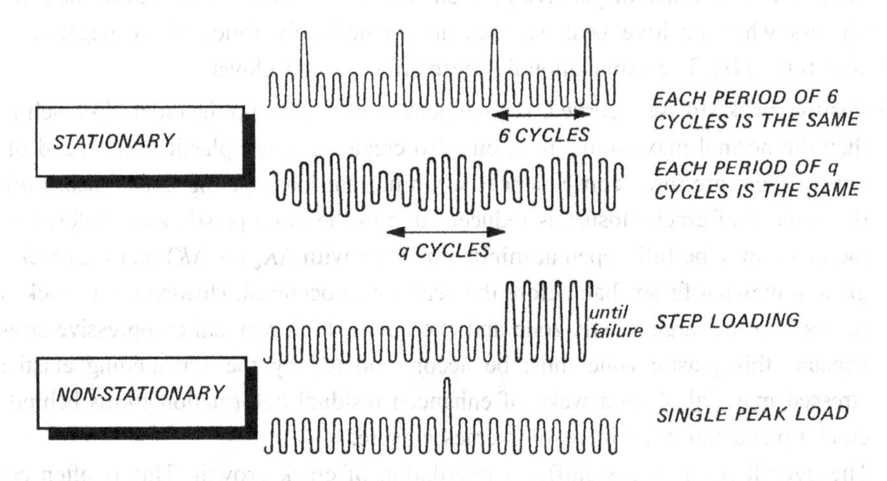

Figure 9.20. Examples of stationary and non-stationary variable amplitude loading.

More Complex Variable Amplitude Loading

For the purpose of this course variable amplitude loading can be placed in two categories:

- stationary variable amplitude loading;
- non-stationary variable amplitude loading.

Examples of both categories are shown in figure 9.20. This shows that repetitions of the same sequence of load cycles are examples of stationary variable amplitude loading. These sequences of load cycles are deterministic, *i.e.* non-random. However, a random load sequence may also be classified as stationary if its statistical description is constant, *i.e.* independent of time.

In practice the loads are often a mixture of deterministic and random loads. For example, a transport aircraft experiences deterministic ground-air-ground transitions (one cycle per flight) and random gust loads (many cycles per flight), figure 9.21. Provided that the statistical description of this mixture of loads is invariant with time, then such a load sequence also may be classified as stationary variable amplitude loading.

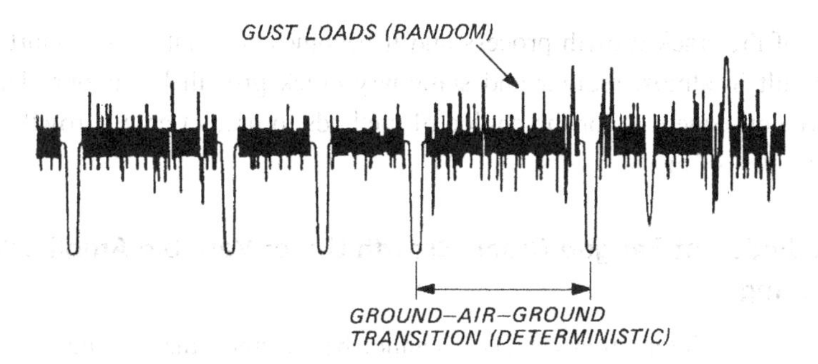

Figure 9.21. Example of load history for a transport aircraft wing.

In the previous part of this section it was shown that positive peak loads can cause crack growth retardation. In turn this retardation may result in crack growth no longer being a regular (*i.e.* stationary) process. To illustrate this, consider the loading histories shown in figure 9.22. The first load history has peak loads with a long recurrence period. Such a load history can result in large discontinuities in the crack growth curve. Curve C in figure 9.19 is a good example. These discontinuities are pronounced because the recurrence period is sufficiently long that each retardation is over before the next peak load occurs.

On the other hand, figure 9.22 shows that the second load history, which has peak loads with a short recurrence period, results in an almost regular crack growth curve. Retardations still occur, but they are superimposed on each other: this is, in fact, highly effective in slowing down the overall crack growth.

It is thus clear that the recurrence periods of peak loads are important for crack growth under variable amplitude loading. Long recurrence periods will disturb the

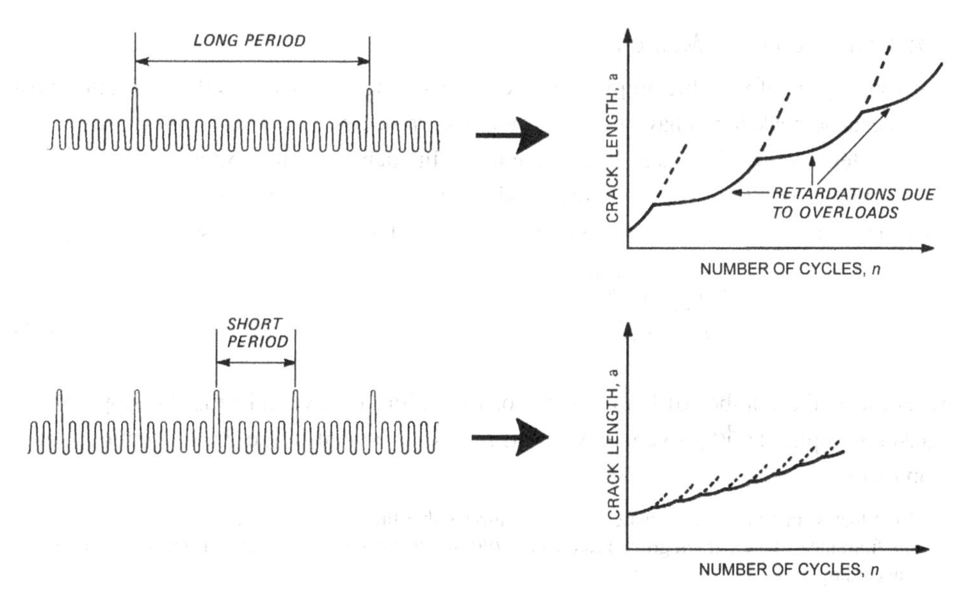

Figure 9.22. Effect of peak load recurrence period on the crack growth curve.

regularity of the crack growth process and it becomes non-stationary. Short recurrence periods result in almost regular and stationary crack growth behaviour. These differences have consequences for the choice of methods to predict crack growth under variable amplitude loading.

9.7 Prediction of Fatigue Crack Growth Under Variable Amplitude Loading

Owing to load interaction effects it cannot be assumed that an increment of crack growth, Δa_i, in cycle i of a variable amplitude load history is directly related to ΔK_i Thus a straightforward summation procedure using constant amplitude crack growth rate data (as in section 9.5) will not give proper results. Nevertheless such a summation is sometimes done. The results are nearly always conservative, often extremely so, since the most important consequence of load interaction effects is crack growth retardation.

An empirical way of accounting for load interaction effects is to determine experimentally whether crack growth rates under various kinds of variable amplitude loading can be correlated by characteristic stress intensity factors, and if so to numerically integrate the crack growth rate curves in the same way as for constant amplitude loading. This method is subject to several limitations, as will be discussed.

Besides this characteristic K method a number of crack growth models have been developed since about 1970 to try and account for load interaction effects and thereby enable predictions of crack growth curves. These models are of two kinds:

1) models based on crack tip plasticity ('first generation' models),
2) models based on crack closure ('second generation' models).

A concise review of these models will be made.

The Characteristic K Method

For some types of variable amplitude load history fatigue crack growth is a regular process because peak loads have either a short recurrence period or only minor effects on crack extension in subsequent load excursions. In such cases it is often possible to correlate crack growth rates by a characteristic stress intensity factor. One possibility is the root mean value of the stress intensity range, ΔK_{rm}. The general expression for ΔK_{rm} is

$$\Delta K_{rm} = \sqrt[m]{\frac{\Sigma(\Delta K_i)^m n_i}{\Sigma n_i}}, \qquad (9.13)$$

where n_i is the number of load cycles corresponding to ΔK_i and m is the slope of the constant amplitude da/dn versus ΔK plot, *i.e.* the exponent in the Paris equation, equation (9.1).

Note that in fact the characteristic K method implies that the regular fatigue crack growth rate da/dn results from the different crack growth rate values $(da/dn)_i$ that each correspond to a stress intensity range ΔK_i, according to

$$\frac{da}{dn} = f_1\left(\frac{da}{dn}\right)_1 + f_2\left(\frac{da}{dn}\right)_2 + \ldots\ldots, \tag{9.14}$$

where f_i is the fraction $n_i/\sum n_i$ of the total number of load cycles with a stress intensity range ΔK_i. It is implicitly assumed that there are no interaction effects between different fractions.

The main problem in the use of equation (9.13) is the derivation of ΔK values. To do this properly it is necessary to account for crack closure and follow a rational procedure of cycle counting. A full discussion of this problem is given in reference 19 of the bibliography, from which the example in figure 9.23 is taken. This figure shows that ΔK_{rm} correlates regularly retarded crack growth under block programme loading with constant amplitude crack growth. The reason is that correct derivation of ΔK_{rm} for block programme loading results in a much smaller value than the ΔK_{rm} (= ΔK_{eff}) for constant amplitude loading, and this successfully accounts for the shift in the block programme crack growth rate curve owing to retardation.

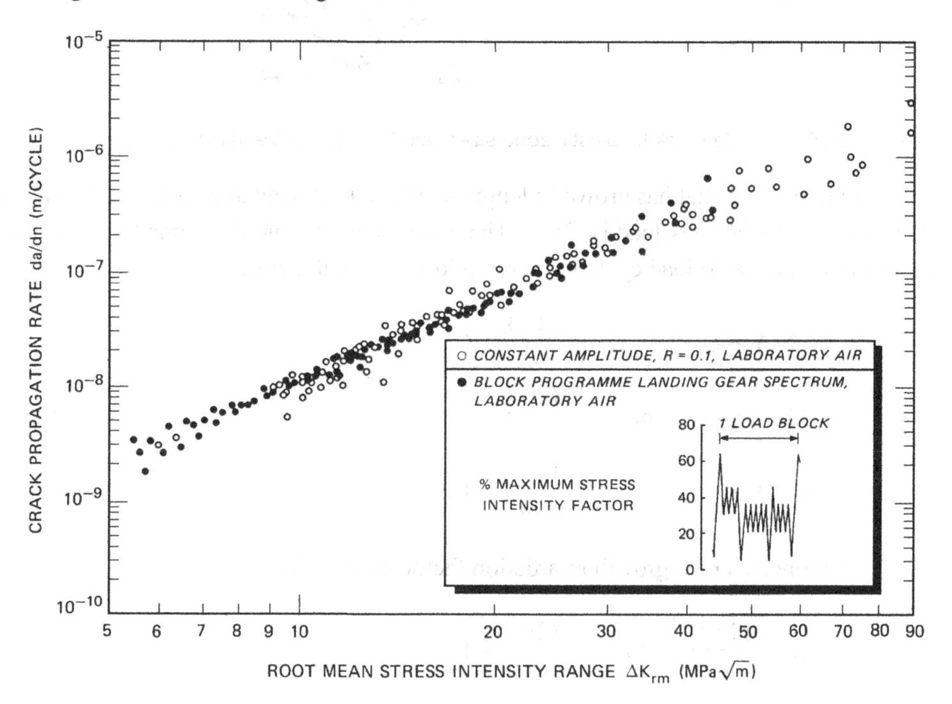

Figure 9.23. Crack growth rates in three ultrahigh strength steels correlated by ΔK_{rm}.

A special case is narrow band random loading of steels, for which m was assumed to be 2 and no account was taken of crack closure, reference 20. When $m = 2$ the stress intensity range becomes the root mean square value, ΔK_{rms}. Since the root mean square is a parameter characterizing a stationary random process it was considered that ΔK_{rms} should correlate crack growth under stationary (narrow band) random loading. In fact, good correlations were obtained. However, such correlations are very empirical and are useful for predicting crack growth only for load histories very similar to those for which the correlations were made.

Crack Growth Models Based on Crack Tip Plasticity

These models assume that after a peak load there will be an interaction effect as long as the crack tip plastic zones for subsequent load cycles are within the plastic zone due to the peak load. The consequences of this assumption will be illustrated using figure 9.24 and a well known crack growth model introduced by Wheeler (reference 21 of the bibliography).

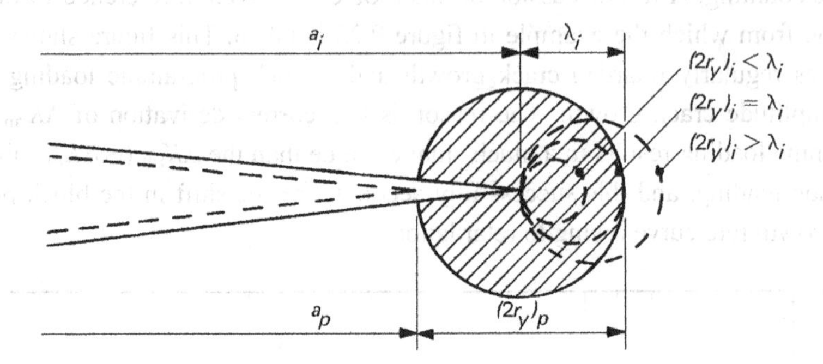

Figure 9.24. Possible crack tip plastic zone sizes due to a load cycle following a peak load.

Consider a crack that has grown to length a_i after a peak load at length a_p. The plastic zone size due to the peak load is $(2r_y)_p$. There are three possible situations for the plastic zone size $(2r_y)_i$ due to load cycle i (which follows the peak load):

$$
\begin{aligned}
&(2r_y)_i < \lambda_i \quad &\text{or} \quad &\frac{(2r_y)_i}{\lambda_i} < 1 \\
&(2r_y)_i = \lambda_i \quad &\text{or} \quad &\frac{(2r_y)_i}{\lambda_i} = 1 \\
&(2r_y)_i > \lambda_i \quad &\text{or} \quad &\frac{(2r_y)_i}{\lambda_i} > 1\,.
\end{aligned}
\tag{9.15}
$$

Wheeler defined a crack growth retardation factor, ϕ, as follows:

$$
\begin{aligned}
&\phi_i = \left[\frac{(2r_y)_i}{\lambda_i}\right]^m \quad &\text{for} \quad &\frac{(2r_y)_i}{\lambda_i} < 1 \\
&\phi_i = 1 \quad &\text{for} \quad &\frac{(2r_y)_i}{\lambda_i} \geq 1\,.
\end{aligned}
\tag{9.16}
$$

The plastic zone sizes $(2r_y)_p$ and $(2r_y)_i$ are given by

$$
2r_y = \frac{1}{\pi}\left(\frac{K_{max}}{C\sigma_{ys}}\right)^2, \tag{9.17}
$$

where C is the plastic constraint factor that accounts for the effect of the stress state on the plastic zone size, *cf.* section 3.5. The increment of crack growth, Δa_i, occurring during load cycle i is

$$\Delta a_i = \phi_i \left(\frac{da}{dn}\right)_{ca}, \tag{9.18}$$

where $(da/dn)_{ca}$ is the constant amplitude crack growth rate appropriate to the stress intensity range, ΔK_i, and stress ratio, R_i, of load cycle i. Note that as soon as $\phi_i = 1$ there is no longer any retardation.

The constant m in equation (9.16), designated by Wheeler as the *retardation shaping exponent*, is empirical. It has to be found by fitting crack growth predictions to test data. For this purpose a fairly simple variable amplitude load history is used. Once m is obtained the crack growth curve for a complex variable amplitude load history is predicted using a step by step counting method for each crack length a_i:

- calculate $(2r_y)_p$ and $(2r_y)_i$ according to equation (9.17),
- calculate λ_i $(= a_p + (2r_y)_p - a_i)$,
- calculate ϕ_i according to equation (9.16) or set $\phi_i = 1$ if $(2r_y)_i \geq \lambda_i$,
- calculate Δa_i according to equation (9.18),
- determine $a_{i+1} = a_i + \Delta a_i$,
- repeat the previous steps for each load cycle.

Note that this procedure is basically different from the numerical integration of constant amplitude data mentioned in section 9.5. For constant amplitude crack growth prediction a crack length increment is taken and the cycles are summed. For variable amplitude crack growth prediction each cycle is taken and the crack length increments are summed.

Crack growth models like that of Wheeler have some success in predicting crack growth under variable amplitude loading. However, the models rely completely on empirical adjustment (*i.e.* of m) and do not incorporate effects such as delayed retardation or the counteracting effect of negative peak loads discussed in the previous section. These limitations make the models unreliable for truly predicting crack growth. In other words they are useful mainly for curve fitting and interpolation of test results, but not for extrapolation to significantly different load histories.

Crack Growth Models Based on Crack Closure

More recent crack growth models incorporate crack closure. This enables delayed retardation and the effect of negative peak loads to be accounted for by the variation in σ_{op} and hence ΔK_{eff}. The main problem is to determine the correct variation in σ_{op} during variable amplitude loading. Discussion of specific solutions to this problem is beyond the scope of this course. Interested readers may consult references 22 and 23 of the bibliography.

Once the variation in σ_{op} has been determined the procedure for estimating the crack growth curve is as follows:

- calculate $(\sigma_{op})_i$ for load cycle i as a function of R and previous positive and negative load cycles (*e.g.* by the method in reference 22),
- determine $(\Delta\sigma_{eff})_i = (\sigma_{max})_i - (\sigma_{op})_i$ and hence $(\Delta K_{eff})_i$,
- calculate $\Delta a = (da/dn)_i = f\{(\Delta K_{eff})_i\}$: for this step constant amplitude crack growth rate data correlated by ΔK_{eff} are necessary, *e.g.* figure 9.12,

- determine $a_{i+1} = a_i + \Delta a_i$,
- repeat the previous steps for each load cycle.

The relatively complex way in which σ_{op} must be determined for each load cycle and the step by step counting method of generating the crack growth curve requires long calculating times. However, the results are very promising in terms of both interpolation of test data and truly predicting crack growth.

Elber (reference 24 of the bibliography) proposed using a constant σ_{op} determined from constant amplitude testing of the same material. This would eliminate the tedious calculation of σ_{op} for each load cycle, but at the same time would limit the use of crack closure based models to variable amplitude load histories that result in fairly regular crack growth.

9.8 Fatigue Crack Initiation

In the introduction to this chapter the importance of the initiation phase in fatigue crack growth was already pointed out. From a fracture mechanics point of view the main difficulty in dealing with this phase is the fact that application of LEFM to microcracks is either impossible, in which case EPFM must be used, or possible only with substantial modifications and restrictions.

Fatigue Limit Stress Range and Fatigue Threshold Stress Intensity Range

In practice the fatigue limit stress range, $\Delta\sigma_e$, is often used for design purposes in a wide variety of engineering materials. $\Delta\sigma_e$ is the limiting value for the stress range that leads to a very long or an infinite fatigue lifetime and thus to some extent characterises initiation behaviour of the material.

Originally the fatigue limit was thought to represent the inability of a material to initiate a crack at the applied stress levels. However, this idea was contradicted by the observation that cracks often develop at stress ranges well below the fatigue limit. At present it is recognised that these so-called *non-propagating cracks* form at a relatively early stage, grow for a short period but then remain dormant, reference 25. In view of this, the fatigue limit should refer to the stress range required for a microcrack to overcome the strongest barrier it will encounter.

During what is considered to be the crack initiation period (*cf.* figure 9.1) fatigue cracks develop during the first load cycles, but depending on the load magnitude become quiescent for some time before they eventually succeed in crossing existing barriers and grow into macrocracks. An example of an important (micro) structural barrier is the grain boundary. A growing microcrack can be retarded or even stop at a grain boundary. This explains why the presence of small grains at the surface can lead to a higher fatigue limit.

In principle the fatigue limit stress range, $\Delta\sigma_e$, is measured on smooth specimens. However, factors such as surface roughness, defects or cracks will lower the measured value (see also under the subheading "Fatigue Strength and Defect Size" in this section). The value of the fatigue limit stress range in the presence of cracks, denoted as $\Delta\sigma_{th}$, can be expected to decrease with crack length. On the other hand, for very short cracks $\Delta\sigma_{th}$ approaches the fatigue limit stress range for smooth specimens, $\Delta\sigma_e$.

On the basis of experiments, Kitagawa and Takahashi (reference 26), put forward the idea that there is a critical crack size, l_o, below which cracks do not affect the fatigue limit and thus

$$\Delta\sigma_{th} = \Delta\sigma_e \quad \text{for } a < l_o . \tag{9.19}$$

This observation is in accordance with the notion discussed above, *i.e.* that small cracks already develop during the first load cycles, and that therefore it is irrelevant whether or not such microcracks are present before the load is applied.

For cracks longer than l_o linear elastic fracture mechanics can be applied and $\Delta\sigma_{th}$ is determined by the threshold stress intensity range for long cracks, ΔK_{th}. Assuming $K_I = \sigma\sqrt{\pi a}$, then

$$\Delta\sigma_{th} = \frac{\Delta K_{th}}{\sqrt{\pi a}} \quad \text{for } a > l_o . \tag{9.20}$$

Figure 9.25 shows an example of a so-called Kitagawa-Takahashi diagram. This is a double logarithmic plot of $\Delta\sigma_{th}$ versus crack length. The two extreme conditions given by equations (9.19) and (9.20) are represented in this plot as straight lines. The crack length at which the two lines intersect is the critical crack size l_o. It follows from equations (9.19) and (9.20) that at this critical crack size

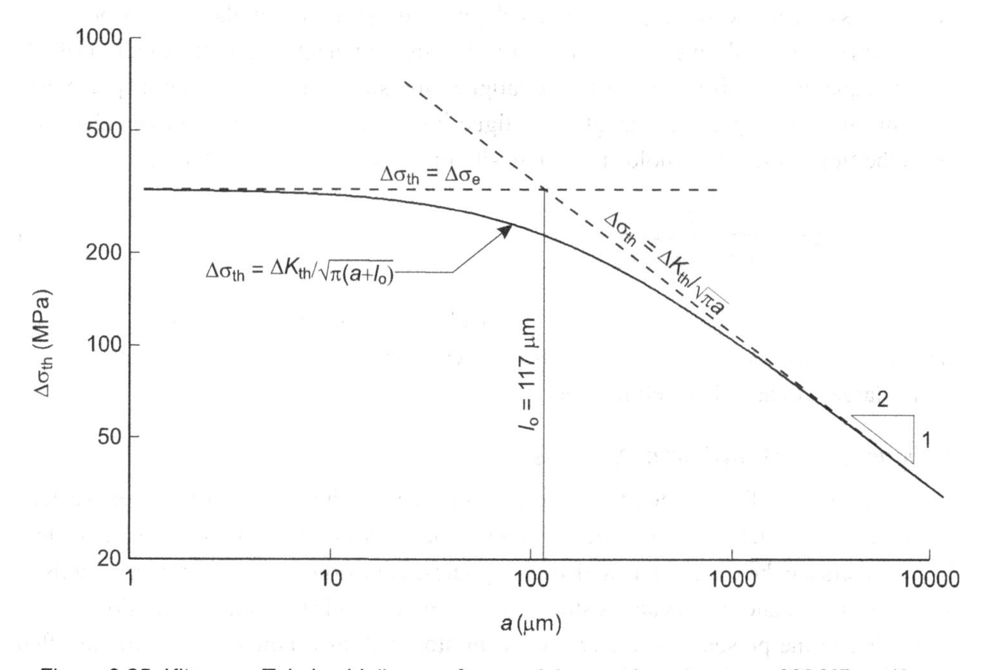

Figure 9.25. Kitagawa-Takahashi diagram for a nodular cast iron, *i.e.* $\Delta\sigma_e$ = 320 MPa, ΔK_{th} = 6.2 MPa$\sqrt{m}$ measured at R = 0.1, see reference 27.

$$\Delta\sigma_e = \frac{\Delta K_{th}}{\sqrt{\pi l_o}}$$

or

$$l_o = \frac{1}{\pi}\left(\frac{\Delta K_{th}}{\Delta\sigma_e}\right)^2 . \tag{9.21}$$

The value of l_o for a certain material can thus be calculated on the basis of the fatigue limit measured on smooth specimens and the fatigue threshold for specimens with long cracks. However, no obvious relation has been established between l_o and the microstructure of different materials.

It should be noted that the 'short crack' behaviour described by the Kitagawa-Takahashi diagram can also be interpreted as a threshold stress intensity range that decreases as soon as the crack length becomes smaller than l_o. However this interpretation is limited by the applicability of LEFM to short cracks.

As might be expected, measurements (reference 26) indicate that the transition between 'short crack' and 'long crack' behaviour is not abrupt. Especially near l_o, both straight lines in figure 9.25 appear to be non-conservative. On empirical grounds El Haddad *et al.*, references 28 and 29, proposed a modified expression for the stress intensity range that accounts for the different behaviour of short cracks as compared to long cracks, *i.e.*

$$\Delta K = C\,\Delta\sigma\sqrt{\pi(a + l_o)} , \tag{9.22}$$

where C is a factor accounting for the crack geometry (≈ 1). Using the expression for l_o, equation (9.21), it follows that when $a \to 0$ the stress intensity range in equation (9.22) will be equal to ΔK_{th} if $C\,\Delta\sigma = \Delta\sigma_e$, the fatigue limit stress range for smooth specimens. This means that for all crack lengths a fatigue limit stress range $\Delta\sigma_{th}$ can be calculated from the 'long crack' threshold stress intensity range ΔK_{th} using the relation

$$\Delta\sigma_{th} = \frac{\Delta K_{th}}{C\sqrt{\pi(a + l_o)}} . \tag{9.23}$$

This relation, also plotted in figure 9.25 for $C = 1$, is found to represent experimental data reasonably well, reference 28. In any case the straight lines non-conservatism seems largely to have been eliminated.

Fatigue Crack Growth from Notches

Cracks preferentially initiate at or near free surfaces, owing to the presence of surface roughness or scratches that can be found even on smooth, highly polished parts (a few micrometers can be enough). But also precipitates, inclusions and other imperfections in the bulk of the material can act as stress concentrators and lead to crack initiation.

Owing to the presence of a stress concentrator such as a notch, crack growth often begins under conditions of local plasticity. The crack then proceeds through the elastic stress-strain field of the notch before it reaches the bulk stress-strain field. This situation

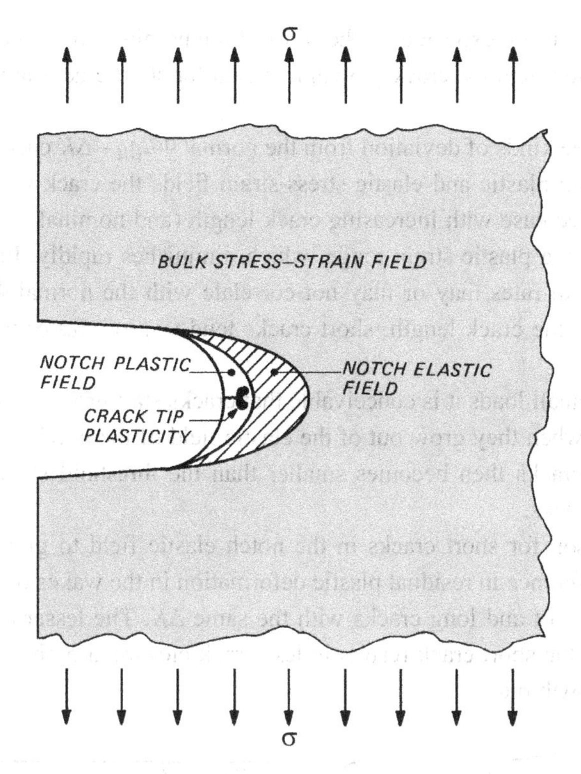

Figure 9.26. Growth of a small fatigue crack at a notch and the associated elastic and plastic stress-strain fields.

is depicted schematically in figure 9.26.

When the crack is propagating in the notch plastic field it is incorrect to use LEFM to characterize crack growth. This is demonstrated by the fact that crack growth rates are

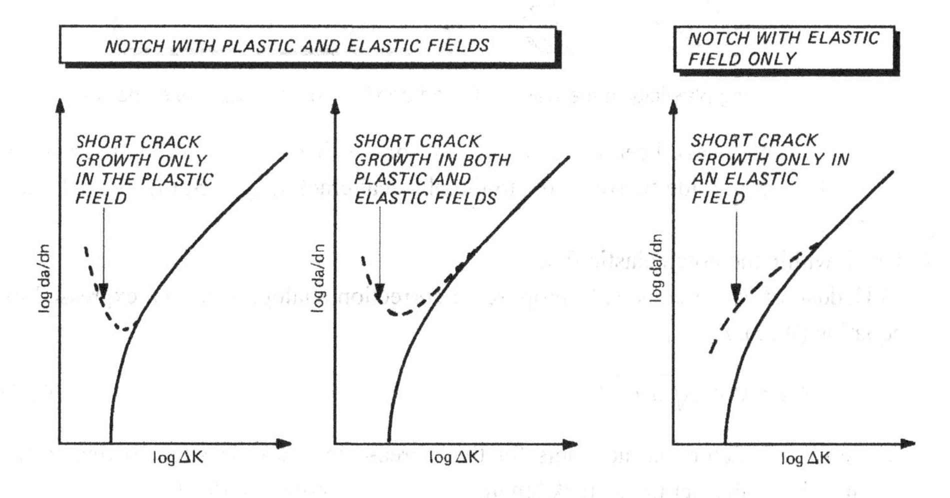

Figure 9.27. Schematic of anomalous crack growth rates for cracks at notches.

much higher than those expected on the basis of the nominal ΔK values. However, even in the notch elastic field the crack growth rate data for short cracks are not correlated by ΔK.

There are three kinds of deviation from the normal $^{da}/_{dn} - \Delta K$ curve, figure 9.27. For notches with both plastic and elastic stress-strain fields the crack growth rates are initially high but decrease with increasing crack length (and nominal ΔK) because growth is controlled by the plastic strain range, which diminishes rapidly. In the notch elastic field crack growth rates may or may not correlate with the normal $^{da}/_{dn} - \Delta K$ curve. This depends on the crack length: short cracks tend to grow faster than long cracks at the same ΔK.

Note that at small loads it is conceivable that cracks start growing at notches, but that they will arrest when they grow out of the elastic field of the notch. The stress intensity range for long cracks then becomes smaller than the threshold stress intensity range, ΔK_{th}, for long cracks.

A major reason for short cracks in the notch elastic field to grow faster than long cracks is the difference in residual plastic deformation in the wakes of the cracks. Figure 9.28 compares short and long cracks with the same ΔK. The lesser amount of residual deformation for the short crack results in less crack closure, a higher ΔK_{eff} and hence a higher crack growth rate.

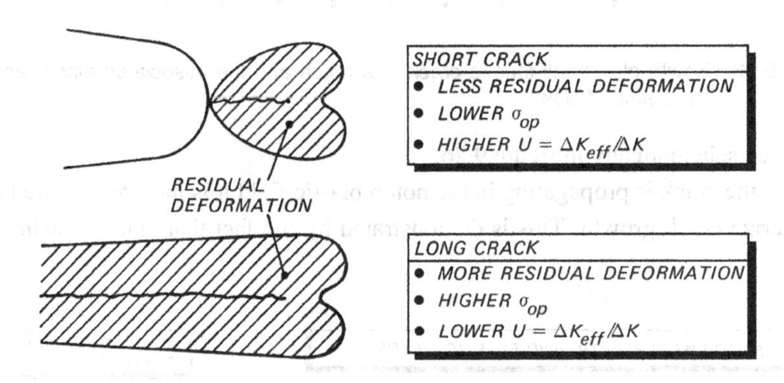

Figure 9.28. Differing plasticity in the wakes of short and long cracks with the same ΔK.

Several attempts have been made to account for the faster growth of short cracks at notches. They all provide 'corrections' to the physical crack length. Some examples are given here.

1) Crack within the notch elastic field

El Haddad *et al.*, reference 29, proposed a correction analogous to that expressed by equation (9.22), *i.e.*

$$\Delta K = k' \, \Delta\sigma\sqrt{\pi(a + l_o)} \,, \qquad (9.24)$$

where k' is a factor that accounts for the increase in crack tip stress owing to the notch and l_o is the fictitious crack length expressed by equation (9.21).

Another useful correction is due to Smith and Miller, reference 30:

$$\text{for } a < 0.13\sqrt{D\rho}\,, \quad \Delta K = \Delta\sigma\sqrt{\pi a}\left(1 + 7.69\sqrt{D/\rho}\right)^{1/2},$$
$$\text{for } a > 0.13\sqrt{D\rho}\,, \quad \Delta K = \Delta\sigma\sqrt{\pi(a + D)}\,, \tag{9.25}$$

where ρ is the actual notch root radius and D is the depth of an elliptical notch with the same root radius.

2) Crack within the notch plastic field
For this case El Haddad *et al.* proposed using either ΔJ for the corrected crack length $(a + l_o)$ or else a strain-based intensity factor, ΔK_ε, defined as

$$\Delta K_\varepsilon = E\,\Delta\varepsilon\sqrt{\pi(a + l_o)}\,, \tag{9.26}$$

where $\Delta\varepsilon$ is the local plastic strain. Equation (9.26) appears simple, but $\Delta\varepsilon$ is not easy to estimate. The same is true of ΔJ. Interested readers should consult references 28 and 29 of the bibliography.

Fatigue Strength and Defect Size

The threshold stress intensity range for crack growth and the fatigue limit stress range are influenced by defects, inclusions and non-homogeneities. A serious reduction in fatigue strength and also a considerable scatter can result. A very useful approach to handle the effect of two-dimensional and three-dimensional defects is proposed by Murakami and Endo (references 31 and 32). They introduced the *square root area parameter* model, in which $\sqrt{\text{area}}$ is defined as the square root of the area obtained by projecting a defect onto the plane perpendicular to the maximum tensile stress.

The use of $\sqrt{\text{area}}$ as the parameter representing the effect of defects on the fatigue properties is suggested by an approximate expression (accurate within 10%) for the maximum stress intensity for surface cracks with widely varying shapes, *i.e.*:

$$K_{\mathrm{I,max}} \approx 0.65\cdot\sigma\cdot\sqrt{\pi\sqrt{\text{area}}}\,. \tag{9.27}$$

For example for a semi-elliptical surface crack one can consider the appropriate K solution (see section 2.8)

$$K_\mathrm{I} = 1.12\frac{\sigma\sqrt{\pi a}}{\frac{3\pi}{8} + \frac{\pi}{8}\left(\frac{a}{c}\right)^2}\left(\sin^2\varphi + \left(\frac{a}{c}\right)^2\cos^2\varphi\right)^{\frac{1}{4}}, \tag{9.28}$$

and express a in terms of the crack area, $\frac{1}{2}\pi ac$, *i.e.*

$$a = \sqrt{a/c}\sqrt{2/\pi}\sqrt{\text{area}}\,. \tag{9.29}$$

In figure 9.29 the ratio $K_\mathrm{I}/\sigma\sqrt{\pi\sqrt{\text{area}}}$ is plotted as a function of a/c for two locations along the crack front, *i.e.* at the crack root ($\varphi = 90°$) and at the intersection with the free surface ($\varphi = 0°$). At these locations K_I is at its maximum for shallow cracks ($a/c < 1$) and deep cracks ($a/c > 1$) respectively. It can be seen that equation (9.27) applies within an accuracy of $\pm$10% for the range $0.2 < a/c < 5$.

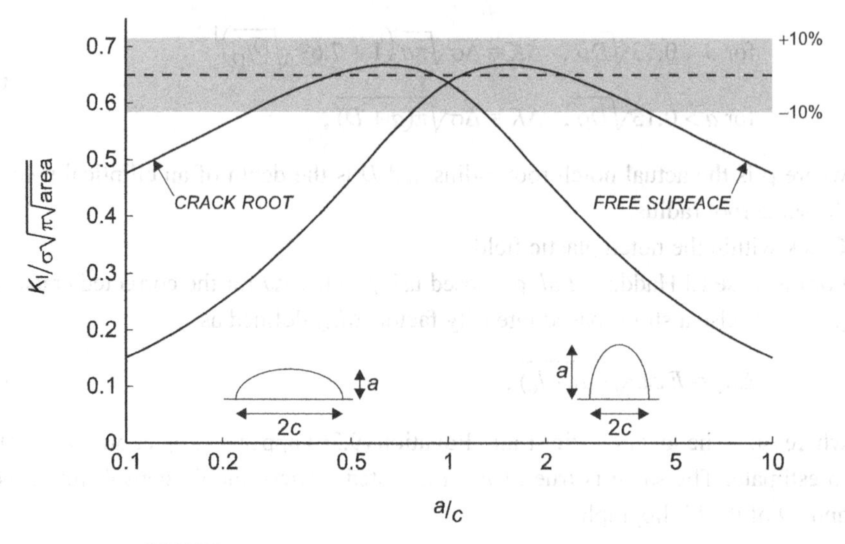

Figure 9.29. $K_I/\sigma\sqrt{\pi\sqrt{\text{area}}}$ for a semi-elliptical surface crack at the free surface and at the crack root as a function of the aspect ratio a/c.

Furthermore, it is argued that for three-dimensional defects, *e.g.* holes, the fatigue properties are determined by small cracks emanating from these defects. The assumption is made that the stress intensity is determined by the projection of the three-dimensional defect shape onto the plane normal to the maximum principal stress. This assumption was experimentally verified by introducing various artificial defects. Results from fatigue experiments in the range $20 < \sqrt{\text{area}} < 1000$ microns (μm) indicated that for a wide range of materials

$$\Delta K_{th} \propto \left(\sqrt{\text{area}}\right)^{1/3} . \tag{9.30}$$

The material behaviour in the $\sqrt{\text{area}}$ model is represented by the Vickers hardness, *HV*, of the material matrix, *i.e.* the matrix without defects. In general ΔK_{th} increases with hardness. However, the stress range below the fatigue limit for which non-propagating cracks are found is larger for softer materials. This behaviour is reflected by the empirical relation

$$\Delta K_{th} \propto (HV + 120) , \tag{9.31}$$

again based on data for a wide range of materials for which $70 < HV < 720$ kgf/mm^2.

For a material like nodular cast iron it is difficult to estimate the hardness of the matrix, because nodules will influence the outcome of the measurement. A simple correction for this is suggested by Endo (reference 33), *i.e.*

$$HV = \frac{HV_n}{1 - f_n} , \tag{9.32}$$

where HV_n is the overall hardness of a matrix containing a volume fraction f_n of porosity. For nodular cast iron f_n is assumed to be the volume fraction of nodules (graphite).

Equations (9.30) and (9.31) ultimately leads to the expression

$$\Delta K_{th} = 3.3 \cdot 10^{-3} \cdot (HV + 120) \cdot \left(\sqrt{\text{area}}\right)^{1/3} \cdot \left(\frac{1-R}{2}\right)^{\alpha} , \quad (9.33)$$

where ΔK_{th} is in MPa$\sqrt{\text{m}}$, HV in kgf/mm^2 and $\sqrt{\text{area}}$ in μm. The last term in the expression accounts for the effect of the load ratio, R; the exponent $\alpha = 0.226 + HV \cdot 10^{-4}$.

The fatigue limit stress range in the presence of surface defects, $\Delta\sigma_{th}$, can now also be calculated. By substituting ΔK_{th} for K_{max} and $\Delta\sigma_{th}$ for σ in equation (9.27) the relation between ΔK_{th} and $\Delta\sigma_{th}$ for surface defects is obtained. In combination with equation (9.33) it follows that

$$\Delta\sigma_{th} = 2.86 \cdot (HV + 120) \cdot \left(\sqrt{\text{area}}\right)^{-1/6} \cdot \left(\frac{1-R}{2}\right)^{\alpha} . \quad (9.34)$$

A K analysis for internal defects with various shapes leads to the approximate equation

$$K_{I,max} \approx 0.5 \cdot \sigma \cdot \sqrt{\pi \sqrt{\text{area}_i}} , \quad (9.35)$$

where $\sqrt{\text{area}_i}$ applies to internal defects. Thus, by comparing this equation with equation (9.27), it can be concluded that an internal defect will yield the same $K_{I,max}$ compared to a surface defect if

$$\sqrt{\text{area}_i} = 1.69 \sqrt{\text{area}} . \quad (9.36)$$

The fatigue limit stress range in the presence of internal defects is now obtained by substituting $\sqrt{\text{area}_i}$ for $\sqrt{\text{area}}$ in equation (9.34), *i.e.*

$$\Delta\sigma_{th} = 3.12 \cdot (HV + 120) \cdot \left(\sqrt{\text{area}_i}\right)^{-1/6} \cdot \left(\frac{1-R}{2}\right)^{\alpha} . \quad (9.37)$$

Equations (9.34) and (9.37) enable the fatigue limit stress range to be estimated without actually doing a fatigue test. An accuracy of better than 15% is claimed for materials with a Vickers hardness in the range of 100 to 740 kgf/mm^2.

The defect size, in terms of $\sqrt{\text{area}}$, must also be within a certain range to apply the $\sqrt{\text{area}}$ model. For very large defects this can be understood by realizing that ΔK_{th} has now become constant and, given the form of both equation (9.27) and (9.35), $\Delta\sigma_{th}$ will be proportional to $\left(\sqrt{\text{area}}\right)^{-1/2}$ instead of $\left(\sqrt{\text{area}}\right)^{-1/6}$. Although the upper limit for $\sqrt{\text{area}}$ cannot be defined exactly, it is estimated to be approximately 1000 μm.

For small defects it is also clear that there must be a limit to the applicability, since the $\sqrt{\text{area}}$ model would predict an infinite fatigue limit for defect-free material. As discussed near the beginning of this section, cracks are assumed to form shortly after the application of a fatigue load. If the applied stress range is below the fatigue limit stress range, $\Delta\sigma_e$, for smooth defect-free specimens, the microcracks can grow somewhat but are then stopped by some microstructural barrier. Defects which are initially present but are smaller than these non-propagating cracks will not affect the fatigue limit. Therefore the lower limit for the applicability of the $\sqrt{area}$ model is related to the maximum size of non-propagating cracks at the fatigue limit stress range $\Delta\sigma_e$. By substituting $\Delta\sigma_e$ for $\Delta\sigma_{th}$ in either equation (9.34) or (9.37) the lower limit of $\sqrt{area}$ can be calculated. Obviously this lower limit depends on the material.

9.9 Bibliography

1. Pook, L.P., *The Role of Crack Growth in Metal Fatigue*, The Metals Society (1983): London.
2. Schijve, J., *Four Lectures on Fatigue Crack Growth, Part II: Fatigue Cracks, Plasticity Effects and Crack Closure*, Engineering Fracture Mechanics, Vol. 11, pp. 167-221 (1979).
3. McEvily, A.J., *On the Quantitative Analysis of Fatigue Crack Propagation*, Fatigue Mechanisms: Advances in Quantitative Measurement of Physical Damage, ASTM STP 811, American Society for Testing and Materials, pp. 283-312 (1983): Philadelphia.
4. Johnson, R., Bretherton, I., Tomkins, B., Scott, P.M. and Silvester, D.R.V., *The Effects of Sea Water Corrosion on Fatigue Crack Propagation in Structural Steel.* European Offshore Steels Research Seminar, The Welding Institute, pp. 387–414 (1980): Cambridge.
5. Crooker, T.W., *The Role of Fracture Toughness in Low Cycle Fatigue-Crack Propagation for High Strength Alloys*, Engineering Fracture Mechanics, Vol. 5 , pp. 35–43 (1973).
6. Van der Veen, J.H., van der Wekken, C.J. and Ewalds, H.L., *Fatigue of Structural Steels in Various Environments*, Delft University of Technology, Department of Metallurgy Report, October 1978.
7. Ewalds, H.L., van Doorn, F.C. and Sloof, W.G., *Influence of the Environment and Specimen Thickness on Fatigue Crack Growth Data Correlation by means of Elber Type Equations*, Corrosion Fatigue, ASTM STP 801, American Society for Testing and Materials, pp. 115–134 (1983): Philadelphia.
8. Barsom, J.M., Imhof, E.J. and Rolfe, S.T., *Fatigue-Crack Propagation in High-Strength Steels*, Engineering Fracture Mechanics, Vol. 2, pp. 301–317 (1971).
9. De Vries, M.I. and Staal, H.U., *Fatigue Crack Propagation of Type 304 Stainless Steel at Room Temperature and 550 °C*, Energy Research Centre Report #RCN 222 (1975): Petten, The Netherlands.
10. Elber, W., The *Significance of Fatigue Crack Closure,* Damage Tolerance in Aircraft Structures, ASTM STP 486, American Society for Testing and Materials, pp. 230–247 (1971): Philadelphia.
11. Schijve, J., *The Stress Ratio Effect on Fatigue Crack Growth in 2024-T3 Alclad and the Relation to Crack Closure*, Delft University of Technology Department of Aerospace Engineering, Memorandum M-336, August 1979.
12. Riemslag A.C., van Kranenburg, C., Benedictus-de Vries, S., Veer, F.A. and Zuidema, J., *Fatigue Crack Growth Predictions in AA 5083 and AA 2024 Using a Simple Geometric Model*, in: Advances in Fracture Research, K. Ravi-Chandar *et al.* (Eds.), Proceedings on CD-rom of 10th International Conference on Fracture (ICF 10), Honolulu, USA, December, 2001, 6 p. (2001).
13. Shih, T.T. and Wei, R.P., *A study of crack closure in fatigue*, Engineering Fracture Mechanics, Vol. 6, pp. 19-32 (1974).
14. Marci, G. and Packman, P.F., *The Effects of the Plastic Wake Zone on the Conditions for Fatigue Crack Propagation*, International Journal of Fracture, Vol. 16, No 2, pp. 133-153 (1980).
15. Zuidema, J. and Krabbe, J.P., *The effect of regular and irregular shear lips on fatigue crack growth in AL 2024*, Fatigue and Fracture of Engineering Materials and Structures, Vol. 20, No. 10, pp. 1413-1422 (1997).
16. Wanhill, R.J.H., *Low Stress Intensity Fatigue Crack Growth in 2024-T3 and T351*, Engineering Fracture Mechanics, Vol. 30, pp. 233-260 (1988).
17. Schmidt, R.A. and Paris, P.C., *Threshold for Fatigue Crack Propagation and the Effects of Load Ratio and Frequency*, Progress in Flaw Growth and Fracture Testing, ASTM STP 536, American Society for Testing and Materials, pp. 79–94 (1973): Philadelphia.
18. Schijve, J. and Broek, D., *Crack Propagation: the Results of a Test Programme Bared on a Gust Spectrum with Variable Amplitude Loading*, Aircraft Engineering, Vol. 34, pp. 314–316 (1962).
19. Wanhill, R.J.H., *Fatigue Fracture in Steel Landing Gear Components*, National Aerospace Laboratory Technical Report 84117, October 1984.

20. Barsom, J.M., *Fatigue Crack Growth Under Variable-Amplitude Loading in Various Bridge Steels*, Fatigue Crack Growth Under Spectrum Loads, ASTM STP 595, American Society for Testing and Materials, pp. 217–235 (1976): Philadelphia.
21. Wheeler, O.E., *Spectrum Loading and Crack Growth,* Journal of Basic Engineering, Transactions ASME, Vol. 94, pp. 181–186 (1972).
22. De Koning, A.U., *A Simple Crack Closure Model for Prediction of Fatigue Crack Growth Rates under Variable-Amplitude Loading*, Fracture Mechanics, ASTM STP 743, American Society for Testing and Materials, pp. 63–85 (1981): Philadelphia.
23. Newman, J.C., Jr., *Prediction of Fatigue Crack Growth under Variable-Amplitude and Spectrum Loading Using a Closure Model*, Design of Fatigue and Fracture Resistant Structures, ASTM STP 761, American Society for Testing and Materials, pp. 255–277 (1982): Philadelphia.
24. Elber, W., *Equivalent Constant-Amplitude Concept for Crack Growth under Spectrum Loading*, Fatigue Crack Growth under Spectrum Loads, ASTM STP 595, American Society for Testing and Materials, pp. 236–250 (1976): Philadelphia.
25. Miller, K.J., *Metal Fatigue - Past, Current and Future*, Institution of Mechanical Engineers, Proc. Instn. Mech. Engrs, Vol. 205, Mech. Eng. Publications Limited, (1991): Suffolk, England.
26. Kitagawa, H. and Takahashi, S., *Application of Fracture Mechanics to Very Small Cracks*, Proc. Int Conf. Mech. Behaviour of Materials (ICM2), pp. 627-631 (1976).
27. Zuidema, J., Wijnmaalen, C.E. and Van Eldijk, C, *Fatigue of Nodular Cast Iron*, Proceedings of the Seventh International Fatigue Congress Fatigue '99, Bejing, P.R. China, June 1999, Volume 3/4, pp. 2071-2076 (1999).
28. El Haddad, M.H., Dowling, N.E., Topper, T.H. and Smith, K.N., J Integral Applications for Short Fatigue Cracks at Notches, International Journal of Fracture, Vol. 16, pp. 15–30 (1980).
29. El Haddad, M.H., Topper, T.H. and Topper, T.N., *Fatigue Life Predictions of Smooth and Notched Specimens Based on Fracture Mechanics*, Journal of Engineering Materials and Technology, Vol. 103, pp. 91–96 (1981).
30. Smith, R.A. and Miller, K.J., *Prediction of Fatigue Regimes in Notched Components*, International Journal of Mechanical Science, Vol. 20, pp. 201–206 (1978).
31. Murakami, Y. and Endo, M., *Effects of Hardness and Crack Geometries on* ΔK_{th} *of Small Cracks Emanating from Small Defects*, The Behaviour of Short Fatigue Cracks, EGF Pub. 1, Mechanical Engineering Publications, pp. 275-293 (1986): London.
32. Murakami, Y. and Endo, M., *Effects of Defects, Inclusions and Inhomogeneities on Fatigue Strength*, International Journal of Fatigue, Vol. 16, pp. 163-182 (1994).
33. Endo, M., *Fatigue Strength Prediction of Nodular Cast Irons Containing Small Defects*, Impact of Improved Material Quality on Properties, Product Performance, and Design, The American Society of Mechanical Engineers, Vol. 28, pp.125-137 (1991): New York.

10 Sustained Load Fracture

10.1 Introduction

Sustained load fracture is a general term for time-dependent crack growth under loading often well below that normally required to cause failure in a tensile or fracture toughness test. Examples of sustained load fracture are:

- creep and creep crack growth
- stress corrosion cracking
- cracking due to embrittlement by internal or external (gaseous) hydrogen
- liquid metal embrittlement.

Creep deformation and cracking constitute a widespread and very important practical problem, particularly in the power generating industry and aircraft gas turbines. However, a good and universally accepted description of creep cracking by fracture mechanics, whether LEFM or EPFM, is as yet unavailable.

The other types of sustained load fracture follow basically similar trends in terms of fracture mechanics. Thus, excluding creep, it is possible to discuss the application of fracture mechanics to sustained load fracture in a general way. As with fatigue crack growth the use of fracture mechanics is mainly limited to LEFM methods, specifically the stress intensity (K) approach.

The use of K to describe sustained load fracture is based mainly on procedures similar to those for fracture toughness testing, including the use of more or less standard specimens with fatigue precracks under nominally plane strain conditions. However, full plane strain is not strictly necessary if it is only required to test specimens with a thickness representative of that for a narrow section component in service.

Experimental methods for fracture mechanics evaluation of sustained load fracture fall into two categories:

1) Time-to-failure (TTF) tests on precracked specimens.
2) Crack growth rate testing.

These methods will be discussed in sections 10.2 and 10.3 respectively, and the experimental problems that arise are dealt with in section 10.4.

The way in which crack growth rate data could be used to predict failure of a structural component in service is treated in section 10.5. However, there are many difficulties that require careful evaluation of the practical significance of test data. This is the subject of section 10.6, which closes the chapter.

10.2 Time-To-Failure (TTF) Tests

For a long time the study of sustained load fracture (principally stress corrosion cracking) relied solely upon TTF tests on smooth specimens. Such tests are still useful, but it is now recognised that fracture mechanics based tests provide an essential supplement. The most striking example concerns titanium alloys, which were thought to be immune to stress corrosion in aqueous solutions until Brown in 1966 tested fatigue precracked cantilever beam specimens with disastrous results (reference 1 of the bibliography to this chapter).

Precracked specimens for TTF tests are configured such that a constant load results in increasing stress intensity with increasing crack length. The specimens are loaded to various initial stress intensity levels, K_{I_i}, and the time to failure is recorded. A representative TTF plot is shown in figure 10.1.

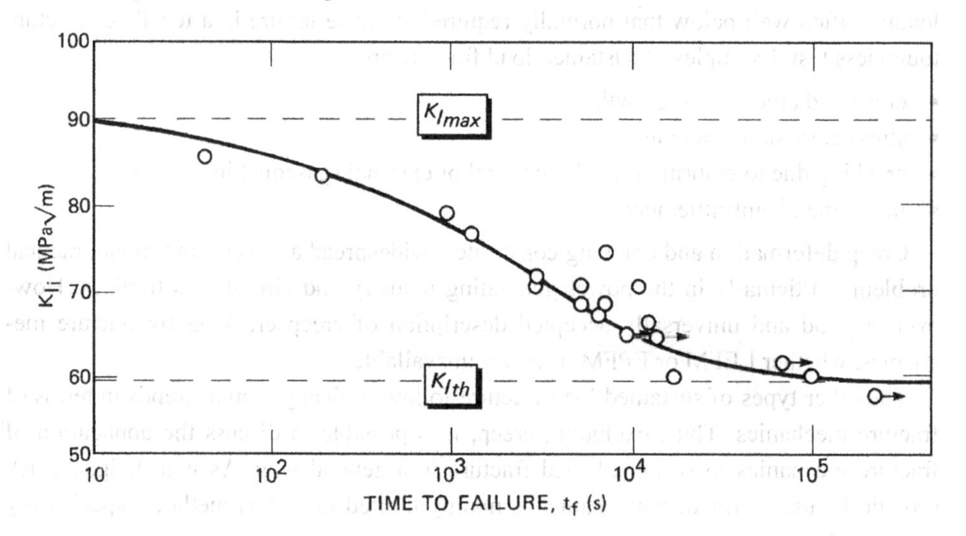

Figure 10.1. Sustained load testing of precracked titanium alloy Ti-6Al-4V specimens in normal air.

The plot shows that two quantities, $K_{I_{max}}$ and $K_{I_{th}}$, can be determined. $K_{I_{max}}$ represents the maximum load carrying ability and will generally be equal to either K_{Ic} or K_Q the valid or invalid fracture toughness values. $K_{I_{th}}$ is the threshold stress intensity, below which there is virtually no crack growth. For stress corrosion cracking the threshold is customarily referred to as $K_{I_{scc}}$.

In some cases $K_{I_{th}}$ may be a true threshold. In general, however, $K_{I_{th}}$ should not be considered a material property. There are two reasons for this. First, the testing time required to establish $K_{I_{th}}$ may be extremely long (years rather than months). Secondly, some materials (notably steels) exhibit long incubation periods before sustained load fracture commences from the fatigue precrack, and the incubation periods increase with decreasing K_{I_i}.

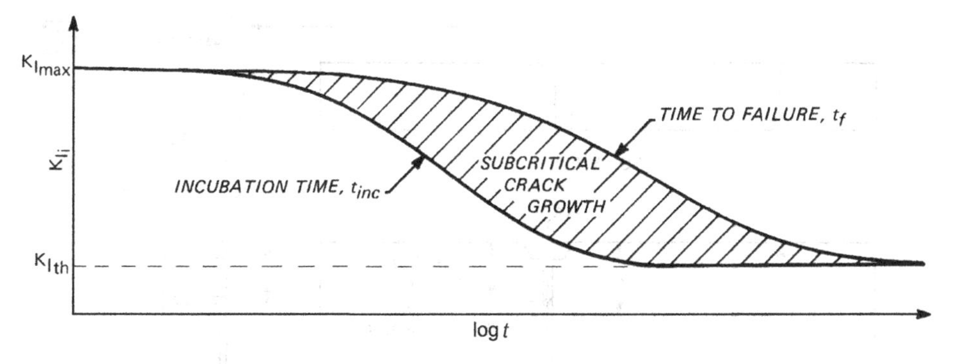

Figure 10.2. Schematic TTF plot.

Figure 10.2 shows schematically that in the regime between $K_{I_{th}}$ and $K_{I_{max}}$ the time to failure, t_f, generally comprises an incubation time, t_{inc}, and a period of subcritical crack growth. The incubation time depends on the material, environment, K_{I_i} and also on the previous loading history (if any) of the specimen. The subcritical crack growth period depends on the specimen configuration, type of loading, amount of crack growth before $K_{I_{max}}$ is reached, and the kinetics of crack growth due to the material-environment interaction.

As stated earlier, precracked specimens for TTF tests are such that a constant load results in increasing stress intensity with increasing crack length. A commonly used specimen is the fatigue precracked cantilever beam specimen developed by Brown and co-workers at the US Naval Research Laboratory. The specimen and loading arrangement are illustrated in figure 10.3.

The cantilever beam specimen usually has side grooves reducing the nominal width from B to B_N. Side grooves are beneficial in obtaining a more even stress distribution through the thickness, since they prevent the occurrence of a plane stress state at the specimen side surfaces. The results are a more uniform crack front and prevention of out-of-plane crack deviation for some specimen types.

Stress intensity factors for cantilever beam specimens can be calculated according to:

$$K_I = \frac{6M}{(B \cdot B_N)^{\frac{1}{2}}(W-a)^{\frac{3}{2}}} \cdot f\left(\frac{a}{W}\right), \tag{10.1}$$

where M is the bending moment, W is the beam depth, and a is the total crack length, see figure 10.3. The geometry factor, $f(a/W)$, is given in the following table.

a/W	$f(a/W)$
0.05	0.36
0.10	0.49
0.20	0.60
0.30	0.66
0.40	0.69
0.50	0.72
≥ 0.60	0.73

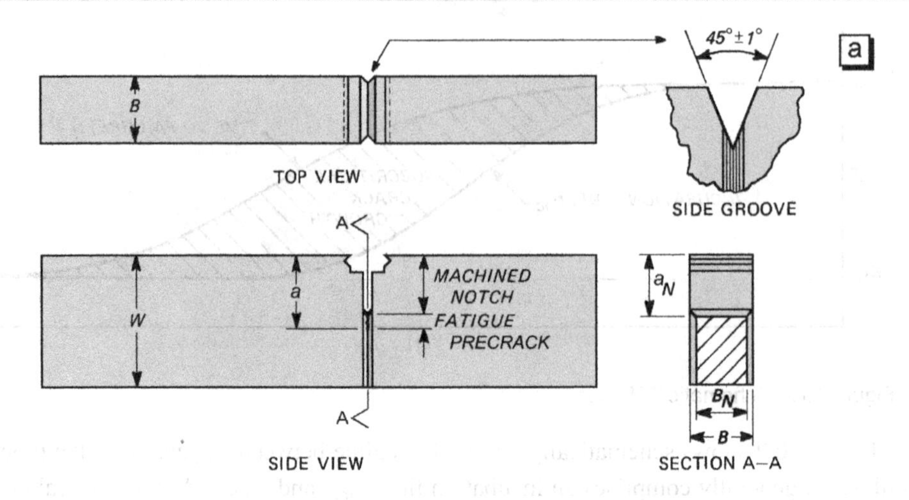

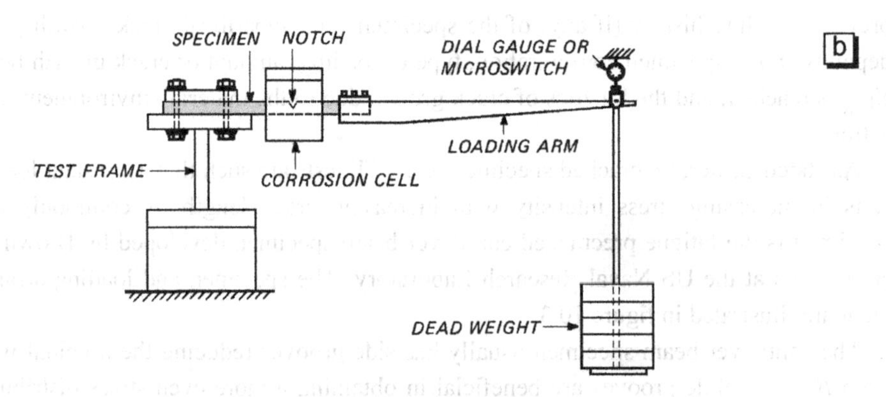

Figure 10.3. (a) Fatigue precracked cantilever beam specimen and (b) loading arrangement for stress corrosion testing.

10.3 Crack Growth Rate Testing

The general features of the dependence of sustained load crack growth on K_I were already mentioned in section 1.10 (figure 1.14) and are shown again in figure 10.4. The crack growth curve consists of three regions. In regions I and III the growth rate da/dt strongly depends on the stress intensity, but in region II the crack growth rate is virtually independent of stress intensity. Regions I and II are the most characteristic, although region II is sometimes a rounded hump rather than a plateau. Region III is often not observed, owing to an abrupt transition from region II to fast fracture. Note that $K_{I_{th}}$ and $K_{I_{max}}$ can be determined, at least in principle, from crack growth rate tests.

Many types of specimen have been proposed for studying sustained load crack growth. Most fall into either the increasing K (constant load) or decreasing K (fixed grip) categories: compare also with section 5.4.

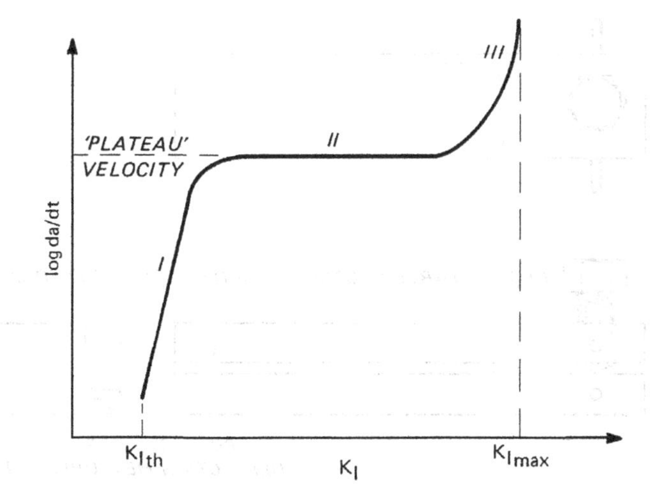

Figure 10.4. Generalised sustained load crack growth behaviour.

Advantages of decreasing K specimens are:

- Fatigue precracking from a sharp notch is not always necessary, since initial mechanical crack growth is stable, or eventually stable in case of a 'pop-in', and usually does not influence subsequent sustained load fracture.
- Virtually the entire crack growth curve can be obtained from one specimen.
- They can be self-stressed (*e.g.* figure 10.5) and therefore portable: for example, they can be exposed outdoors.
- Steady state conditions for crack growth and arrest at $K_{I_{th}}$ are more readily achieved.

A disadvantage of using decreasing K specimens is the occurrence of corrosion product wedging for some material-environment combinations. For a decreasing K specimen the initial displacement is fixed, so that the crack tip tends to narrow during propagation. Corrosion products may form a wedge between the crack surfaces and lead to a higher crack growth rate at a given nominal K_I and to apparently lower $K_{I_{th}}$ values. These ef-

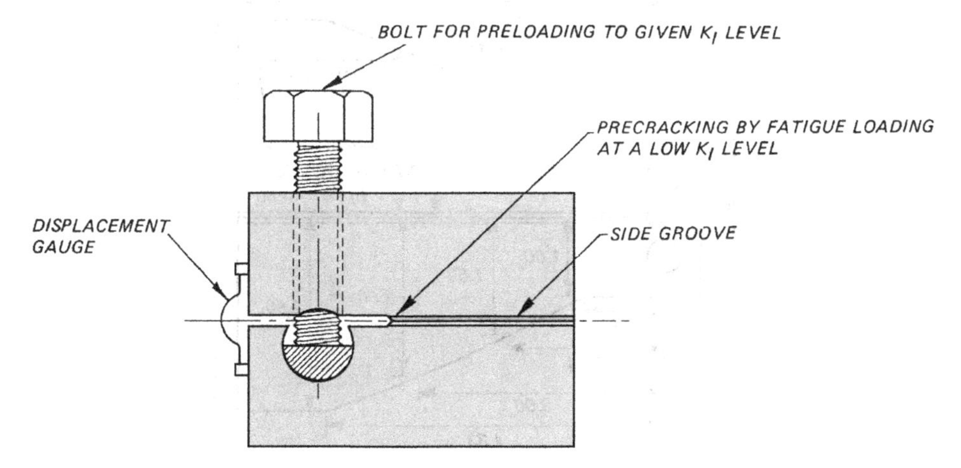

Figure 10.5. Modified crack-line wedge-loaded specimen (CLWL).

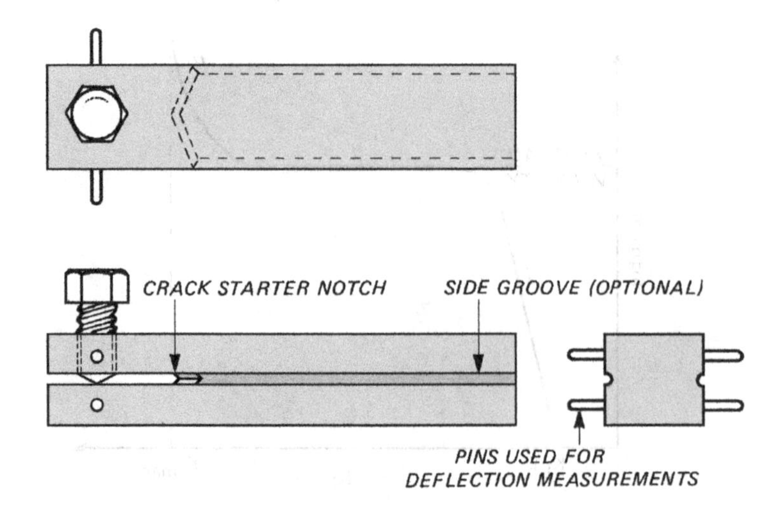

Figure 10.6. Bolt loaded double cantilever beam specimen (DCB).

fects are very difficult to assess quantitatively.

Two types of commonly used decreasing K specimens, the modified crack-line wedge-loaded (CLWL) and the double cantilever beam (DCB) specimens, are shown in figures 10.5 and 10.6. These types of specimens (though not self-stressed) have already been mentioned in sections 5.4 and 4.4 respectively. Both types require determination of the elastic compliance in order to calculate stress intensity factors, and the expressions for K_I are unfortunately — cumbersome, equations (5.2) and (4.25). Some more guidance for the calculation of stress intensity factors in CLWL specimens is given in reference 2.

There is a third category of specimens, namely those giving a constant K under constant load. Such specimens have not been so widely used, but for detailed studies of

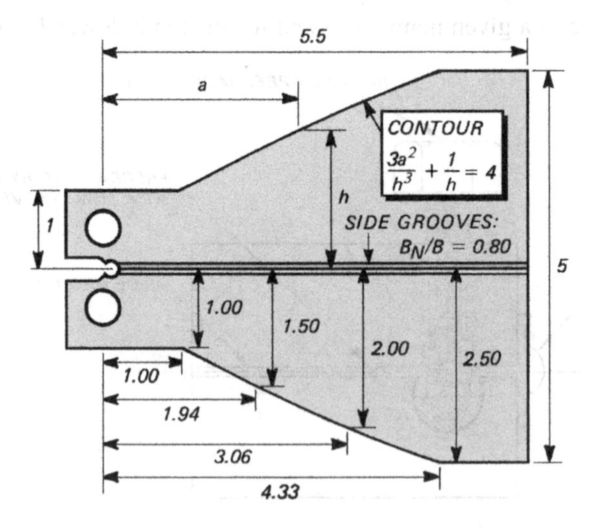

Figure 10.7. Tapered double cantilever beam specimen (TDCB).

sustained load fracture they provide a valuable adjunct to increasing and decreasing K tests. An example is the tapered double cantilever beam (TDCB) specimen, figure 10.7. As mentioned in section 4.4, for this specimen $(3a^2 + h^2)/h^3$ is constant. This results in a linear increase in compliance with crack length, *i.e.* dC/da is constant. For a given load P the stress intensity factor under plane strain conditions is given by

$$K_I = P\sqrt{\frac{E}{2B_N(1-\nu^2)} \cdot \frac{dC}{da}} \quad (10.2)$$

Hence K_I is constant when dC/da is constant. In practice the contour in figure 10.7 is often approximated by a straight line in order to facilitate specimen manufacture.

Difference in Behaviour for Increasing and Decreasing K Specimens

A schematic of the difference in behaviour of increasing K (constant load) and decreasing K (fixed grip) specimens is given in figure 10.8. Note that the ordinate is now K_I, not K_{I_i} as in the case of TTF plots.

10.4 Experimental Problems

There are a number of experimental difficulties associated with sustained load fracture testing, including

- the incubation time, t_{inc}
- non-steady state crack growth

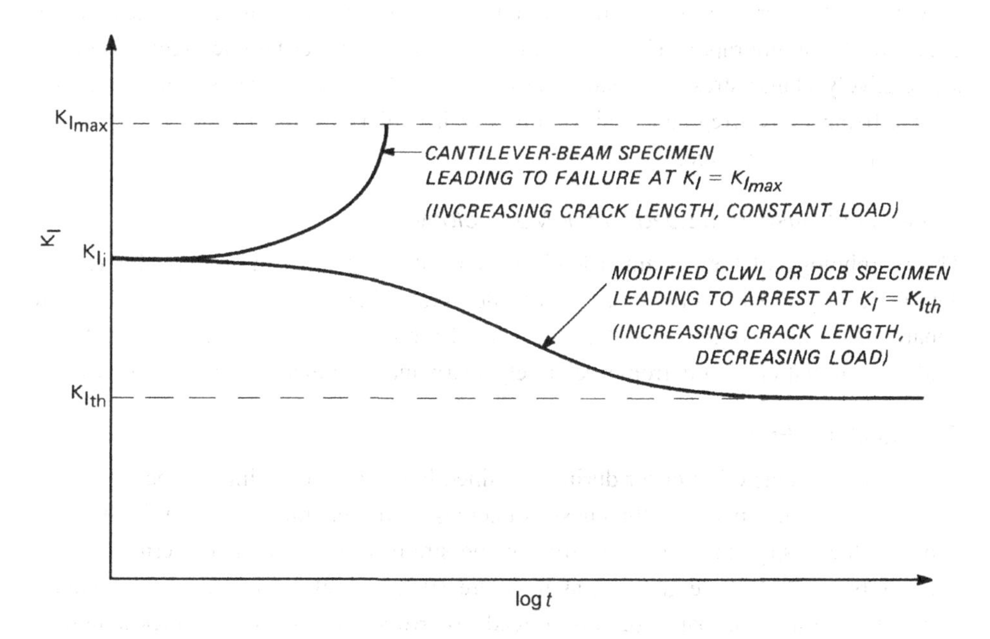

Figure 10.8. Difference in behaviour for increasing K (cantilever beam) and decreasing K (modified CLWL or DCB) specimens.

- effect of precrack morphology
- influence of sustained load crack morphology
- curved crack fronts
- corrosion product wedging.

Incubation Time, t_{inc}

As mentioned already in section 10.2, the incubation time, t_{inc}, depends on K_{I_i} and the previous loading history of the specimen. The loading history stems from the fatigue precracking process and can include effects of plastic deformation. Exposure to an aggressive environment before sustained loading can also affect t_{inc}. Furthermore, t_{inc} — or rather: the apparent incubation time — strongly depends on the methods of detecting crack growth, *e.g.* visual inspection, compliance measurements or acoustic emission.

Non-Steady State Crack Growth

A high K_{I_i} applied to a decreasing K specimen or a rapid increase in stress intensity for an increasing K specimen can lead to crack growth rates initially lower than those predicted by the da/dt-K_I diagram. This non-steady state crack growth is caused by changes in the electrochemical reactions at the crack tip. (Further discussion of this interesting problem is beyond the scope of this course.)

Effect of Precrack Morphology

Stress corrosion tests on aluminium alloys with DCB specimens have shown that mechanical 'pop-in' precracks result in displacement of the crack growth curve (particularly region I, figure 10.4) to higher K_I values as compared to fatigue precracks, reference 3 of the bibliography. This was explained by the fact that fatigue precracks generate relatively planar stress corrosion cracks with uniform crack fronts, whereas pop-in cracks, being more irregular, lead to similarly irregular stress corrosion cracks which require more driving force.

Influence of Sustained Load Crack Morphology

The morphology of the sustained load crack front is especially important for tests to determine $K_{I_{th}}$. Crack tip blunting and/or microbranching in decreasing K tests lead to apparent $K_{I_{th}}$ values higher than those obtained from increasing K tests, where sustained load crack initiation occurs from a relatively sharp and unbranched fatigue precrack.

Curved Crack Fronts

Curved crack fronts often occur during sustained load fracture testing and become more pronounced as the specimen thickness is decreased. In general the reason for this curvature is the change in stress state from plane strain in the specimen interior to plane stress at the surface. Side grooves, as in figure 10.3, are helpful in reducing curvature, which is a nuisance primarily because it leads to errors in using surface crack lengths to calculate crack growth rates.

Corrosion Product Wedging

Corrosion product wedging has been mentioned already in section 10.3 as a disadvantage of using decreasing K specimens, since the wedging action leads to higher crack growth rates at a nominal K_I and to apparently lower $K_{I_{th}}$ values. To determine whether corrosion product wedging has influenced crack growth the specimen can be unloaded after testing and the deflection at the load line can be remeasured and compared to that at the beginning of the test. If the deflections are nearly the same, no substantial amount of corrosion products has accumulated in the crack and therefore wedging has not occurred.

10.5 Method of Predicting Failure of a Structural Component

As shown in figure 10.2 the time to failure, t_f, generally comprises an incubation time, t_{inc}, and a period of subcritical crack growth. Therefore in the first instance the prediction of failure of a structural component is a twofold problem:

1) Prediction of t_{inc}.
2) Prediction of the crack growth period.

As a reminder, the incubation time depends on the material, environment. initial stress intensity (K_{I_i}) and prior loading history. A convenient way of determining t_{inc} for a certain material environment combination is to use decreasing K (fixed grip) specimens and load them to various K_{I_i} values. By starting at high K_{I_i} levels the long incubation times can be avoided and yet the locus of t_{inc} can be found, as shown schematically in figure 10.9. Note that, as in figure 10.8, the ordinate is K_I, not K_{I_i}.

Crack growth rates, da/dt, depend on the material, environment and stress intensity factor K_I. Thus the crack growth period will depend strongly on specimen (or compo-

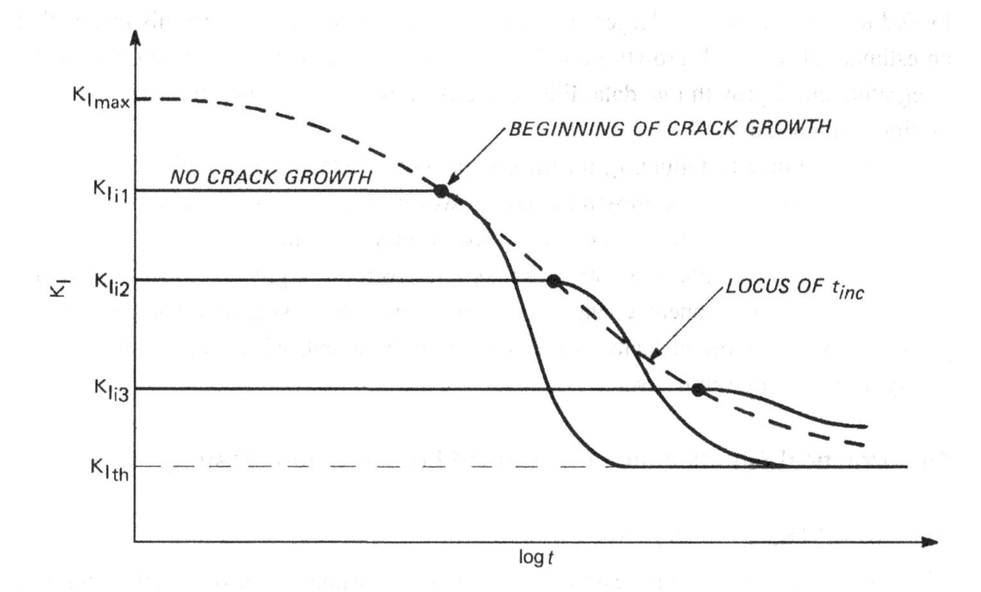

Figure 10.9. Schematic determination of t_{inc} using decreasing K specimens (reference 4).

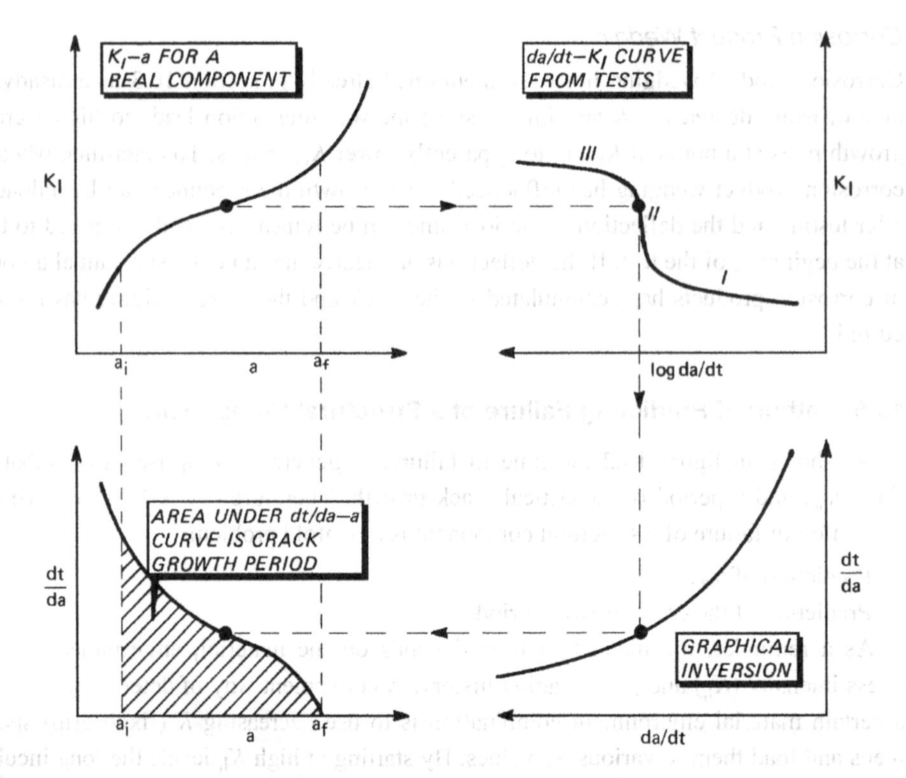

Figure 10.10. Schematic of integrating crack growth rate data from tests in order to predict the crack growth period for a real component.

nent) geometry and size, since these parameters control the variation of K with crack length. For example, if large and small specimens which are geometrically similar are loaded to the same K_{I_i}, the larger specimen takes longer to fail. In turn this means that an estimate of the crack growth period for a structural component must be obtained by integrating crack growth rate data. Figure 10.10 shows schematically how such an integration could be done.

Finally, the time to failure, t_f, for the component is obtained at least in theory — by summing t_{inc} and the crack growth period. However, in practice there may be great difficulties. For example, the component geometry may be complex and stress intensity factors difficult to obtain. Also, stresses may relax as a crack grows, or else may redistribute owing to permanent changes in displacement (*e.g.* slip of bolted or riveted joints). Another important factor is the service environment, which may differ considerably from test environments.

10.6 Practical Significance of Sustained Load Fracture Testing

Threshold Stress Intensity, $K_{I_{th}}$

Most sustained load fracture tests are done with the primary aim of determining $K_{I_{th}}$. Although useful for specific applications, actual values of $K_{I_{th}}$ are not used as general design criteria because their significance for service performance has not been established. This is not surprising in view of the fact that $K_{I_{th}}$ depends on the material, environment, temperature, stress state (plane strain or plane stress) and often on very long testing times: the longer the testing times, the lower the apparent value of $K_{I_{th}}$.

Despite these problems $K_{I_{th}}$ is useful as a guide to material selection and design. For example, figure 10.11 shows stress corrosion $K_{I_{scc}}$ data for a number of commercial high strength steels. In the very high strength regime $K_{I_{scc}}$ is low, irrespective of alloy type, composition, microstructure or heat treatment. This trend, in addition to the fact that stress corrosion crack growth rates are high, indicates that stress corrosion cracking cannot be tolerated by components made from very high strength steels. A fracture mechanics approach to design cannot be used. Instead, the components must be rigorously inspected or proof loaded to ensure that they are free from macroscopic flaws, and thorough protection against corrosion must be applied (this is good practice, whatever the material).

Proof loading is the application of a high load in the knowledge or expectation that if the component does not fail, it will then survive in service a given time, after which the proof load may be repeated to establish a further safe period of operation. There is always the possibility that the component will fail during proof loading. Understandably, therefore, this method of component verification is not very popular.

As a development of the foregoing, an important contribution to interpretation of $K_{I_{th}}$ testing has been made by the US Naval Research Laboratory, reference 6 of the bibliography. This is the concept of Ratio Analysis Diagrams (RADs) in which a grid of lines

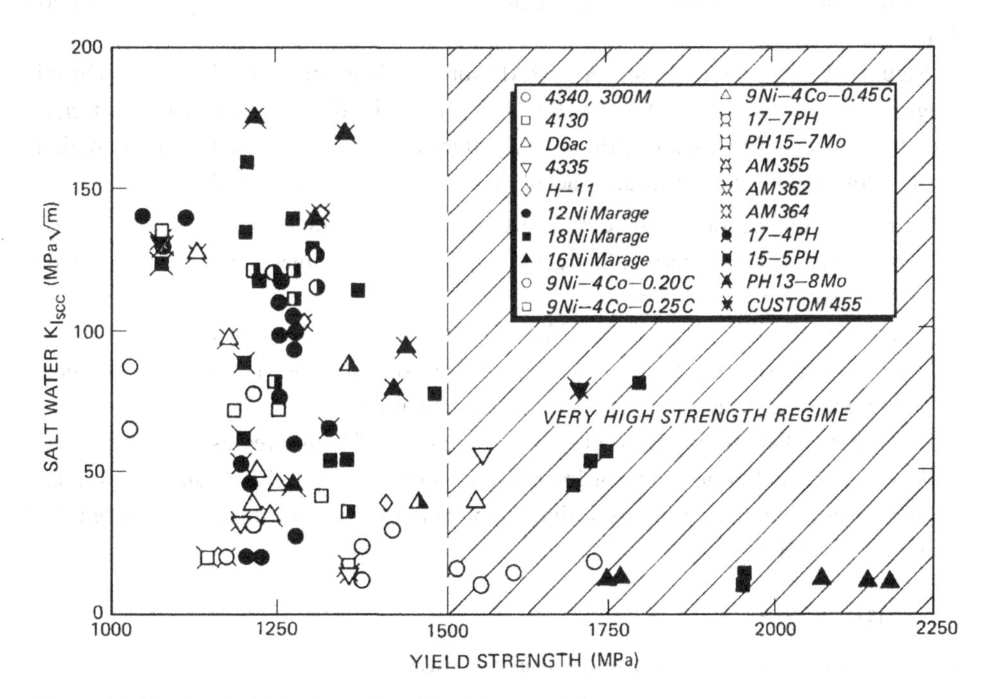

Figure 10.11. $K_{I_{scc}}$ for high strength steels, reference 5.

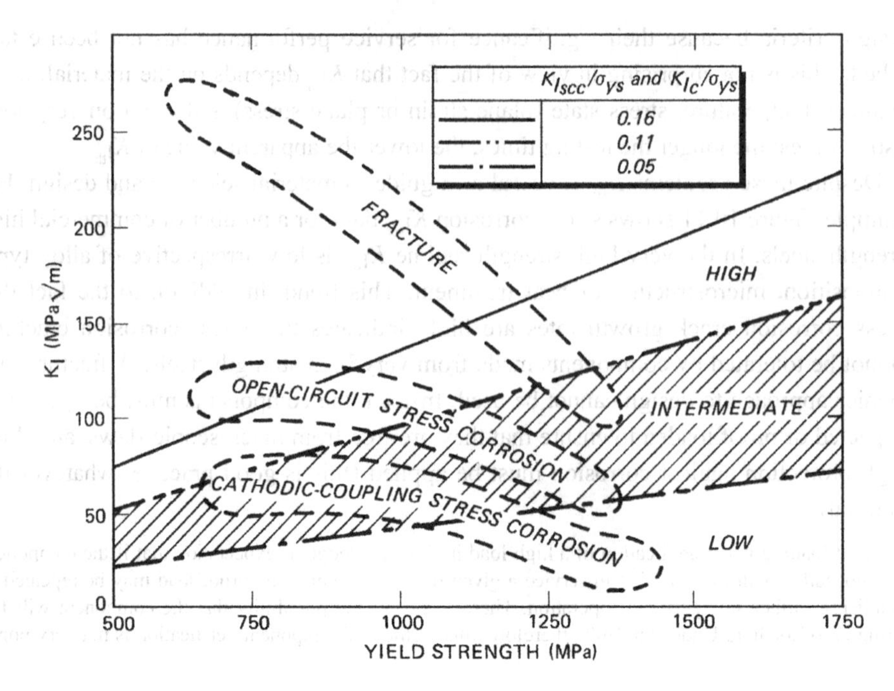

Figure 10.12. Ratio Analysis Diagram (RAP) showing the fracture (K_{Ic}) and stress corrosion ($K_{I_{scc}}$) properties of high strength stainless steels, reference 6.

of constant K_I/σ_{ys} are superimposed on K_{Ic} and $K_{I_{th}}$ data plotted against σ_{ys}. An example of such a diagram for practical interpretation of K_{Ic} has already been given in section 5.6.

Figure 10.12 shows a K_{Ic} and $K_{I_{scc}}$ RAD for a high strength stainless steel. On this diagram the data zones for fast fracture (K_{Ic}) and initiation of stress corrosion crack growth ($K_{I_{scc}}$ are plotted with a grid of K_I/σ_{ys} lines that separate the diagram into regions of high, low and intermediate ratios according to the following rationale:

1) The high ratio region is where high stresses and large cracks are necessary to cause fast fracture or stress corrosion crack growth. A fracture mechanics approach to design is relatively straightforward.
2) The low ratio region is where fast fracture or stress corrosion crack growth can initiate from very small defects at moderate to low stress levels. This region is where a fracture mechanics approach to design cannot be used.
3) The intermediate region is where the combination of high stresses and small flaws, low stresses and large flaws, or intermediate stress levels and flaw sizes are critical. This region is where a highly refined application of fracture mechanics is required for adequate design.

Crack Growth Rate Tests

Crack growth rate data can be useful in several ways:

- estimating $K_{I_{th}}$ from decreasing K tests, section 10.3

- predicting the service lives of components, provided that the incubation time t_{inc}, non-steady state crack growth, and the practical difficulties mentioned at the end of section 10.5 are absent or can be accounted for
- deciding whether a period of safe crack growth exists, and if so, determining inspection intervals for parts assumed or known to contain flaws.

For some material-environment combinations the crack growth rates measured in tests are so high that even though $K_{I_{max}}$ may be much higher than $K_{I_{th}}$, it is clear that there is no possibility of setting reasonable inspection intervals for a period of safe crack growth. Examples are liquid metal embrittlement of various materials and aqueous stress corrosion of high strength steels and titanium alloys. In these cases either the material must have a high $K_{I_{th}}$ value, or else sustained load fracture cannot be tolerated, as mentioned earlier.

On the other hand, important problems like stress corrosion cracking of high strength aluminium alloys can be better understood by analysis of crack growth rate data. Figure 10.13 shows crack growth rate data for several 7000 series aluminium alloys exposed to an outdoors environment of intermittent rain and moderate humidity. The data cover a range of possibilities for region II crack growth:

a) Fast crack growth in 7079-T651.

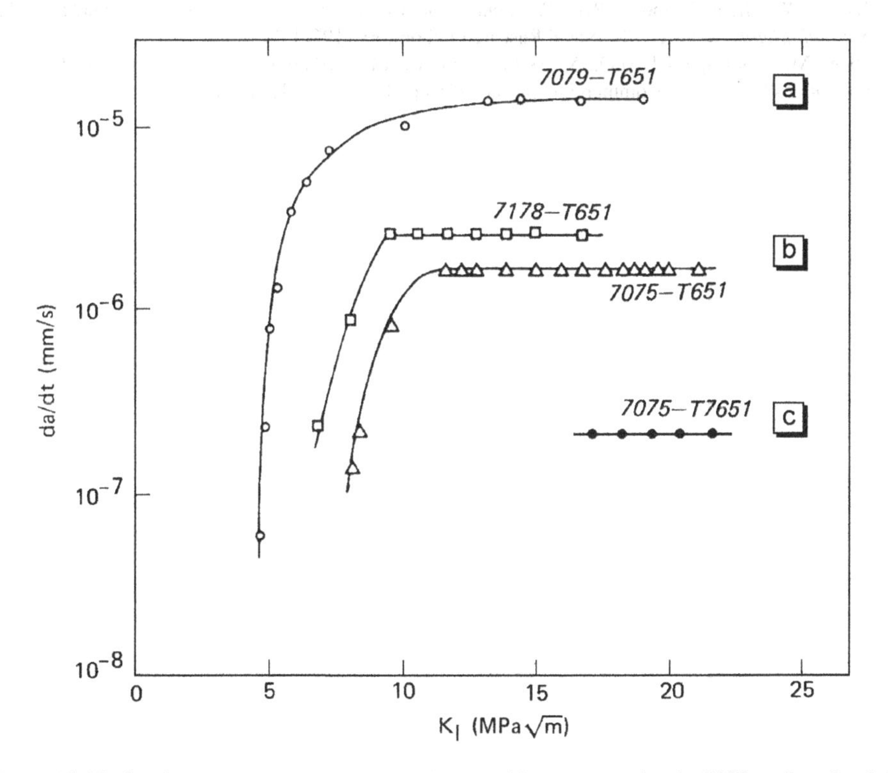

Figure 10.13. Outdoor exposure stress corrosion cracking propagation in 7000 series aluminium alloy plate, reference 7.

b) Fairly slow crack growth in 7075-T651 and 7178-T651.
c) Very slow crack growth in 7075-T7651.

As an example, consider components made from these alloys and installed in structures with feasible inspection intervals of a few months. Only in case (b) is the region II crack growth rate useful for estimating an inspection period of this time scale. In case (a) the region II crack growth rate is too high. In case (c) crack growth is so slow that inspection for cracks every few months is a waste of time.

10.7 Bibliography

1. Brown, B.F., *A New Stress-Corrosion Cracking Test for High Strength Alloys*, Materials Research and Standards, Vol. 6, pp. 129-133, (1966).
2. Chell, G.G. and Harrison, R.P., *Stress Intensity Factors for Cracks in Some Fracture Mechanics Test Specimens under Displacement Control*, Engineering Fracture Mechanics, Vol. 7, pp. 193-203, (1975).
3. Dorward, R.C. and Hasse, K.R., *Flaw Growth of 7075, 7475, 7050 and 7049 Aluminium Plate in Stress Corrosion Environments*, Final Report Contract NAS8-30890, Kaiser Aluminium and Chemical Corporation Centre for Technology (1976): Pleasanton California.
4. Wei, R.P., Novak, S.R. and Williams, D.P., *Some Important Considerations in the Development of Stress Corrosion Cracking Test Methods*, Materials Research and Standards, Vol. 12. pp. 25-30 (1972).
5. Wanhill, R.J.H., *Microstructural Influences on Fatigue and Fracture Resistance in High Strength Structural Materials*, Engineering Fracture Mechanics, Vol. 10, pp. 337-357 (1978).
6. Judy, R.W., Jr. and Goode, R.J., *Prevention and Control of Subcritical Crack Growth in High-Strength Metals*, Report 7780, Naval Research Laboratory (1970): Washington, D.C.
7. Hyatt, M.V. and Speidel, M.O., *Stress Corrosion Cracking of High-Strength Aluminium Alloys*, Report D6-24840, Boeing Commercial Airplane Group (1970): Seattle, Washington.

11 Dynamic Crack Growth and Arrest

11.1 Introduction

In this concluding chapter on fracture mechanics and crack growth the behaviour of statically loaded cracks growing beyond instability will be discussed.

The onset of instability usually means failure of a component or structure. As such, either instability cannot be allowed to occur, or else a partial (component) failure can be tolerated until its detection by regular inspection. In the latter case the component must be repaired or replaced if possible. Otherwise the structure has to be retired from service.

Certain structures belong to a category for which instability is always a possibility and failure would be intolerable. Examples are pipelines, nuclear reactor pressure vessels and liquid natural gas containers. For this category of structures the design and construction must include features to ensure crack arrest.

The foregoing examples illustrate the importance of unstable, *i.e.* dynamic, crack growth and its arrest. However, the study of this phenomenon is a highly specialised one, and is not amenable to detailed treatment in a basic course on fracture mechanics. In what follows only general remarks about the different concepts will be made.

In section 11.2 two basic aspects of dynamic crack growth will be described. First the velocities of fast fractures and the fact that they are finite. Second, as a consequence of finite crack velocity it is possible for crack branching to occur. Then in section 11.3 the conditions for crack arrest will be given together with their practical significance.

Section 11.4 is an overview of the widely used fracture mechanics methods of analysing dynamic crack growth and arrest in pipelines and thick-walled pressure vessels, and includes the concept of dynamic fracture toughness.

In sections 11.5 and 11.6 some information will be given about experimental determination of dynamic fracture toughness and the instantaneous value of the dynamic stress intensity factor.

Finally, section 11.7 reports about more recent progress made in the field of elastic-plastic fracture dynamics.

11.2 Basic Aspects of Dynamic Crack Growth

There are two basic aspects of dynamic crack growth:

1) Finite velocities of crack propagation.
2) Crack branching: this falls in the category of macrobranching rather than microbranching (which commonly occurs during all kinds of crack growth).

Finite Velocities of Crack Propagation

Dynamic crack growth may be considered in terms of an energy balance. This will be shown with the help of figure 11.1, which is a simplified crack resistance (G, $R - a$) diagram. After initiation of unstable crack extension there is excess energy which increases during crack growth. By the time the crack has reached a length a_i the total excess energy has amounted approximately to the shaded area in figure 11.1. (In practice G does not have to increase linearly with increasing crack length. Nor is it necessarily valid that R remains constant during dynamic crack growth.)

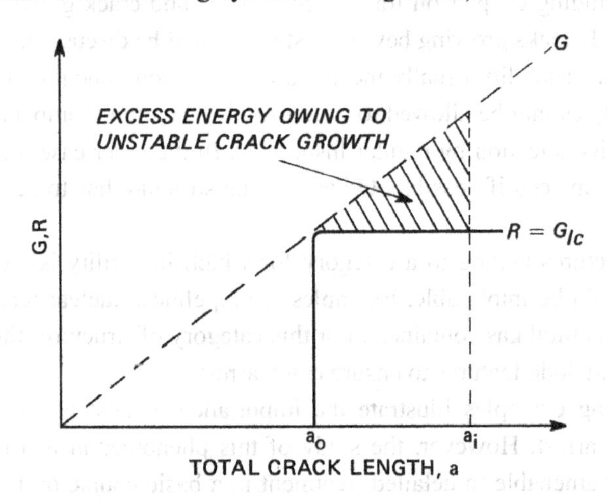

Figure 11.1. $G,R - a$ diagram showing the excess in energy some time after initiation of unstable crack extension under plane strain conditions.

The approximation in figure 11.1 is, however, convenient and adequate for further analysis to indicate that crack velocities are finite. The analysis which follows was first published by Mott, reference 1 of the bibliography to this chapter, as long ago as 1948.

Assuming that the formulation for G is the same at the onset of crack extension and during unstable fracture, the excess energy in figure 11.1 can be expressed by[1]

$$U_{\text{excess}} = \int_{a_o}^{a_i} (G - R)\,\mathrm{d}a = -R(a_i - a_o) + \int_{a_o}^{a_i} \frac{\pi\sigma^2 a}{E'}\,\mathrm{d}a\,. \tag{11.1}$$

Note that since $R = G_{Ic}$ is assumed to be constant there is a condition of plane strain. Thus $E' = E/(1 - \nu^2)$. Also, R is given by $R = \pi\sigma^2 a_o/E'$. Substituting in equation (11.1)

[1] As before, all energy quantities are defined per unit thickness.

we obtain

$$U_{\text{excess}} = -\frac{\pi\sigma^2 a_o}{E'}(a_i - a_o) + \frac{\pi\sigma^2}{2E'}(a_i^2 - a_o^2)$$

$$= \frac{\pi\sigma^2}{2E'}(a_i - a_o)^2 . \qquad (11.2)$$

Mott argued that for a propagating crack the excess energy is stored as kinetic energy. Some 25 years later this postulate was experimentally confirmed by Hahn *et al.*, reference 2.

A simple expression for the stored kinetic energy is obtainable from the opening displacement of the crack flanks. From section 2.3 the displacement v in the y direction is

$$v = \frac{2\sigma}{E'}\sqrt{a^2 - x^2} . \qquad (11.3)$$

If x is expressed as a fraction of a, *i.e.* $x = Ca$ where $0 < C < 1$, then

$$v = \frac{2\sigma}{E'}\sqrt{a^2(1 - C^2)} = C_1\frac{\sigma a}{E'} . \qquad (11.4)$$

As the crack propagates the displacement v will change with time. Denoting the rate of change $^{\mathrm{d}}/_{\mathrm{d}t}$ with '˙', we may write

$$\dot{v} = \frac{C_1\sigma\dot{a}}{E'} . \qquad (11.5)$$

The material adjacent to the crack flanks is displaced with a velocity $\dot{v}$. The kinetic energy in the displaced material is $T = \frac{1}{2}m\dot{v}^2$. For a material of density ρ and per unit thickness

$$T = \tfrac{1}{2}\rho\cdot\text{area}\cdot\dot{v}^2 = \tfrac{1}{2}\rho\iint \dot{v}^2\,\mathrm{d}x\,\mathrm{d}y = \tfrac{1}{2}\rho\,\dot{a}^2\frac{\sigma^2}{E'^2}\iint C_1^2\,\mathrm{d}x\,\mathrm{d}y . \qquad (11.6)$$

The solution of the integral in equation (11.6) will have the dimension [LENGTH]^2. For a semi-infinite plate the only significant length is the crack length a. Thus the integral can be expressed as ka^2, and for a crack of length a_i (see figure 11.1)

$$T = \tfrac{1}{2}\rho\,\dot{a}^2 k a_i^2\frac{\sigma^2}{E'^2} . \qquad (11.7)$$

If all the excess energy owing to unstable crack growth is converted into kinetic energy, then U_{excess} in equation (11.2) will equal T. Thus

$$\frac{\pi\sigma^2}{2E'}(a_i - a_o)^2 = \frac{\rho k\sigma^2}{2E'^2}(\dot{a}\,a_i)^2$$

so $$\dot{a}^2 = \frac{\pi}{k}\frac{E'}{\rho}\left(\frac{a_i - a_o}{a_i}\right)^2$$

and $$\dot{a} = \sqrt{\frac{\pi}{k}}\sqrt{\frac{E'}{\rho}}\left(1 - \frac{a_o}{a_i}\right). \tag{11.8}$$

The quantity $\sqrt{E'/\rho}$ is the velocity, V_g, of a longitudinal wave in a material. Consequently equation (11.8) can be rewritten as

$$\frac{\dot{a}}{V_g} = \sqrt{\frac{\pi}{k}}\left(1 - \frac{a_o}{a_i}\right). \tag{11.9}$$

For long propagating cracks $a_i >> a_o$ and equation (11.9) has a limit value of $\sqrt{\pi/k}$. The limit value has been calculated to be much less than unity, and experimental measurements have provided confirmation of this. Thus it can be stated that the maximum velocity of a propagating crack will always be a fraction of the longitudinal wave speed.

More elaborate analyses have shown that for a brittle material the theoretical maximum crack velocity is equal to the velocity of a surface wave.

Crack Branching

Dynamic crack growth may be accompanied by multiple branching of the crack. Some authors have attempted to explain branching in terms of kinetic energy. But experiments using high speed cameras have shown that branching does not alter the crack velocity. This calls into question any analysis based on kinetic energy, since if kinetic energy were used for crack branching the velocity would decrease.

Such experiments have shown, however, that crack branching under mode I loading occurs only when a specific stress intensity factor is exceeded. An illustration of this is given in figure 11.2 for six glass plates containing crack starters in the form of notches of increasing sharpness in the order a – f. The blunter the notch the higher the stress required to initiate fracture. This means that at any crack length the stress intensity factor for the propagating crack is highest in specimen (a) and lowest in specimen (f). The re-

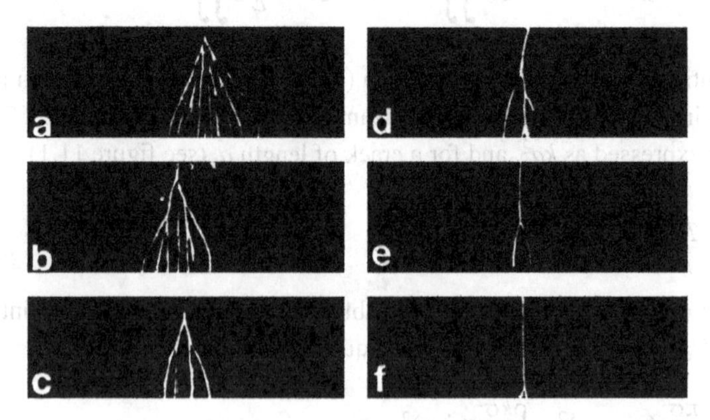

Figure 11.2. Crack branching in glass plates. Courtesy J.E. Field, Cambridge University.

sult, as figure 11.2 shows, is that crack branching is greatest in specimen (a) and least in specimen (f).

The dependence of crack branching on the stress intensity factor can be explained qualitatively. When a specific stress intensity factor is exceeded new cracks initiate ahead of the main crack. At first such cracks initiate so close to the main crack that they are overtaken by it. Eventually the initiation of new cracks is sufficiently far ahead of the main crack that they can accelerate to a velocity whereby they are no longer overtaken. Branching then occurs.

Evidence for this explanation has been provided by experiments on brittle materials. In particular, the observation of roughening of the surfaces of the main crack shortly before branching is evidence that new cracks initiate ahead of the main crack and are at first overtaken by it.

Crack branching can be promoted by the finite geometry of specimens. This effect has its origin in the maximum crack velocity being less than the velocities of waves in the material. Shock waves caused by impact loading and/or dynamic crack growth travel faster than the crack. The shock waves reflect from the back surface of a specimen. On arriving back at the crack tip the shock waves suddenly increase the stress intensity factor. This often causes multiple branching: excellent examples are sometimes provided by broken window panes.

11.3 Basic Principles of Crack Arrest

Like dynamic crack growth, crack arrest can be considered in terms of an energy balance. In the first instance one might consider the problem of crack arrest as an energy rate balance, such that the crack stops if G somehow decreases below R. If R is constant this criterion is exactly the reverse of that for unstable crack extension (*cf.* section 4.5). The situation for plane strain conditions is depicted schematically in figure 11.3.a.

However, for some materials it is very probable that R is not constant but depends on the crack velocity. The reason is as follows. The material in front of a fast running crack will be loaded at very high strain rates, and for strain rate sensitive materials the yield strength increases with increasing strain rate. In turn, a higher yield strength decreases the amount of crack tip plasticity and hence R, which is mainly plastic energy. In this case the energy rate balance criterion for crack arrest is as depicted in figure 11.3.b.

The energy rate balance criterion is in fact an oversimplification. In section 11.2 it was argued that excess energy after instability is converted to kinetic energy. If this kinetic energy can be used for crack propagation the situations given in figure 11.3 are no longer valid, and the problem must be solved by equating the total amounts of energy and not just energy rates. This is shown schematically in figure 11.4 for plane strain conditions. Note that for a strain rate sensitive material (situation in figure 11.4.b) R will increase as the arrest point is approached. This is because the decrease in kinetic energy will be accompanied by a decrease in crack velocity and hence lower strain rates and a lower yield strength ahead of the crack.

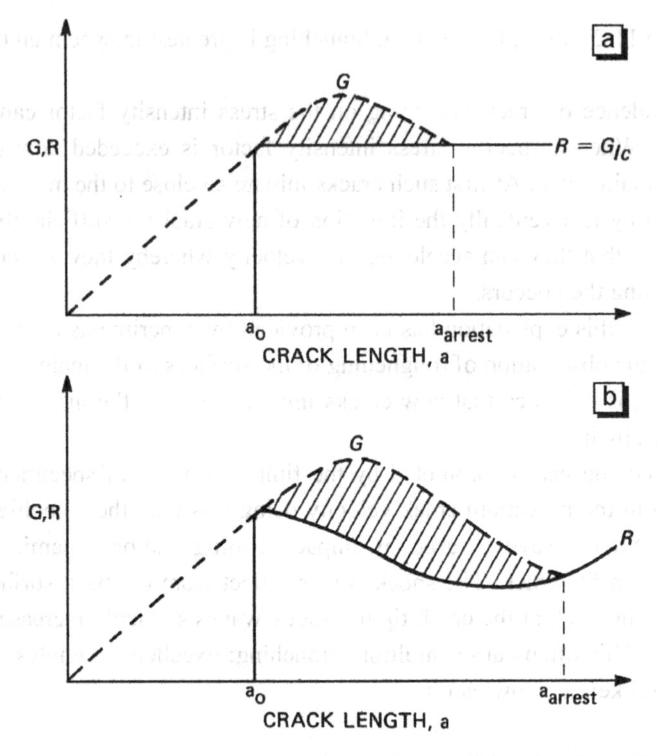

Figure 11.3. Crack arrest in terms of an energy rate balance ($G = R$ at arrest) for plane strain conditions with R (a) insensitive and (b) sensitive to crack velocity.

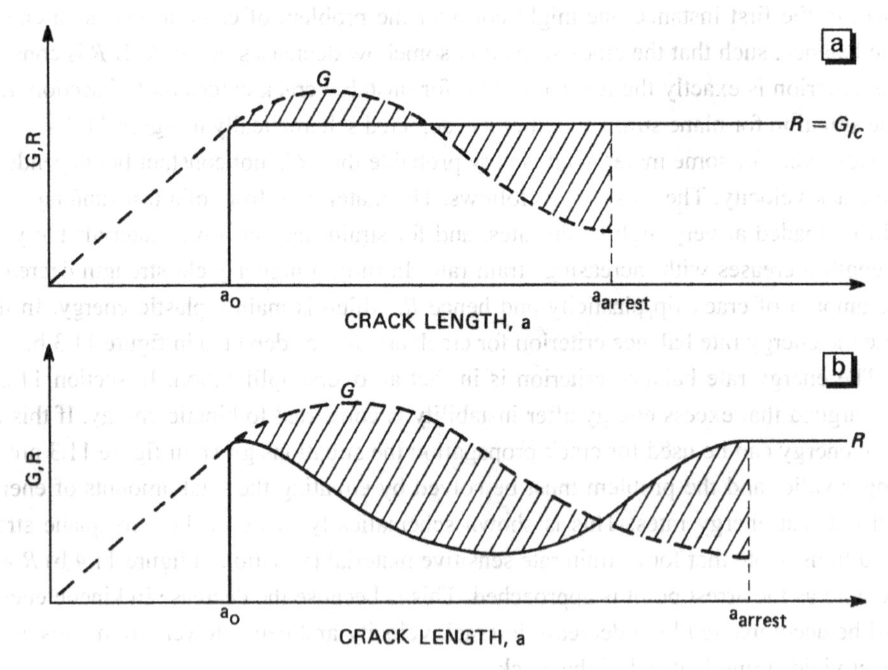

Figure 11.4. Crack arrest in terms of a balance of total energies for plane strain conditions with R (a) insensitive and (b) sensitive to crack velocity.

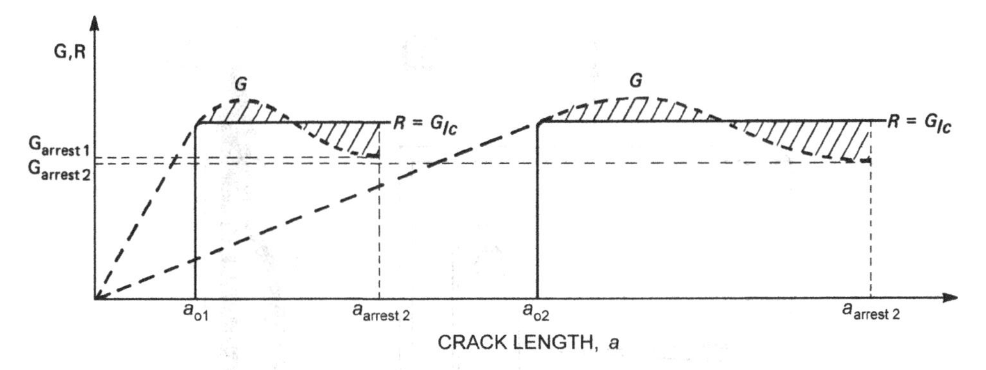

Figure 11.5. Effect of initial crack length on G_{arrest} for a strain rate insensitive material under plane strain conditions.

Although it is inconceivable that all the kinetic energy of a fast running crack is available for crack propagation, the situations shown in figure 11.4 are generally considered representative of actual crack-arrest behaviour in many materials, since they are qualitatively consistent with experimental results.

A potentially important consequence of crack arrest depending (approximately) on a balance of total energies is that the value of G at arrest is not a material constant, since it will depend on the variation of both G and R with crack length and velocity. This is exemplified in figure 11.5, which shows that even for the same maximum value of G and a constant R, the value of G at arrest can be clearly different for different initial (and hence final) crack lengths. This problem will be mentioned again in section 11.4 in terms of the static stress intensity factor at crack arrest.

The Practical Significance of Crack Arrest

The foregoing considerations, even though qualitative, are sufficient to demonstrate that crack arrest in a structure can occur if the energy release rate, G, decreases and/or the crack resistance, R, increases.

A decrease in G is obtained if

- the crack grows into a decreasing stress field, as in the case of wedge loading (see section 5.4);
- the load causing instability is transient and decreases with time;
- part of the load on the cracked element is taken up and transmitted by other structural elements: this is also referred to as load shedding by the cracked element.

A simple example of the load being taken up and transmitted by another structural element is given in figure 11.6. This shows a cracked plate with a bolted-on arrest strip. As the crack approaches, the arrest strip increasingly resists the normal displacement of the crack flanks. Consequently, part of the load on the plate is taken up and transmitted by the arrest strip, as indicated schematically in figure 11.6.b. This principle is much used (in a more sophisticated way) in aircraft structures, and serves the dual purpose of arresting fast running cracks and slowing down fatigue crack growth. More information on this subject is given in reference 3 of the bibliography to this chapter.

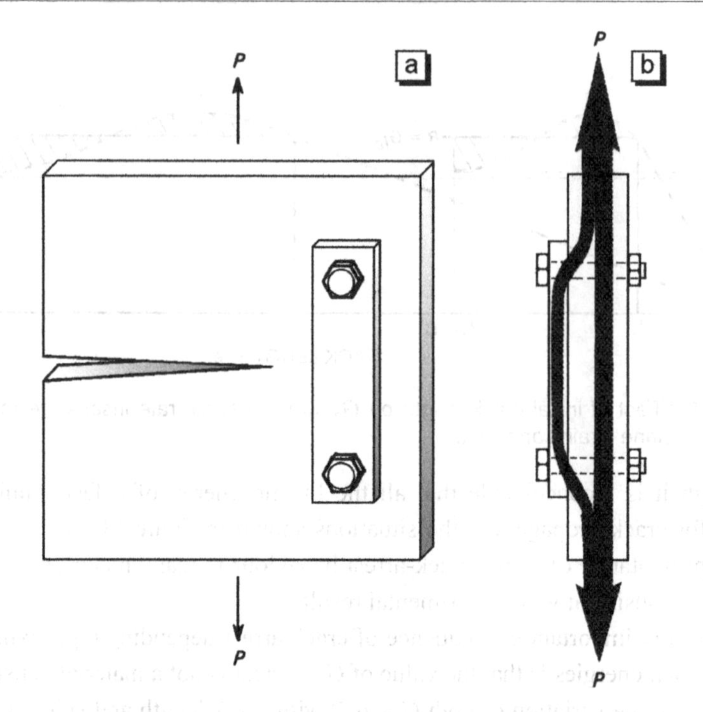

Figure 11.6. Cracked plate with bolted on arrest strip.

Another crack-arrest configuration, this time for a pipeline, is shown in figure 11.7. The arrest of fast running cracks in pipelines is extremely important. It has been known for fast fractures in gas pipelines to run for kilometres, with disastrous consequences. Pipeline fracture is further discussed in section 11.4.

An increase in a material's crack resistance, *R*, is not easily obtained. The only possibility that has practical significance is to dimension the structure such that instability would be accompanied by a change from plane strain to plane stress conditions. This would result in a rapidly rising *R*-curve (see section 4.6) which could soon cause crack arrest even if *G* continued to increase. This is shown schematically in figure 11.8.

There are other possibilities of increasing *R*. They have very limited applicability. Examples are direct insertion of strips of laminated material or strips of a material with much higher fracture toughness (and usually a much lower yield strength). These possi-

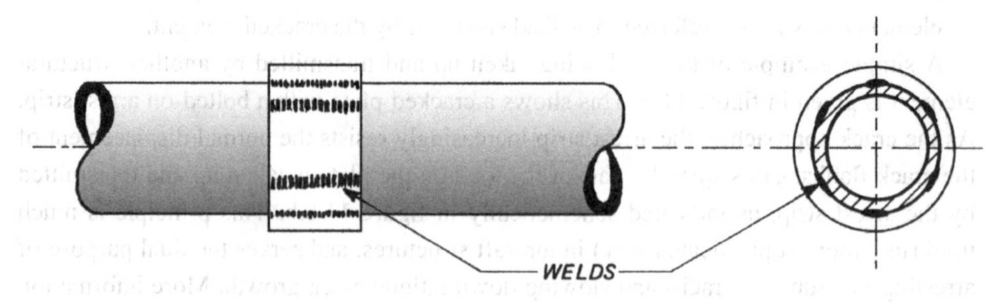

Figure 11.7. Pipeline with welded on crack-arrest ring.

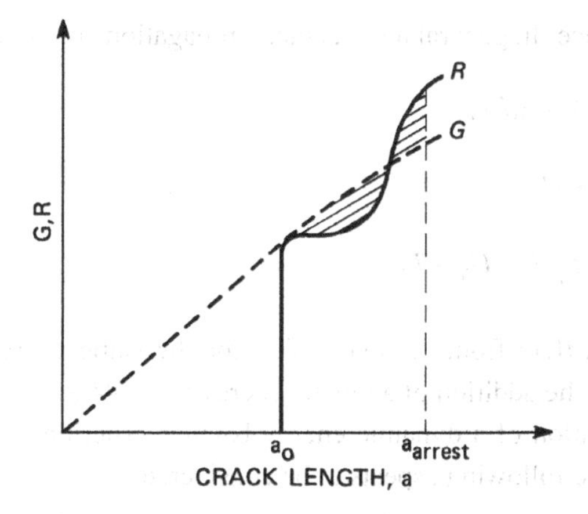

Figure 11.8. Increase in crack resistance owing to a transition from plane strain to plane stress.

bilities are feasible for weldable materials, but apart from technical difficulties there is a major manufacturing problem, since to be truly effective the strips must be spaced fairly close together.

11.4 Fracture Mechanics Analysis of Fast Fracture and Crack Arrest

Fracture mechanics analysis of fast fracture and crack arrest is a highly specialised topic, and a comprehensive treatment of the subject is beyond the scope of this course. Instead we shall make some general remarks concerning the two main problem areas, pipelines and thick-walled pressure vessels, for which dynamic fracture and crack arrest must be considered.

Pipelines

As mentioned in the previous section, the arrest of fast running cracks in pipelines is extremely important, since fast fractures have been known to run for kilometres. The problem is especially severe for gas pipelines, for the following reason. When a gas pipeline fractures the gas depressurises rapidly to cause a decompression shock wave travelling at the speed of sound ($\approx$ 400 m/s) in the gas. If the fast fracture in the pipe is able to travel faster than the decompression shock wave (as is quite possible) the crack tip continues to run in fully loaded material and there is no chance, without crack-arrest rings, for crack arrest to occur.

On the other hand, if a liquid-filled pipeline breaks open the pressure will drop faster, since the decompression shock wave travels at a higher speed (*e.g.* 1500 m/s). The drop in pressure will cause a decrease in the load acting on the pipe wall and consequently a decrease in G of the crack. This decrease may well be sufficient for a running crack to arrest.

Early studies of the fast fracture problem in pipelines were based on the assumption that the crack arrest will occur at a characteristic stress intensity (see reference 4 for more detailed information). However, the problem is currently viewed in terms of a dynamic energy balance. In general terms, crack propagation continues as long as

$$G_{\mathrm{dyn}}(a,t) > R(\dot{a}) , \tag{11.10}$$

where t denotes time and

$$G_{\mathrm{dyn}} = \frac{\mathrm{d}}{\mathrm{d}a}(F - U_{\mathrm{a}} - T) . \tag{11.11}$$

Equation (11.11) differs from the expression for the static energy release rate, G, *cf.* equation (4.17), by the addition of a kinetic energy term, $^{\mathrm{d}T}/_{\mathrm{d}a}$.

Practical application of a dynamic energy balance criterion to fast fracture of a gas pipeline requires the following aspects to be considered:

- the work done by the pressurised gas on the pipe walls as they crack;
- the contribution of kinetic energy to crack growth;
- the inertia of the pipe walls that have already cracked;
- the decrease of gas pressure owing to leakage;
- the effect of the large amount of plastic deformation of the pipe walls behind the crack tip (the pipe walls bend outwards, see figure 11.9);
- possible constraint of the plastic deformation of the pipe walls owing to the pipe being covered by soil (generally called "backfill").

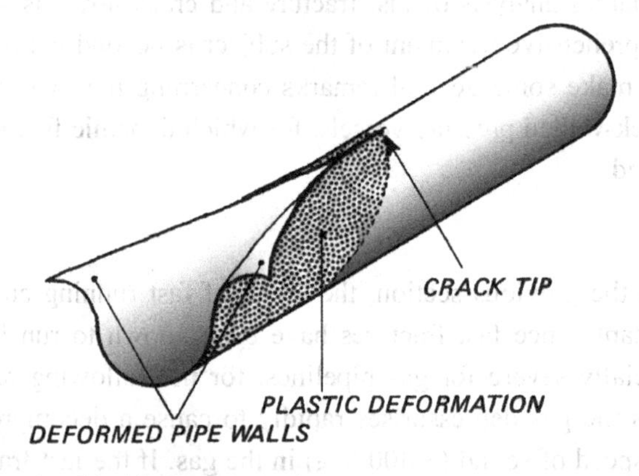

Figure 11.9. Schematic of fast fracture in a pipe.

In reference 5 a simplified (but still complex) expression for the dynamic energy release rate was obtained with the assumption of 'steady state' crack propagation, *i.e.* G_{dyn} independent of crack length and dependent only on time and hence crack velocity. The variation of G_{dyn} with crack velocity was calculated using equation (11.11). An example is given in figure 11.10.a, showing that G_{dyn} strongly depends on crack velocity. For this particular example G_{dyn} reaches a maximum value $\approx$ 6.4 $^{\mathrm{MJ}}/_{\mathrm{m}^2}$ when there is no soil coverage. Thus if the pipe material has a minimum crack resistance, R_{min}, of 6.5 $^{\mathrm{MJ}}/_{\mathrm{m}^2}$ it is not possible for fast fracture to achieve a steady state of continuing crack propagation: *i.e.* if fast fracture somehow initiates it will arrest, even when crack-arrest rings are not used.

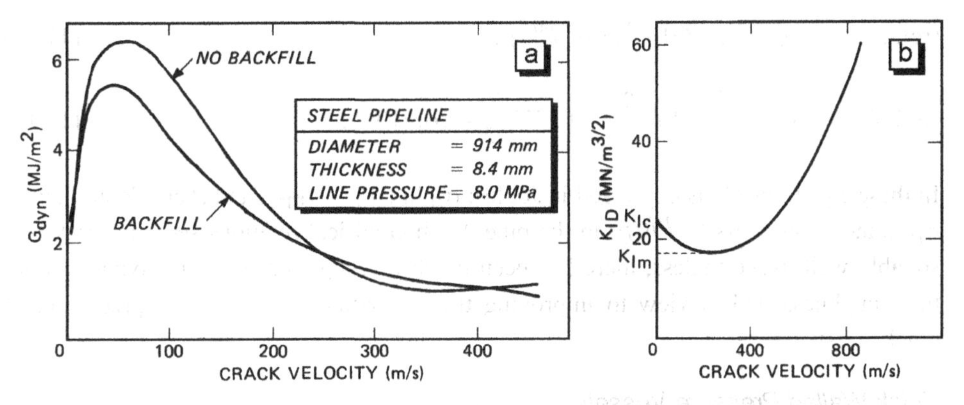

Figure 11.10. a) G_{dyn} calculated as a function of crack velocity, after reference 5;
b) K_{ID} measured as a function of crack velocity for a pressure vessel steel.

If it is at all possible to calculate G_{dyn} for a fast fracture situation, then it is evidently important to obtain data for the dynamic crack resistance, $R(\dot{a})$, or the equivalent dynamic fracture toughness, $K_{ID}(\dot{a})$. Note that as in the static case (section 4.8) we can express the dynamic crack resistance curve in terms of stress intensity factors. The criterion for crack arrest is that G_{dyn} becomes less than R_{min} or that K_I^{dyn} becomes less than the minimum in the variation of K_{ID} with crack velocity, usually denoted as K_{Im}. Figure 11.10.b shows that K_{ID} and hence R can strongly depend on crack velocity. Estimates of K_{ID} are possible by assuming that during steady state crack propagation $G_{dyn} = R(\dot{a})$ and that

$$\left(\frac{K_{ID}}{K_{Ic}}\right)^2 = \frac{R(\dot{a})}{G_{Ic}} = \frac{G_{dyn}}{G_{Ic}}, \tag{11.12}$$

where G_{dyn} follows from equation (11.11), the solution of which, however, requires a complete dynamic analysis of the fracture problem. Alternatively, instead of a cumbersome evaluation of K_{ID} for different crack velocities, a measure for K_{Im} can be determined directly by experiment. This is discussed in section 11.5.

In fact, for steel pipelines the problem is generally solved in a more empirical way. The energy release rate, G_{dyn}, is correlated with the energy, C_V, of conventional Charpy impact tests carried out above the brittle-to-ductile transition temperature (figure 1.2). The minimum crack resistance required for a pipeline steel is then expressed as a minimum required Charpy energy, $C_{V_{min}}$.

Correlations between G_{dyn} and C_V have no physical background, but they are used because the Charpy Impact test is still the most common method of determining a steel's resistance to brittle fracture. The correlations are carried a step further by establishing empirical relations between $C_{V_{min}}$ and the main parameters, besides the crack velocity, that determine the maximum value of G_{dyn}. These parameters are the pipe radius, R, wall thickness, B, and line pressure, P.

Several empirical relations for $C_{V_{min}}$ exist. Examples are (ref. 5) $$C_{V_{min}} = 3.36 \cdot 10^{-4} (\sigma_H^{1.5} \cdot R^{0.5}) , \qquad (11.13)$$

(ref. 6) $$C_{V_{min}} = \left(1.713 \frac{R}{B^{0.5}} - 0.2753 \frac{R^{1.25}}{B^{0.75}}\right) \sigma_H \cdot 10^{-3} . \qquad (11.14)$$

In these equations C_V is expressed in Joules; σ_H in Pa (= N/m^2); and R and B are in mm. σ_H is the hoop stress (= PR/B) in the pipe. Such empirical relations seem to work reasonably well. Nevertheless, there has been much investigation of actual dynamic fracture toughness with a view to improving the understanding and hence prediction of crack arrest.

Thick-Walled Pressure Vessels

Besides pipelines, dynamic fracture and crack arrest are also of concern for thick-walled (nuclear) pressure vessels, but with major differences. The critical cracks to be considered are part-through cracks growing from the inside surface of the pressure vessel wall. An example has been given in section 2.8. Cracks may become unstable owing to thermal shock (unusually rapid cooling within the pressure vessel) whereby high tensile stresses are induced at the inside surface of the wall.

It is essential that the cracks stop soon after instability so that wall penetration does not occur. The main reason why crack arrest is possible is the steep negative stress gradient through the wall to the outside surface. An increase in fracture toughness due to a lesser amount of radiation damage can be an additional factor in a nuclear pressure vessel.

Since there is only a very limited amount of permissible dynamic crack growth, the cracks will not have much kinetic energy. This fact is reflected in one of the methods developed to analyse dynamic fracture in pressure vessels steels, the "crack-arrest toughness" approach of Crosley and Ripling (reference 7).

Crosley and Ripling assert that when the amount of unstable crack growth is very limited the dynamic effects on stress intensity can be neglected. If this is the case, then the calculated static stress intensity factor, K_{Ia}^{stat} for a crack that has just arrested should be a reasonable approximation to the actual stress intensity at crack arrest. For this concept to be useful it has to be demonstrated that K_{Ia}^{stat} is approximately constant within the appropriate range of crack lengths and velocities, and that there is in fact little difference between K_{Ia}^{stat} and the value of K_I^{dyn} at crack arrest, K_{Ia}^{dyn}.

For several years there has been considerable controversy over the appropriateness of the static approach advocated by Crosley and Ripling. In particular, Kanninen and co-workers (*e.g.* reference 2) suggest that a dynamic energy balance approach should be used. The controversy has led to much effort in determining dynamic fracture toughness in order to compare the static and dynamic approaches to crack arrest.

Kalthoff *et al.* (reference 8) have made a summary comparing the results of the static and dynamic approaches in a three-dimensional graph, shown in slightly modified form in figure 11.11. The K_I versus crack length curves are schematic for the situation of crack arrest in a double cantilever beam specimen (DCB), but this does not detract from

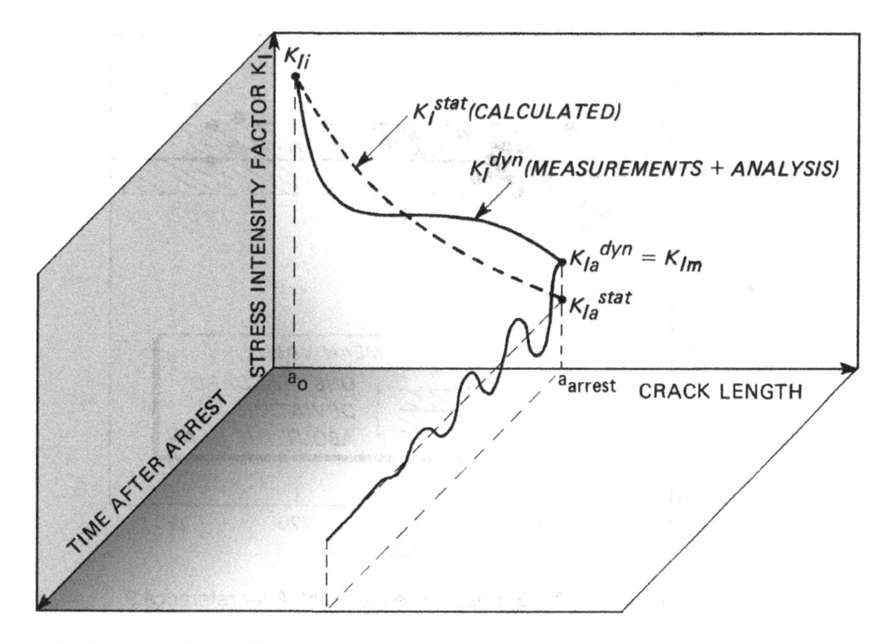

Figure 11.11. Comparison of the static and dynamic approaches to crack arrest in a double cantilever beam specimen (DCB).

the generality of comparison. During instability $K_{\mathrm{I}}^{\mathrm{dyn}}$ is at first lower than $K_{\mathrm{I}}^{\mathrm{stat}}$. However, as the arrest point is approached $K_{\mathrm{I}}^{\mathrm{dyn}}$ becomes greater than $K_{\mathrm{I}}^{\mathrm{stat}}$. At arrest $K_{\mathrm{Ia}}^{\mathrm{dyn}}$ is equal to the minimum value for the dynamic fracture toughness, K_{Im}, and is greater than $K_{\mathrm{Ia}}^{\mathrm{stat}}$. Note that after arrest $K_{\mathrm{I}}^{\mathrm{dyn}}$ oscillates with time and eventually becomes $K_{\mathrm{Ia}}^{\mathrm{stat}}$. At arrest $K_{\mathrm{Ia}}^{\mathrm{stat}}$ is a good approximation of $K_{\mathrm{Ia}}^{\mathrm{dyn}}$ only if the oscillations are limited and damp out quickly.

The difference between $K_{\mathrm{Ia}}^{\mathrm{dyn}}$ and $K_{\mathrm{Ia}}^{\mathrm{stat}}$ increases with increasing amount of unstable crack growth, $a_{\mathrm{arrest}} - a_{\mathrm{o}}$. Also, this difference depends on the specimen geometry: for DCB specimens the difference is significant, but for crack-line wedge-loaded specimens (CLWL) the difference is usually small.

There are a number of reasons why $K_{\mathrm{Ia}}^{\mathrm{stat}}$ is adequate for predicting crack arrest in thick-walled pressure vessels. The two main reasons are:

1) Many investigations support the $K_{\mathrm{Ia}}^{\mathrm{stat}}$ approach when the amount of unstable crack growth is very limited, as it has to be in pressure vessels. (Thus the potentially important variation in G_{arrest}, and hence $K_{\mathrm{Ia}}^{\mathrm{stat}}$, illustrated in figure 11.5 is not significant for unstable crack growth and arrest in pressure vessels.)
2) Results of a co-operative test programme carried out by ASTM members between 1977 and 1980 showed that the data could be described fairly well by a constant $K_{\mathrm{Ia}}^{\mathrm{stat}}$ value, figure 11.12, despite wide variation in the initial stress intensity, K_{Ii}, causing instability. A higher K_{Ii} means a higher initial crack driving force. This results in a higher crack velocity and more kinetic energy, so that if dynamic effects had been significant the constancy of $K_{\mathrm{Ia}}^{\mathrm{stat}}$ would not have been observed.

Note that given the reasonable degree of success for the static approach, it is obviously

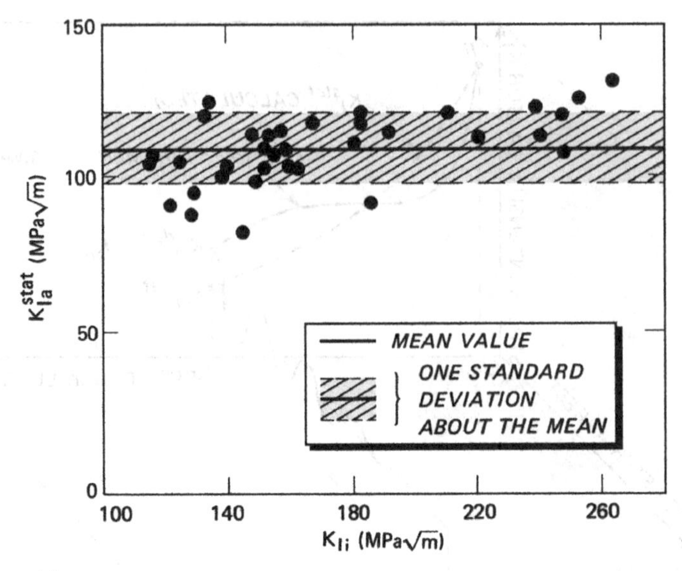

Figure 11.12. K_{Ia}^{stat} versus K_{Ii} for SA533 pressure vessel steel. After reference 9.

much simpler to use a static value of the stress intensity factor than to carry out a dynamic analysis.

11.5 Determination of the Crack-Arrest Toughness

To determine the crack-arrest toughness of a material a running crack must be created that experiences either an increasing fracture toughness, a decreasing stress intensity or a combination of both during its growth. In the past several types of specimen have been used, including double cantilever beams with and without taper (DCB and TDCB), single edge notched plates (SEN) and crack-line wedge-loaded specimens (CLWL). Extensive experimental and analytical work resulted in a preference for the CLWL arrangement shown in figure 11.13, for the following reasons:

- wedge loading, rather than pin loading, limits dynamic energy exchanges between specimen and testing machine;
- the CLWL specimen is relatively economical of material: this is an important consideration for high toughness materials like pressure vessel steels, since large specimens are normally required;
- the amount of side grooving to ensure an in-plane fracture path is much less than that for DCB specimens.

Even though the CLWL specimen is relatively economical of material, the requirement of an initial stress intensity factor well above the arrest value would mean impossibly large specimens for high toughness materials. This problem has been approached in two ways, illustrated in figure 11.14. In one case a CLWL specimen similar in shape to that shown in figure 5.17 has a brittle weld bead at the tip of the crack starter slot (see reference 9). The other specimen type is duplex, with crack initiation occurring in a hardened steel starter section welded onto the test material (see reference 10). Both

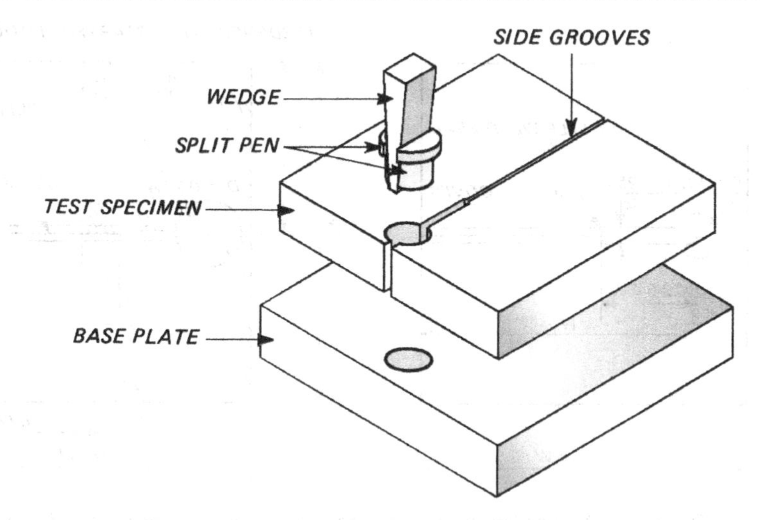

Figure 11.13. Schematic of the crack-line wedge-loading arrangement for crack-arrest toughness testing.

types of specimen have a blunt starter notch, so that the stress intensity to initiate cracking will be higher than the arrest value. Furthermore, side grooves are applied in order to prevent the formation of shear lips (see section 3.6) and restrict crack branching.

Test Procedure

In 1988 ASTM published a standard test procedure, designated E 1221, for determining crack-arrest toughness for ferritic steel, see reference 11. Only an outline of the test procedure and analysis will be given here. This determination follows a static approach, *i.e.* the crack-arrest toughness, now conveniently denoted as K_{Ia}, is in fact $K_{\mathrm{Ia}}^{\mathrm{stat}}$ (see section 11.4), since it is calculated using the load and crack length immediately after arrest.

The procedure is based on the arrangement shown in figure 11.13 and uses the specimen shown on the left-hand side of figure 11.14 (although the specimen on the right-hand side may also be used). The specimen thickness, B, is either equal to the thickness used in the application or sufficient to create a plane strain condition (see below). The side grooves on each side should be $B/8$ deep.

The load exerted by the wedge on the specimen cannot be measured directly. Instead, both the load applied to the wedge as well as the crack-mouth opening displacement are monitored during the test. Wedge-loading can be done at a relatively slow rate, so that recording the load signal from the testing machine and the displacement from a clip gauge mounted on the specimen is not a problem.

To initiate unstable growth followed by arrest a cyclic loading procedure is used: this requires some special attention. First the specimen is loaded to a pre-determined displacement. Assuming the crack has not yet initiated, the specimen is then unloaded. This leads to a certain zero-load displacement offset, which can be due to local cracking

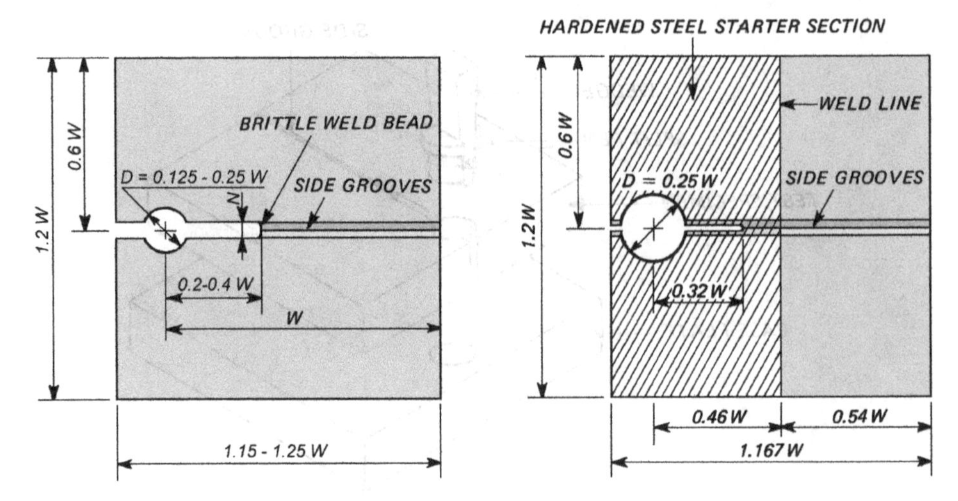

Figure 11.14. Examples of crack-line wedge-loaded specimens for crack-arrest fracture-toughness testing.

in the weld, notch-tip plasticity and/or seating of fixtures or clip gauge. Subsequently the specimen is reloaded to a somewhat higher displacement value. This whole sequence of loading and unloading is repeated until the crack actually shows temporarily unstable growth.

The calculation of both the initial stress intensity, K_{Ii}, and the crack arrest value, K_{Ia}, (described below) are based on the displacement excluding 50% of the zero-load displacement offset accumulated during the load cycles. This percentage is more or less arbitrary, since the effect of these phenomena on K_I are not completely clear. For example, seating of fixtures and local cracking should not be included. On the other hand, residual stresses due to plasticity may to some extent provide a driving force once the crack has grown through the plastic zone.

After the test the crack length at arrest, a_{arrest}, must be measured. Since the specimens are fairly thick ($B \approx 50$ mm) the through-thickness crack length must be measured and averaged. This can be done via heat tinting. The cracked but unbroken specimens are heated in a furnace to discolour the fracture surfaces by oxidation. Subsequently the specimens are broken open and the crack lengths measured directly.

For the calculation of K_{Ii} the initial crack length, a_o, and the displacement at initiation are used, while a candidate value for K_{Ia}, K_{Qa}, is evaluated using a_{arrest} and the displacement at arrest. The latter will generally be slightly larger than the displacement at initiation and should be measured within 100 milliseconds after crack arrest. This is to avoid measuring a slow increase in displacement due to time-dependent behaviour.

Both K_{Ii} and K_{Qa} are now calculated using the relation given in the ASTM standard, *i.e.*

$$K_I = \frac{E\delta\sqrt{B/B_N}}{\sqrt{W}} \cdot f\left(\frac{a}{W}\right), \qquad (11.15)$$

where δ = crack mouth opening displacement,

B_N = net specimen thickness at side grooves,

$$f\left(\frac{a}{W}\right)=\sqrt{1-\frac{a}{W}}\left\{0.748-2.176\left(\frac{a}{W}\right)+3.56\left(\frac{a}{W}\right)^2-2.55\left(\frac{a}{W}\right)^3+0.62\left(\frac{a}{W}\right)^4\right\}.$$

For the candidate arrest value to be qualified as a linear elastic plane strain value, K_{Ia}, a number of checks must be made. The unstable crack growth must be larger than twice the slot width, N (*cf.* left-hand side of figure 11.14), and exceed the plane-stress plastic zone size at initial loading, *i.e.*

$$a_{arrest}-a_o \geq \frac{1}{2\pi}\left(\frac{K_{Ii}}{\sigma_{ys}}\right)^2 . \tag{11.16}$$

Furthermore, the unbroken ligament must be larger than 0.15 W and satisfy

$$W-a_{arrest} \geq 1.25\left(\frac{K_{Qa}}{\sigma_{yd}}\right)^2 , \tag{11.17}$$

where σ_{yd} is an assumed value for the yield strength for appropriate loading times and temperature and set equal to σ_{ys} + 205 MPa. Finally, the specimen thickness must be larger than

$$B \geq 1.0\left(\frac{K_{Qa}}{\sigma_{yd}}\right)^2 . \tag{11.18}$$

11.6 Determination of Dynamic Stress Intensity Factors

Calculation of dynamic stress intensity factors, K_I^{dyn}, is a very difficult problem that has had only limited success. Consequently, actual determination usually involves a combination of experimental measurements and analysis. Either a direct or indirect method of determination can be used.

In a direct method the crack tip characterising parameters are measured during fast crack propagation and require high speed photography to record instantaneous positions of fast running cracks. Two such methods suitable for metallic materials are

- the shadow optical method of caustics in reflection, reference 8,
- dynamic photo-elasticity of birefringent coatings on specimens, reference 12.

An indirect method uses information on crack growth history obtained from measurements, for example by applying timing wires broken by the advancing crack, see reference 13. Afterwards an analysis of the advancing crack using numerical techniques, for example a finite element method, can be used to calculate fracture characteristics such as the dynamic stress intensity factor.

The Shadow Optical Method of Caustics

The physical basis of this method is illustrated in figure 11.15 for the case of a transparent cracked specimen illuminated by parallel light. Owing to the stress concentration at

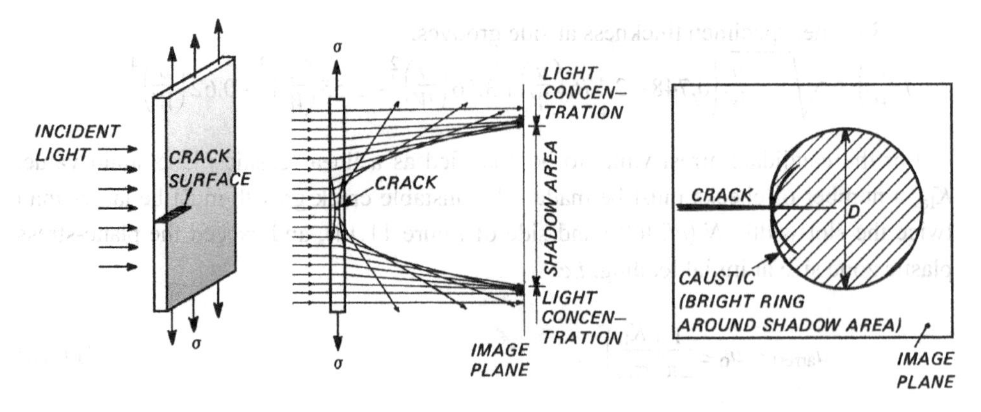

Figure 11.15. Schematic of the shadow optical method of caustics for transmitted light.

the crack tip the specimen will locally contract and the refractive index will change. As a consequence the incident light will be deflected outwards during transmission. The overall effect is to produce a shadow spot bounded by a bright ring (the caustic) on an image plane at any distance behind the specimen.

A similar effect is obtainable from a non-transparent (*e.g.* metallic) specimen acting as a mirror. In this case the light being reflected forms an analogous shadow spot and caustic when observed in a virtual image plane behind the specimen.

In both transmission and reflection arrangements the shadow spots for fast running cracks can be recorded by a high-speed camera focused on the appropriate image plane. The instantaneous dynamic stress intensity factor for a running crack is then obtained from

$$K_{\mathrm{I}}^{\mathrm{dyn}} = F(\dot{a})\,M\,D^{5/2}\,, \tag{11.19}$$

where D is the diameter of the caustic; $F(\dot{a})$ is a factor accounting for the crack velocity dependence of the shadow spot and is ≈ 1 for typical fast fractures; and M accounts for specimen thickness, elastic and optical properties, and the distance between the specimen and the image plane. More details on the values of $F(\dot{a})$ and M are given, for example, in reference 14.

Using this technique Kalthoff *et al.* (reference 8) have obtained $K_{\mathrm{I}}^{\mathrm{dyn}}$ versus crack length curves for a number of materials, including a high strength steel. The results enabled them to indicate the general trend shown in figure 11.11.

A variation of the shadow optical method of caustics is the method of projection on the focal plane. This method was first proposed by Kim, see reference 15, and has the advantage relative to the original method that it does not require high speed photography for monitoring the stressed state. The initially parallel light rays transmitted through a transparent fracture specimen (or reflected from the surface of an opaque specimen) are focused by a converging lens. The real-time stress intensity factor variation, $K_{\mathrm{I}}(t)$, of a moving crack tip can be measured using a single, stationary photodetector. The principle of the method is based on the fact that any variation in $K_{\mathrm{I}}(t)$ leads to a change of the light intensity $I(t)$ impinging on the photodetector located on the focal plane. The following relation is derived:

$$I(t) = B\,K_I(t)^{4/3} \tag{11.20}$$

in which B depends on the speed of the crack for dynamic K_I measurements. For details the reader is referred to reference 15.

Dynamic Photo-elasticity of Coatings on Specimens

This method is an extension of the study of dynamic crack growth in transparent birefringent polymers. It has its basis in the fact that cracks and notches in birefringent polymers give rise to characteristic patterns of isochromatic fringe loops, whose size and shape can be analysed to derive the stress intensity factor when the loading conditions and optical properties of the polymers are known. This is a highly specialised topic: the interested reader is referred to *e.g.* reference 16 for information on such analyses.

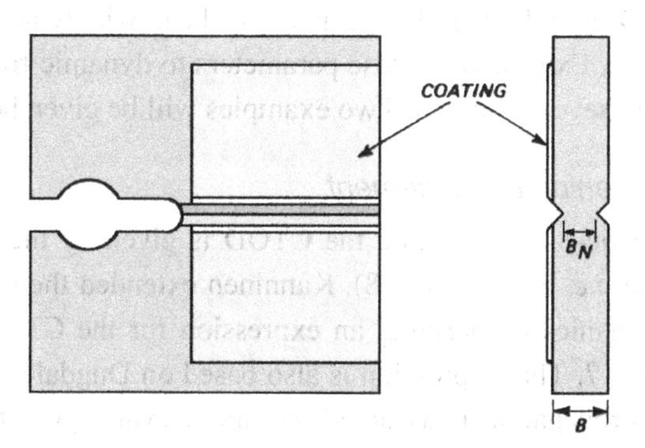

Figure 11.16. Schematic of birefringent coating on a crack-line wedge-loaded specimen (CLWL). After reference 12.

To use the method for studying fast fracture in metallic specimens a birefringent coating is bonded on to a side surface as shown in figure 11.16 for a CLWL specimen. (Note that the coating is not continuous. This eliminates uncertainty as to whether the isochromatic pattern in the coating is influenced by fracture of the coating itself.) A dynamic fracture toughness test is then carried out in synchronisation with high-speed photographic recording of the instantaneous crack length in the specimen and the associated isochromatic fringe loops in the coating.

For any recorded instantaneous crack length in the metallic specimen the dynamic stress intensity factor is

$$(K_I^{dyn})_m = \left(\frac{E_m}{E_c}\right)\left(\frac{1+\nu_c}{1+\nu_m}\right)(K_I^{dyn})_c\sqrt{\frac{B}{B_N}}, \tag{11.21}$$

where the subscripts c and m refer to coating and metallic specimen respectively, and $(K_I^{dyn})_c$ for the coating has to be obtained by the rather complex analysis of the isochromatic fringe loops. The factor $\sqrt{B/B_N}$ is included in equation (11.21) because of the use of side grooves to ensure an in-plane fracture path in the specimen.

Indirect Method for Determination of K_I^{dyn}

This method of determination of the dynamic stress intensity factor uses a combined approach in which the crack propagation is measured and subsequently used in a finite element analysis to determine K_I^{dyn}. For this analysis it may be necessary to use special techniques to describe the singularity at the crack tip and the fact that the tip is moving. Furthermore, it can suffice to assume elastic material behaviour. However, for an adequate description of the material behaviour it may also prove essential to include plasticity in the analysis and, since strain rates become very high for fast running cracks, to include the effect of strain rate on yield behaviour.

11.7 Approaches in Elastic-Plastic Dynamic Fracture Mechanics

As was stated in chapter 6 on Elastic-Plastic Fracture Mechanics, the Crack Opening Displacement (COD) and J integral concepts have been widely accepted as crack characterising parameters. Extension of these parameters to dynamic fracture mechanics has been investigated by several authors. Two examples will be given here.

Dynamic Crack Opening Displacement

For quasi-static elastic-plastic fracture the CTOD is given by the equation derived by Burdekin and Stone, *i.e.* equation (6.28). Kanninen extended the COD approach to dynamic fracture mechanics by deriving an expression for the CTOD for a propagating crack, see reference 17. This expression is also based on Dugdale's strip yield model. A steady-state crack propagation is assumed, *i.e.* all relevant quantities, such as stresses and displacements, are independent of time for a constant position relative to the moving crack tip. For a crack propagating with a speed $\dot{a}$ the static CTOD value, equation (6.28), is multiplied by the function $L(\dot{a})$ given by

$$L(\dot{a}) = \frac{1+\nu}{D(\dot{a})}\left(\frac{\dot{a}}{C_2}\right)^2 \left\{\left(1-\frac{\dot{a}}{C_1}\right)^2\right\}, \tag{11.22}$$

where C_1 = longitudinal wave velocity in a plate assuming plane stress,
C_2 = shear wave velocity,

$$D(\dot{a}) = 4\sqrt{1-\left(\frac{\dot{a}}{C_1}\right)^2}\sqrt{1-\left(\frac{\dot{a}}{C_2}\right)^2} - \left\{2-\left(\frac{\dot{a}}{C_2}\right)^2\right\}^2.$$

In the limit of $\dot{a} \to 0$ the function $L(\dot{a}) = 1$, while it monotonically increases with $\dot{a}$ until it reaches infinity at the speed of Rayleigh waves.

Assuming a constant (critical) CTOD for a propagating crack, Kanninen used equation (11.22) to predict the crack velocity as a function of crack length. A reasonable agreement was found with experiments on steel foil in tension.

Dynamic J Integral

The J integral has found widespread application in quasi-static elastic-plastic fracture mechanics. This path-independent contour integral actually represents the energy release rate for cracks in nonlinear elastic material. Elastic-plastic material behaviour can be adequately described only when unloading is absent or limited. The use of J is therefore restricted to crack growth initiation (chapter 7) or to a limited amount of stable growth (chapter 8).

Obviously J cannot be used to describe dynamic crack propagation in elastic-plastic material since considerable unloading can be expected to occur. However, it will still prove meaningful to consider the integral expression for the dynamic J in nonlinear elastic material. Since a detailed derivation is beyond the scope of this course, only the result will be given here.

For a crack propagating in the x_1 direction with a speed $\dot{a}$ one can derive J as follows, see *e.g.* reference 17:

$$J = \int_{\Gamma} \left\{ \left(W + \tfrac{1}{2}\rho\,\dot{a}^2 \frac{\partial u_i}{\partial x_1}\frac{\partial u_i}{\partial x_1} \right) n_1 - T_i \frac{\partial u_i}{\partial x_1} \right\} \mathrm{d}s + \int_{\Omega} \rho \left(\ddot{u}_i - \dot{a}^2 \frac{\partial^2 u_i}{\partial x_1^2} \right) \frac{\partial u_i}{\partial x_1} \mathrm{d}A\,, \qquad (11.23)$$

where Γ = arbitrary contour surrounding the crack tip,
W = strain energy density,
ρ = mass density,
u_i, $\ddot{u}_i$ = displacement and acceleration respectively,
n_i = outward-directed unit vector normal to Γ,
T_i = traction acting on Γ,
Ω = area bounded by Γ, but excluding an infinitely small area containing the crack tip.

Note that for the static case, when both $\dot{a}$ and $\ddot{u}_i$ are zero, this expression reduces to equation (6.29). The most striking difference compared with the static equation, however, is the fact that J in equation (11.23) is no longer expressed by a path-independent contour integral only.

This can be understood by realising that in dynamic situations stress waves will be travelling to and from the crack tip. If a wave front has traversed one contour but not another, integrals evaluated along these contours will generally not yield identical values. Actually J can be written in the form of a contour integral only, but this contour must then bound an infinitely small volume around the crack tip. Obviously, such a formulation is impractical to perform calculations with, *e.g.* using the finite element method.

For the special case of steady-state crack propagation (see under the previous subheading) $\dot{u}_i = -\dot{a}\,\partial u_i/\partial x_1$. Since now $\ddot{u}_i = -\dot{a}\,\partial \dot{u}_i/\partial x_1 = \dot{a}^2\,\partial^2 u_i/\partial x_1^2$ the area integral in equation (11.23) vanishes and J can be conveniently evaluated only by using data at some distance from the running crack tip, *i.e.* along Γ.

Although the dynamic J cannot be used to describe crack propagation in elastic-plastic material, it can be used up to the point of initiating crack growth. By equating $\dot{a}$

to zero, equation (11.23) reduces to

$$J = \int_{\Gamma} \left(W n_1 - T_i \frac{\partial u_i}{\partial x_1} \right) ds + \int_{\Omega} \rho \ddot{u}_i \frac{\partial u_i}{\partial x_1} dA \,. \tag{11.24}$$

Thus besides the usual contour integral expression, *cf.* equation (6.29), an integral over the area included by the contour must also be considered to account for the dynamic effects when evaluating J up to the point of initiating crack growth.

Note that a standard test method which actually uses the dynamic J integral has not yet been developed.

11.8 Bibliography

1. Mott, N.F., *Fracture of Metals: Theoretical Considerations*, Engineering, Vol. 165, pp. 16-18 (1948).
2. Hahn, G.T., Hoagland, R.G., Kanninen, M.F. and Rosenfield, A.R., *The Characterization of Fracture Arrest in a Structural Steel, Pressure Vessel Technology*, American Society of Mechanical Engineers, Part II, pp. 981-994 (1973): New York.
3. Broek, D., *Elementary Engineering Fracture Mechanics*, Martinus Nijhoff (1982): The Hague.
4. Maxey, W.A., Kiefner, J.F., Eiber, R.J. and Duffy, A.R., *Ductile Fracture Initiation, Propagation and Arrest in cylindrical Vessels*, Fracture Toughness, ASTM STP 514, American Society for Testing and Materials, pp. 70-81(1972): Philadelphia.
5. *Running Shear Fracture in Line Pipe*, AISI Technical Report, American Iron and Steel Institute, September 1974.
6. Fearnehough, G.D., *Fracture Propagation Control in Gas Pipelines: a Survey of Relevant Studies*, International Journal of Pressure Vessels and Piping, Vol. 2, pp. 257-281 (1974).
7. Crosley, P.B. and Ripling, E.J., *Characteristics of a Run Arrest Segment of Crack Extension*, Fast Fracture and Crack Arrest, ASTM STP 627, American Society for Testing and Materials, pp. 203-227 (1977): Philadelphia.
8. Kalthoff, J.F., Beinert, J., Winkler, S. and Klemm, W., *Experimental Analysis of Dynamic Effects in Different Crack Arrest Test Specimens*, Crack Arrest Methodology and Applications, ASTM STP 711, American Society for Testing and Materials, pp. 109-127 (1980): Philadelphia.
9. Crosley, P.B. and Ripling, F.J., *Significance of Crack Arrest Toughness* (K_{Ia}) *Testing*, Crack Arrest Methodology and Applications, ASTM STP 711, American Society for Testing and Materials, pp. 321-337 (1980): Philadelphia.
10. Hoagland, R.G., Rosenfield, A.R., Gehlen, P.C. and Hahn, G.T., *A Crack Arrest Measuring Procedure for K_{Im}, K_{ID} and K_{Ia} Properties*, Fast Fracture and Crack Arrest, ASTM STP 627, American Society for Testing and Materials, pp. 177-202 (1977): Philadelphia.
11. ASTM Standard E 1221-96, *Standard Test Method for Determining Plane-Strain Crack-Arrest Fracture Toughness, K_{Ia}, of Ferritic Steels*, ASTM Standards on Disc, Vol. 03.01 (2001): West Conshohocken, Philadelphia.
12. Kobayashi, T. and Dally, J.W., *Dynamic Photo-elastic Determination of the* á –K *Relation for 4340 Alloy Steel*, Crack Arrest Methodology and Applications, ASTM STP 711, American Society for Testing and Materials, pp. 189-210 (1980): Philadelphia.
13. Kanninen, M.F., Gehlen, P.C., Barnes, C.R., Hoagland, R.G., Hahn, G.T. and Popelar, C.H., *Dynamic Crack Propagation Under Impact Loading*, in: Nonlinear and Dynamic Fracture Mechanics, Perrone, N. and Atluri, S.N. (Eds.), ASME AMD Vol. 35, pp. 185-200 (1979).
14. Beinert, J., Kalthoff, J.F. and Maier, M., *Neuere Ergebnisse zur Anwendung des Schattenfleckverfahrens auf stehende und schnell-laufende Brüche*, 6th International Conference on Experimental Stress Analysis, VDI-Verlag GmbH, pp. 791-798 (1978): Düsseldorf.
15. Kim, K.S., *A Stress Intensity Factor Tracer*, Journal of Applied Mechanics, Vol. 52, No.2, pp. 291-297 (1985).

16. Kobayashi, A.S., *Photo-elasticity Techniques, Experimental Techniques in Fracture Mechanics*, Society for Experimental Stress Analysis, Iowa State Press, Chapter 6, Monograph No. 1, pp. 126-145 (1973): Ames, Iowa.
17. Kanninen, M.F. and Popelar, C.H., *Advanced Fracture Mechanics*, Oxford University Press (1985): New York.

Part V
Mechanisms of Fracture in Actual Materials

12 Mechanisms of Fracture in Metallic Materials

12.1 Introduction

Since World War II there has been great progress in understanding the ways in which materials fracture. Such knowledge has proved essential to better formulation of fracture mechanisms. Nevertheless, it is still not possible to use this knowledge, together with other material properties, for predicting fracture behaviour in engineering terms with a high degree of confidence.

Some insight into the problems involved is given in chapter 13, and it is the intention of the present chapter to provide the necessary background information on fracture mechanisms.

Metallic materials, especially structural engineering alloys, are highly complex. An

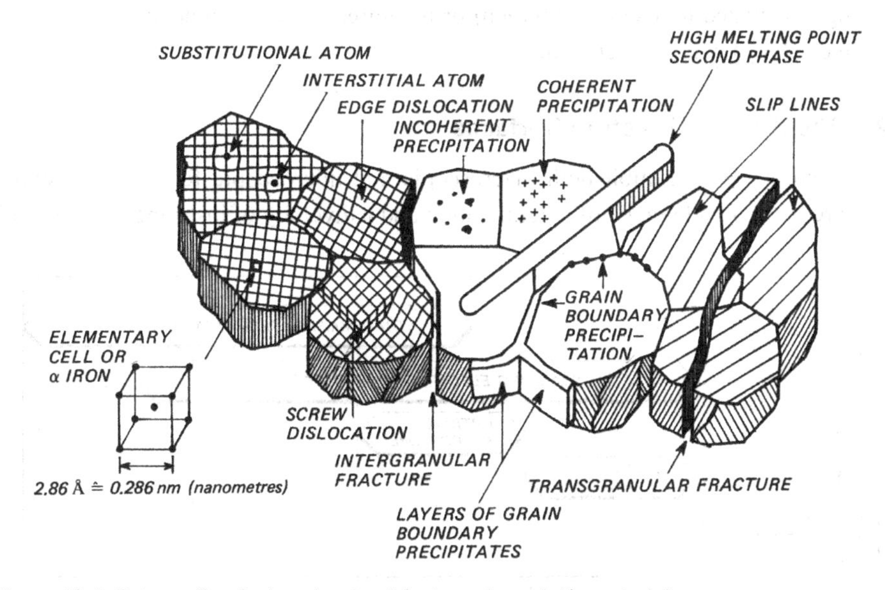

Figure 12.1. Schematic of microstructural features in metallic materials.
Courtesy Gerling Institut für Schadenforschung und Schadenverhütung, Cologne, FRG.

indication of this complexity is given by figure 12.1, which shows various microstructural features (not all of which need be present in a particular material) and also the two main types of fracture path, transgranular and intergranular fracture. Of fundamental importance is the fact that almost all structural materials are polycrystalline, *i.e.* they consist of aggregates of grains, each of which has a particular crystal orientation. The only exceptions are single crystal turbine blades for high performance jet engines.

Before the various mechanisms of fracture are discussed some information will be given in sections 12.2 and 12.3 on the following topics:

1) The instruments used in fractography, which is the study of fracture surfaces. In particular, the use of electron microscopes will be mentioned.
2) The concept of dislocations (see figure 12.1). The nucleation and movement of dislocations causes shear, *i.e.* slip, on certain sets of crystal planes, and the overall effect of slip is plastic de formation.

Sections 12.4–12.7 cover specific aspects of transgranular and intergranular fracture, namely

- transgranular fracture:
  - ductile fracture by microvoid coalescence
  - brittle fracture (cleavage)
  - fatigue crack initiation and growth
- intergranular fracture:
  - grain boundary separation with microvoid coalescence
  - grain boundary separation without microvoid coalescence.

Finally, in section 12.8 some of the types of fracture that can occur owing to sustained loading at elevated temperatures (creep) or in aggressive environments (*e.g.* stress corrosion cracking) will be discussed.

12.2 The Study of Fracture Surfaces

Fracture surfaces exhibit both macroscopic and microscopic features, the study of which requires a wide range of magnification and a diversity of instruments, figure 12.2.

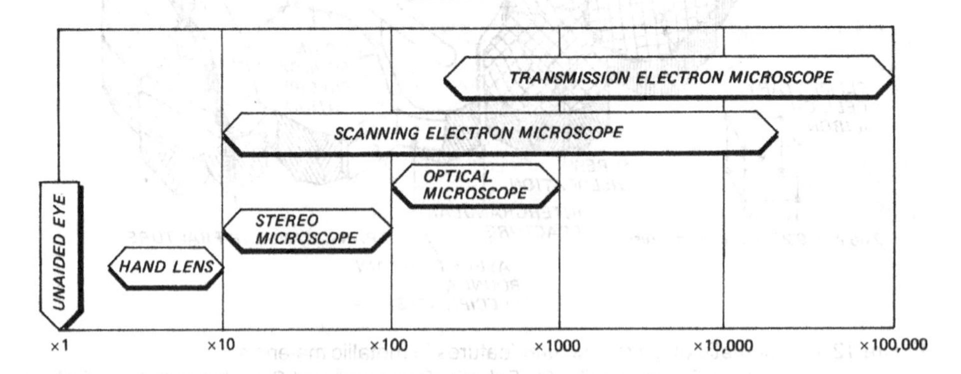

Figure 12.2. Approximate ranges of magnification for instruments used to study fracture surfaces and microstructure.

Macroscopic examination should always be done first. This can be done with the unaided eye or a hand lens, and is often sufficient to indicate the directions in which cracks have grown and the locations of crack origins. Also, it is sometimes possible to distinguish immediately between fatigue and overload failures and whether the fracture is relatively recent. Older fracture surfaces tend to be discoloured owing to corrosion.

If the location of a crack origin is known, a stereo microscope is most useful for seeing whether there are special features associated with the origin. When this has been done the electron microscopes have to be used, especially the scanning electron microscope, which has a large depth of field and can be used from low to high magnifications. The scanning electron microscope has, in fact, virtually replaced the optical microscope for direct examination of fracture surfaces. However, the optical microscope is indispensable for metallography, including polished and etched cross-sections through fracture surfaces and cracks. An example is given in figure 12.3.

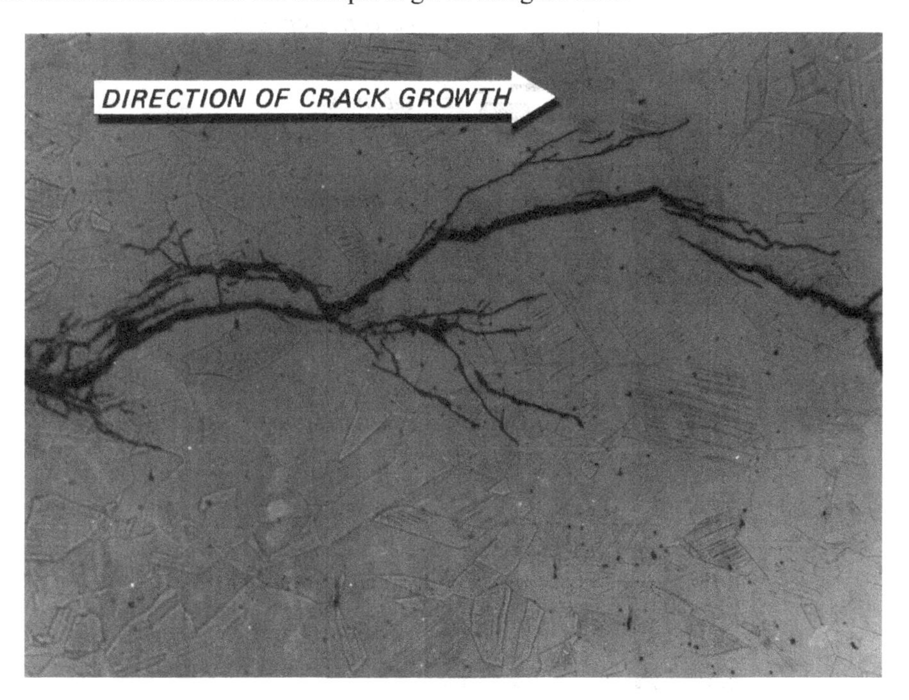

Figure 12.3. Transgranular stress corrosion crack in AISI 304 stainless steel. Optical metallograph, ×320.

The use of electron microscopy is essential for determining the types of fracture with certainty. This is because characteristic, identifying features are often revealed only at magnifications $\approx \times 1000$ or higher. Since the use of electron microscopes is somewhat specialised, the principles of their operation will be discussed briefly here.

The Transmission Electron Microscope (TEM)

The transmission electron microscope came into routine use in the 1950s. Broadly speaking, it can be compared to a photographic enlarger, using an electron beam and

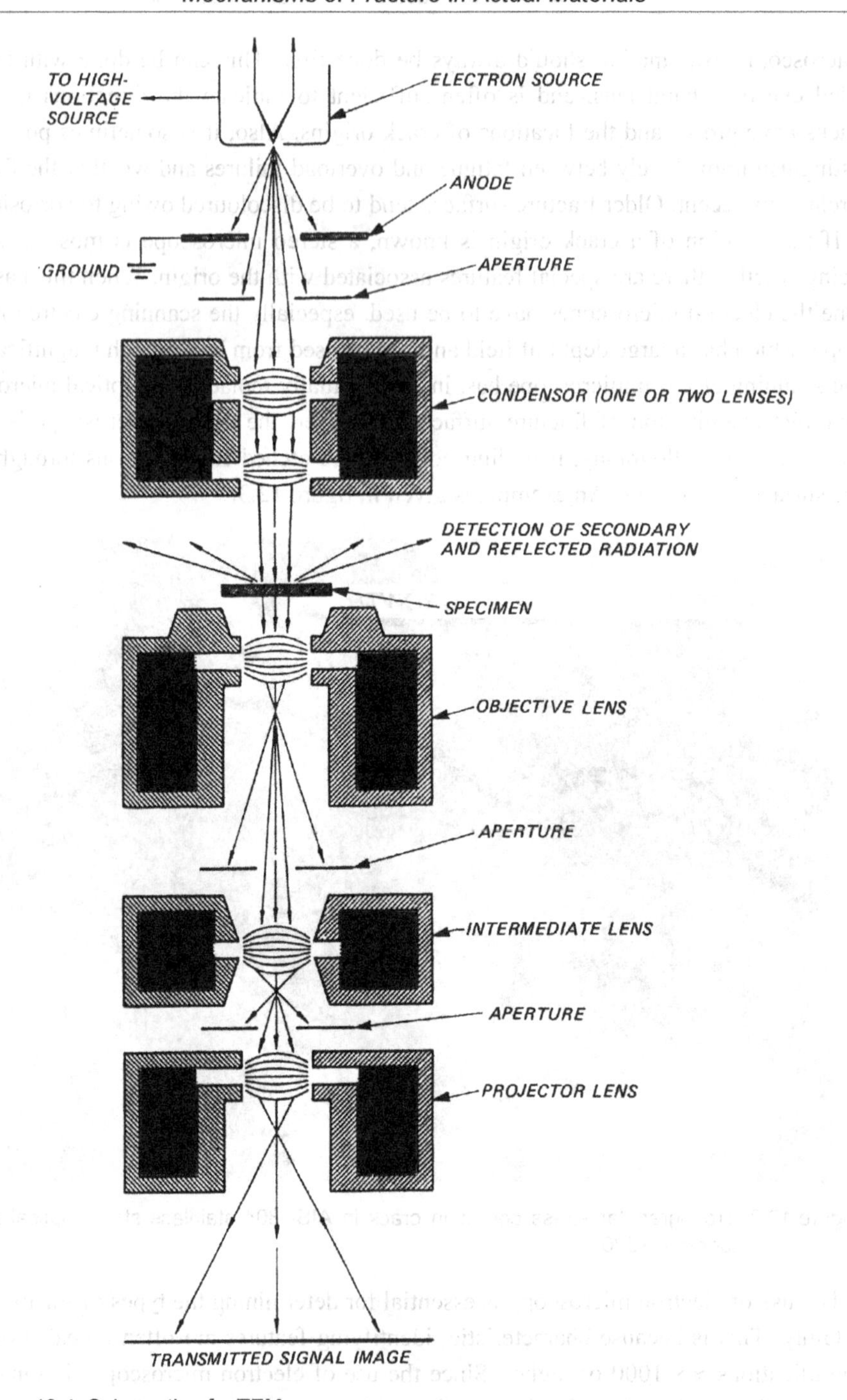

Figure 12.4. Schematic of a TEM.

electromagnetic lenses instead of light and optical lenses.

Figure 12.4 gives a schematic of a TEM. Note that the electron beam passes through the specimen in the microscope (like light through the negative in an enlarger). Electrons can pass through a material thickness of only a few tens of nanometres (1 nm =

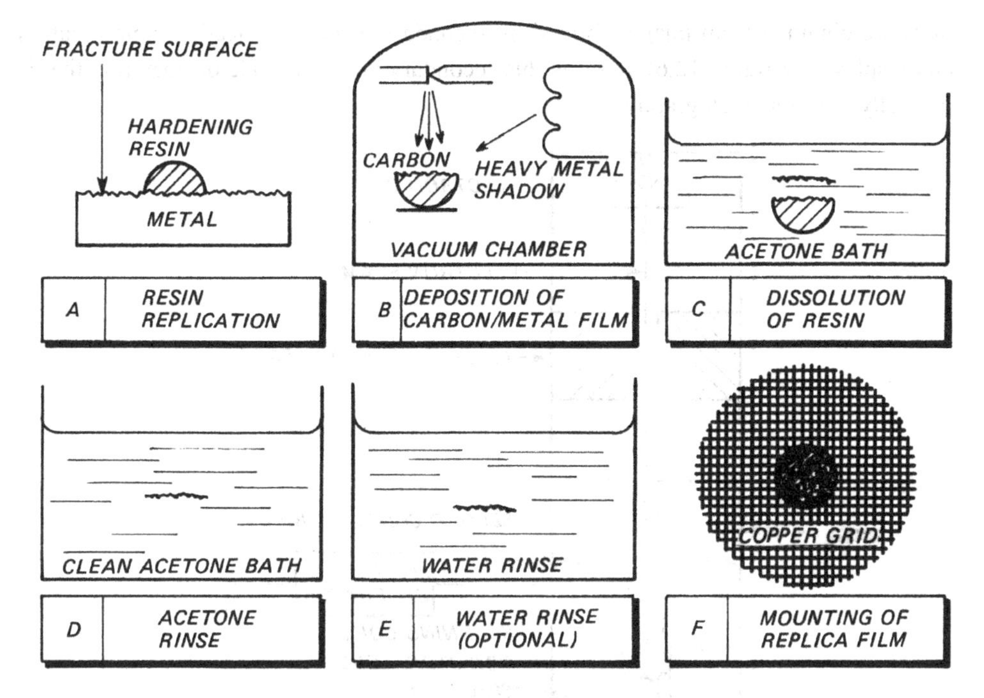

Figure 12.5. Schematic of preparation of a fracture surface replica.

10^{-9} m). This means that the specimen must be very thin. For fractography it is therefore necessary to make a thin film-type replicate specimen of the actual fracture surface. A technique for preparing fracture surface replicas is illustrated in figure 12.5. The relatively thick copper grid is required to support the replica and enable its insertion in the TEM.

The advantages of the TEM for fractography are that fracture surface details can be studied at very high magnifications (see figure 12.2) with good resolution and high contrast. Disadvantages include the time and skill required to prepare good replicas, the obscuring of about 50% of the replica by the supporting copper grid, and the fact that the replicas cannot be larger than a few millimetres in any direction.

The Scanning Electron Microscope (SEM)

The scanning electron microscope is a later development than the TEM and came into routine use in the late 1960s. The operating principle of a SEM is completely different from that of a TEM.

Figure 12.6 gives a schematic of a SEM. The electron beam is highly focused by two condenser lenses and impacts the specimen as a small spot. The impact results in the generation of various forms of radiation, figure 12.7, of which the backscattered and secondary electrons are important for image forming.

The backscattered and secondary electrons come from different zones near the surface of the specimen, as shown in figure 12.8. The secondary electron emission zone is much smaller than the zone for backscattered electrons. Consequently, higher resolutions are obtained from images formed by secondary electron collection, amplification and display (see figure 12.6) and it is this secondary electron mode of operation that is normally used for fractography.

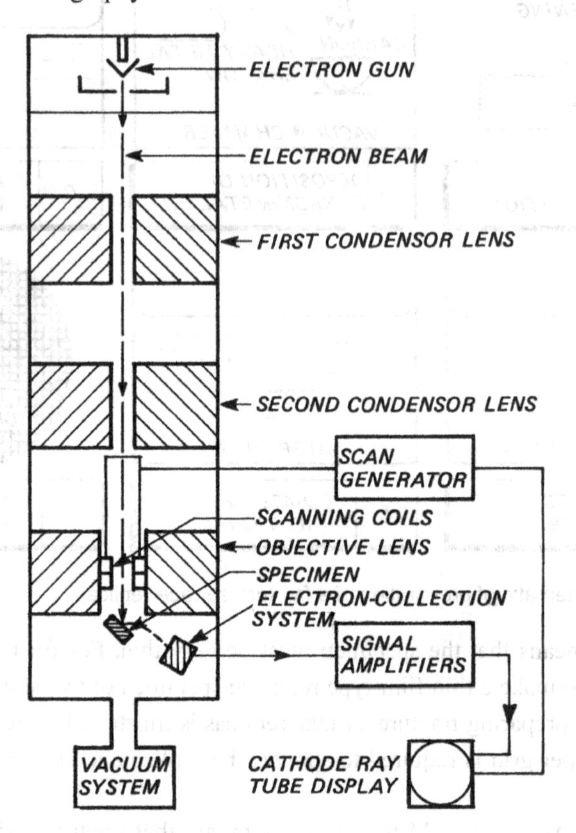

Figure 12.6. Schematic of a SEM.

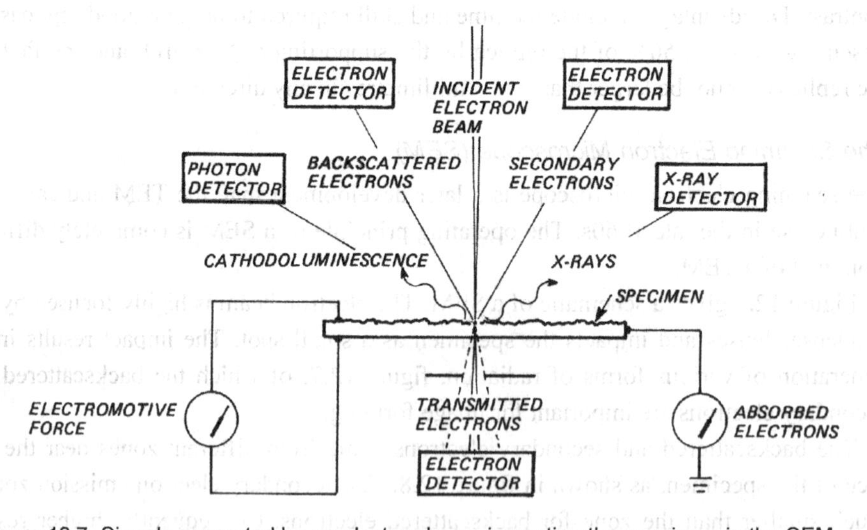

Figure 12.7. Signals generated by an electron beam impacting the specimen in the SEM.

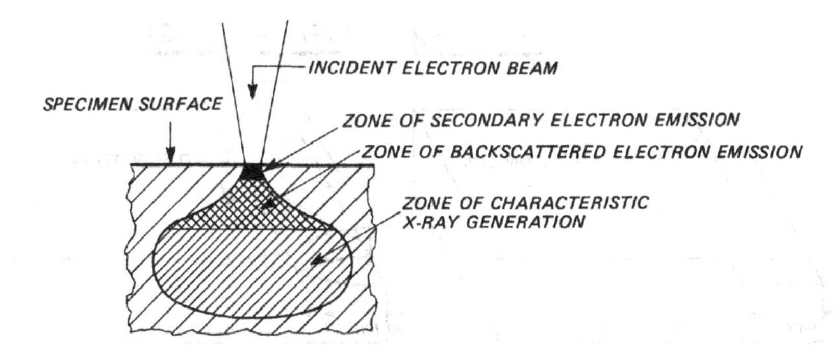

Figure 12.8. Zones in a specimen that are sources for signals generated by an electron beam impacting the specimen.

Actual operation requires the electron beam to scan rapidly back and forth in a systematic manner over the specimen surface. This is achieved using scanning coils in an objective lens, see figure 12.6. The scanning coils are controlled by a raster scan generator, which simultaneously controls the deflection coils of the cathode ray tube display. In this way the signal from the electron collector is displayed in the same pattern as it was generated by the electron beam impacting the specimen.

The relative intensities of the generated secondary electrons form an image of the specimen. Magnification in the SEM is changed by keeping the display size constant and changing the area of specimen scanned. Typical magnifications for fractographic studies range from × 10 to × 20,000, see figure 12.2.

The main advantages of using a SEM for fractography are that the specimen is observed directly and the image has a realistic three-dimensional appearance. Also, modern microscopes can accommodate fairly large specimens with maximum dimensions 100 mm. These advantages have made the SEM the preferred instrument for fractography, despite the better resolution of the TEM. However, both instruments should be used for fracture surface examination, especially when service failures are involved, in order to maximise the chance of identifying characteristic features of different types of fracture.

12.3 Slip, Plastic Deformation and Dislocations

The most common form of plastic deformation is slip. This is the shearing of whole blocks of crystal over one another, see figure 12.9. Thus plastic deformation is intrinsi-

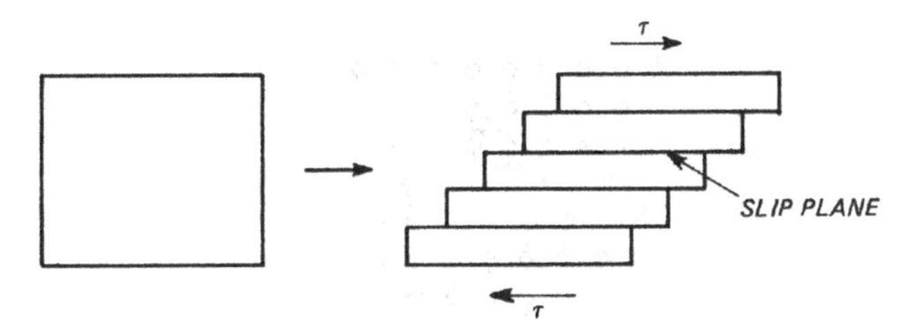

Figure 12.9. Schematic of plastic deformation (shear) by slip.

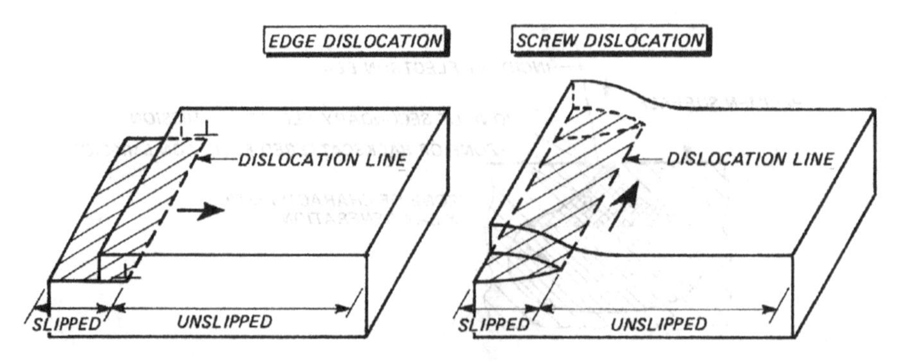

Figure 12.10. Slip owing to movement of pure edge and pure screw dislocations.

cally inhomogeneous. Slip always occurs by displacement of blocks of crystal in specific crystallographic directions on particular sets of crystal planes called slip planes. When slip occurs on several sets of crystal planes in each grain of a polycrystalline material the overall result is plastic deformation that is effectively homogeneous. This is why fracture mechanics concepts, which are based on continuum behaviour, can be used to describe many aspects of crack extension and fracture in structural materials.

The displacement of crystal blocks does not occur simultaneously over the entire slip plane. Instead it occurs consecutively, beginning in a very small region on the slip plane and spreading outwards. The boundary between the regions where slip has and where slip has not occurred is called a dislocation and is commonly represented as a line in the slip plane. The two basic types of dislocation and slip displacements associated with them are shown in figure 12.10. When the displacement is perpendicular to the dislocation line it is called an edge dislocation. If the displacement is parallel to the dislocation line it is of the screw type. In practice, most dislocation lines are neither pure edge nor pure screw but are mixtures of the two.

Note the symbol ⊥ is used for an edge dislocation. This symbol and its inverse represent the fact that an edge dislocation can be depicted as an extra half plane of atoms that lies above (⊥) or below (⊤) the slip plane and moves parallel to it. Figure 12.11 is an idealised atomistic picture of an edge dislocation with the extra half plane of atoms above the slip plane. The displacement b is called the Burgers vector. It is constant along the dislocation line, both for edge and screw dislocations. From figure 12.10 it follows that b is perpendicular to an edge dislocation line and parallel to a screw dislocation line.

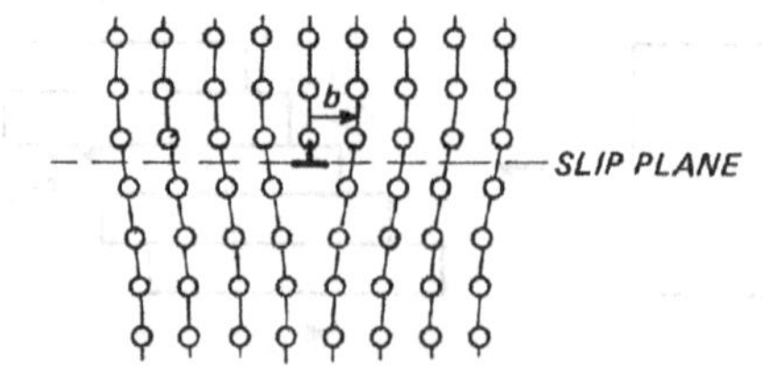

Figure 12.11. Idealised representation of an edge dislocation.

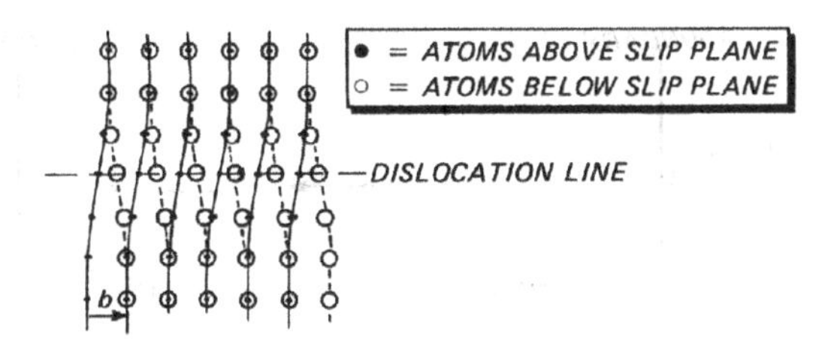

Figure 12.12. A screw dislocation in a simple cubic lattice. After reference 1.

A screw dislocation does not have an extra half plane of atoms associated with it. The representation of screw dislocations is more difficult than that for an edge dislocation: a simplified example is shown in figure 12.12.

As mentioned earlier, most dislocation lines are neither pure edge nor pure screw. Furthermore, the boundary between slipped and unslipped regions often takes the form of an enclosed loop, which must be partly edge and partly screw. Such loops are illustrated in idealised form in figure 12.13. These loops are identical, and show that the relation between a dislocation and its Burgers vector is not entirely unique. The Burgers vector is ambiguous in sign unless the slipped region is specified. Also note that the screw and edge components of each loop are of opposite sign. This is because b is everywhere the same for each loop, while the screw and edge components move in opposite directions to each other. For example, the material at A is compressed (extra half plane of atoms above the slip plane, $\perp$) but at B the material is extended (missing a half plane of atoms above the slip plane, which is equivalent to an extra half plane of atoms below the slip plane, $\top$).

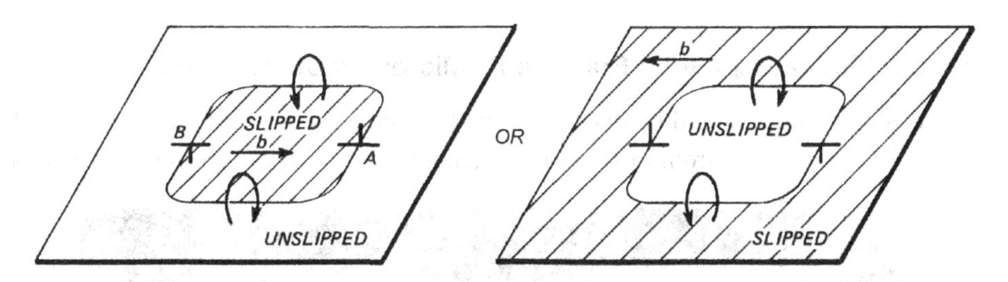

Figure 12.13. Identical dislocation lines (loops) bounding slipped and unslipped areas.

There are stress fields associated with dislocations, and they behave as if subject to a line tension. This tension contracts a dislocation line and may collapse a loop unless there is an applied stress large enough to keep it open or even expand it, thereby causing slip.

The stress required to move a dislocation depends on a number of factors, including its character (edge and screw components), its detailed configuration in the crystal lattice, the presence of other dislocations, each with its own stress field, and barriers such as intermetallic particles and precipitates which do not fit into the crystal lattice. In polycrystalline materials grain boundaries are major barriers to dislocation movement. This is because the slip directions in neighbouring grains are generally different, and

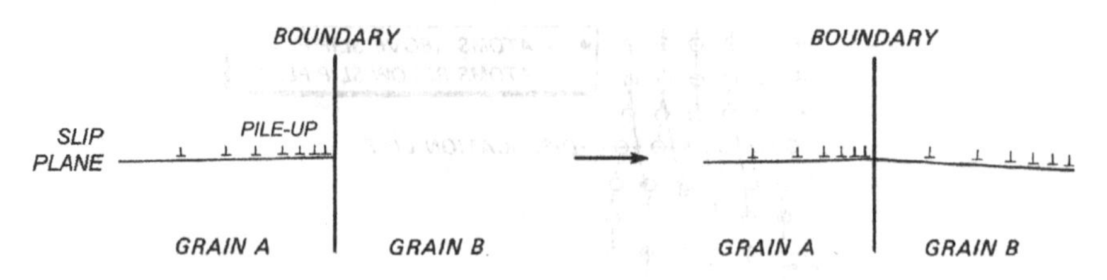

Figure 12.14. Spreading of slip from one grain to another.

then a dislocation cannot pass from one grain to another without changing its Burgers vector, which is a high energy process.

Because there are barriers to dislocation movement, the spreading of slip within a grain and from grain to grain requires either that the dislocations cut through or bypass the barriers, or else that other dislocations are activated. In fact, both processes can occur within a grain, but it is the activation of other dislocations that enables slip to spread from grain to grain. The way in which this occurs is shown in figure 12.14. Dislocations nucleated by some distant source (usually another dislocation) in a slip plane of grain A are obstructed by the grain boundary, resulting in a dislocation pile-up. The pile-up pushes the lead dislocation hard against the boundary, such that there is a high stress concentration. Eventually the concentration of stress is sufficient to activate a dislocation source for slip in grain B, which plastically deforms to alleviate the stress concentration.

The foregoing discussion is a very brief introduction to the concept of dislocations. For readers interested in a more extensive treatment there are several excellent books, including the classic by Cottrell, reference 1 of the bibliography to this chapter.

12.4 Ductile Transgranular Fracture by Microvoid Coalescence

Ductile fracture is caused by overload and, depending on the constraint, can often be recognised immediately from macroscopic examination of a failed specimen or compo-

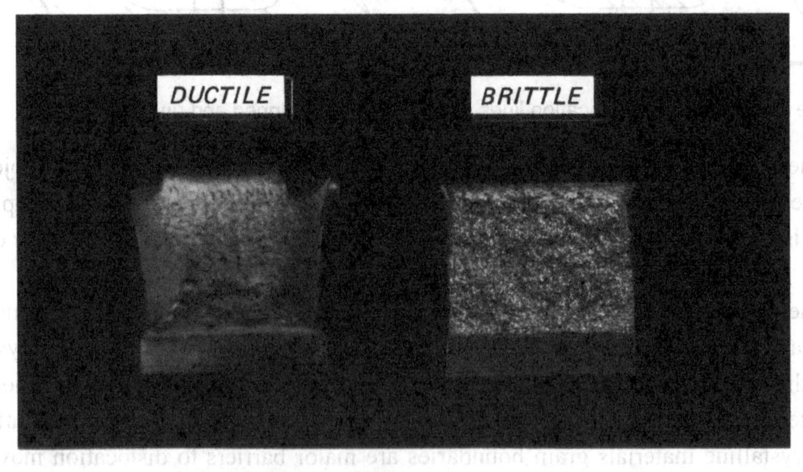

Figure 12.15. Ductile and brittle fractures of impact test specimens.

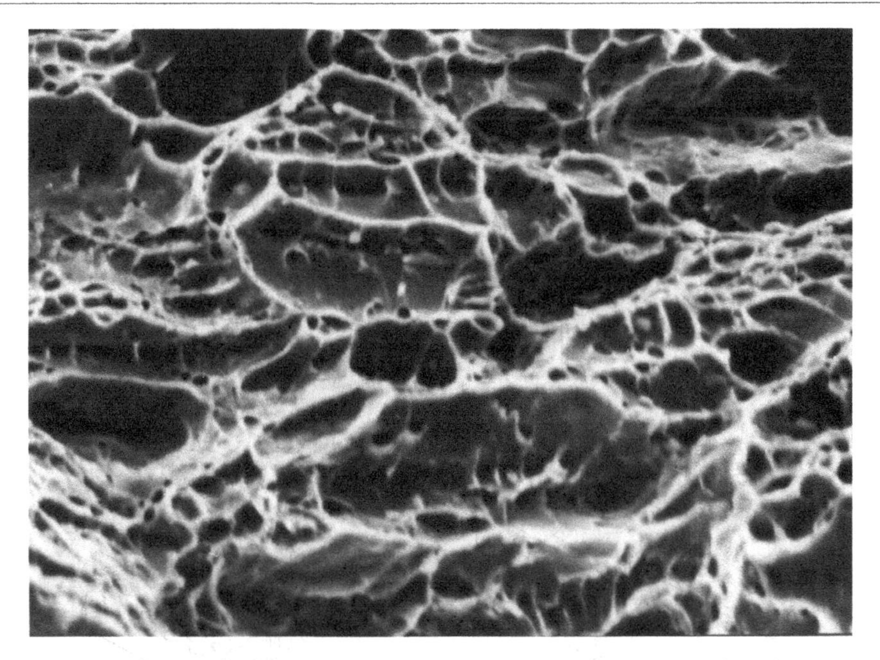

Figure 12.16. Microvoid coalescence (dimpled rupture) in a structural steel. SEM fractograph, × 1600.

nent. If there is very little constraint there will be a significant amount of contraction before failure occurs. Figure 12.15 shows two fractured specimens after Charpy Impact testing. One failed in a ductile manner, the other was brittle. The difference is easily observed owing to the lack of constraint in this type of specimen.

However, when there is high constraint (*e.g.* thick sections) a ductile fracture may occur without noticeable contraction. In such cases the only macroscopic difference is the reflectivity of the fracture surface, which tends to be dull for a ductile fracture and shiny and faceted for a brittle fracture.

On a microscopic scale most structural materials fail by a process known as microvoid coalescence, which results in a dimpled appearance on the fracture surface. An example is given in figure 12.16, which shows both small and large dimples.

Dimple shape is strongly influenced by the type of loading. This is illustrated in figure 12.17. Fracture under local uniaxial tensile loading usually results in formation of equiaxed dimples. Failures caused by shear will produce elongated or parabolic shaped dimples that point in opposite directions on matching fracture surfaces. And tensile tearing produces elongated dimples that point in the same direction on matching fracture surfaces.

The microvoids that form dimples nucleate at various internal discontinuities, the most important of which are intermetallic particles and precipitates and grain boundaries. As the local stress increases the microvoids grow, coalesce and eventually form a continuous fracture surface. A schematic of how a crack extends by microvoid formation at particles is given in figure 12.18. Note that voids can initiate both at matrix/particle interfaces and as a result of particle fracture. Also, large particles are nucleation sites for large voids, and small particles result in small voids.

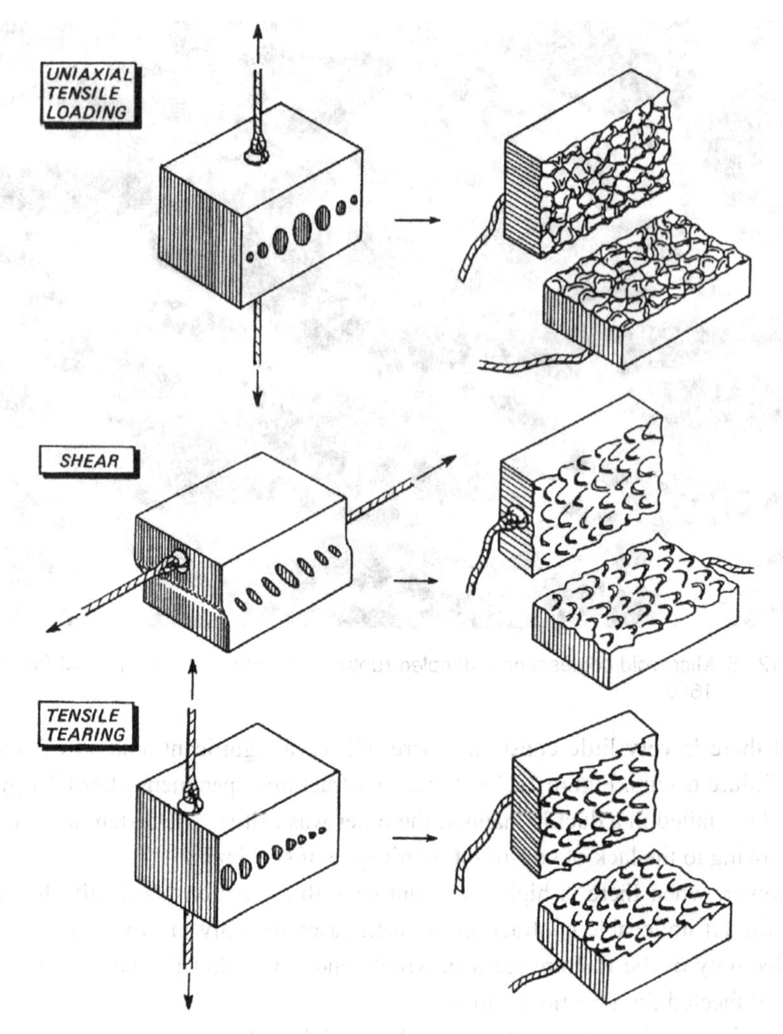

Figure 12.17. Dimple formation owing to uniaxial tensile loading, shear and tensile tearing.

There are several dislocation models for void nucleation and growth. A model by Broek (reference 2) is illustrated in figure 12.19. The first step is the generation of dislocation loops around particles. The edge and screw components of such loops are of opposite sign. This is obvious for the screw components, which surround the particle by moving in opposite directions. However, it is less obvious for the edge components. The reasoning is as follows. Being of opposite sign the screw components will attract each other, and it is possible for them to link up and reform the original edge dislocation (⊥, extra half plane of atoms above the slip plane) if a segment of edge dislocation of opposite sign (⊤, extra half plane of atoms below the slip plane) is left to complete the loop.

As the number of piled up loops increases, the leading loops are pushed to the matrix/particle interface and a void is nucleated. Since there are actually many pile-ups on different slip planes, once a void is nucleated it can grow by assimilating dislocations from these pile-ups.

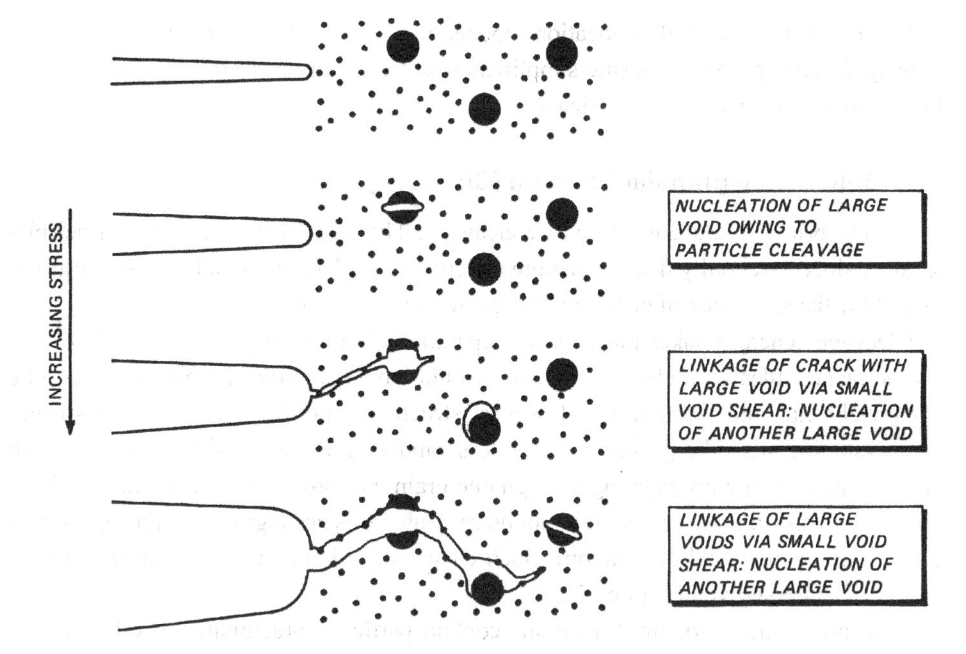

Figure 12.18. Schematic of crack extension by transgranular microvoid coalescence.

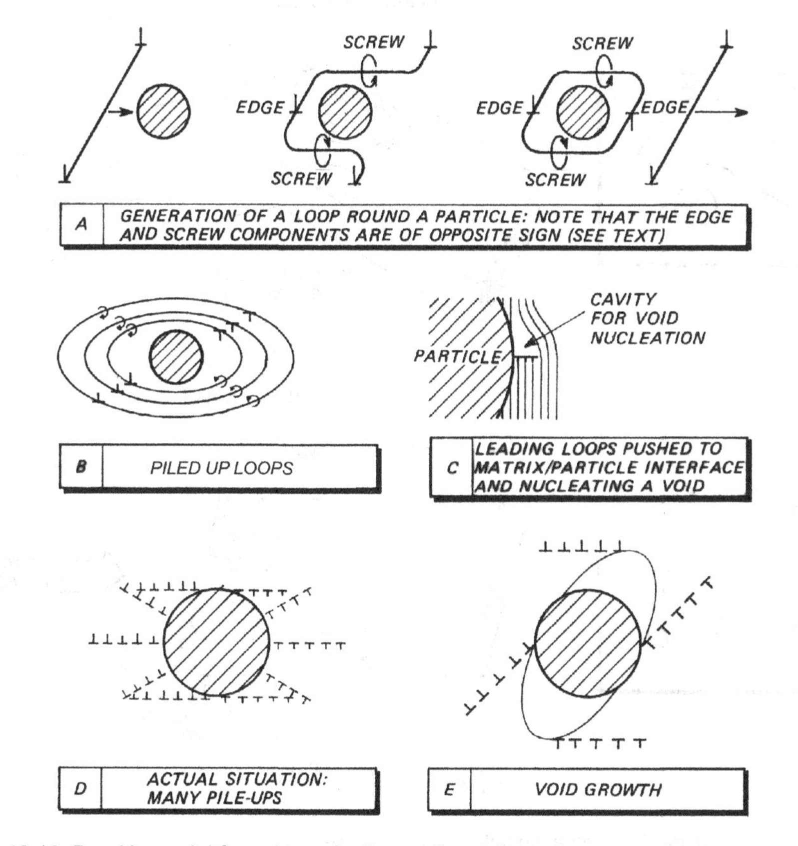

Figure 12.19. Broek's model for void nucleation and growth.

Quantitative analysis of dislocation models of void nucleation and growth is an extremely difficult problem. Drastic simplifications are usually necessary, so that the usefulness of the models is very restricted.

12.5 Brittle Transgranular Fracture (Cleavage)

A truly brittle fracture is caused by cleavage. The term brittle fracture can be misleading, since essentially ductile fracture (microvoid coalescence) under high constraint may show the same lack of contraction expected for cleavage.

Cleavage generally takes place by the separation of atomic bonds along well-defined crystal planes. Ideally, a cleavage fracture would have perfectly matching faces and be completely flat and featureless. However, structural materials are characteristically polycrystalline with the grains more or less randomly oriented with respect to each other. Thus cleavage propagating through one grain will probably have to change direction as it crosses a grain or subgrain boundary (subgrains are regions within a grain that differ slightly in crystal orientation). Such changes in direction resulted in the faceted fracture surface shown in figure 12.15.

In addition, most structural materials contain particles, precipitates or other imper-

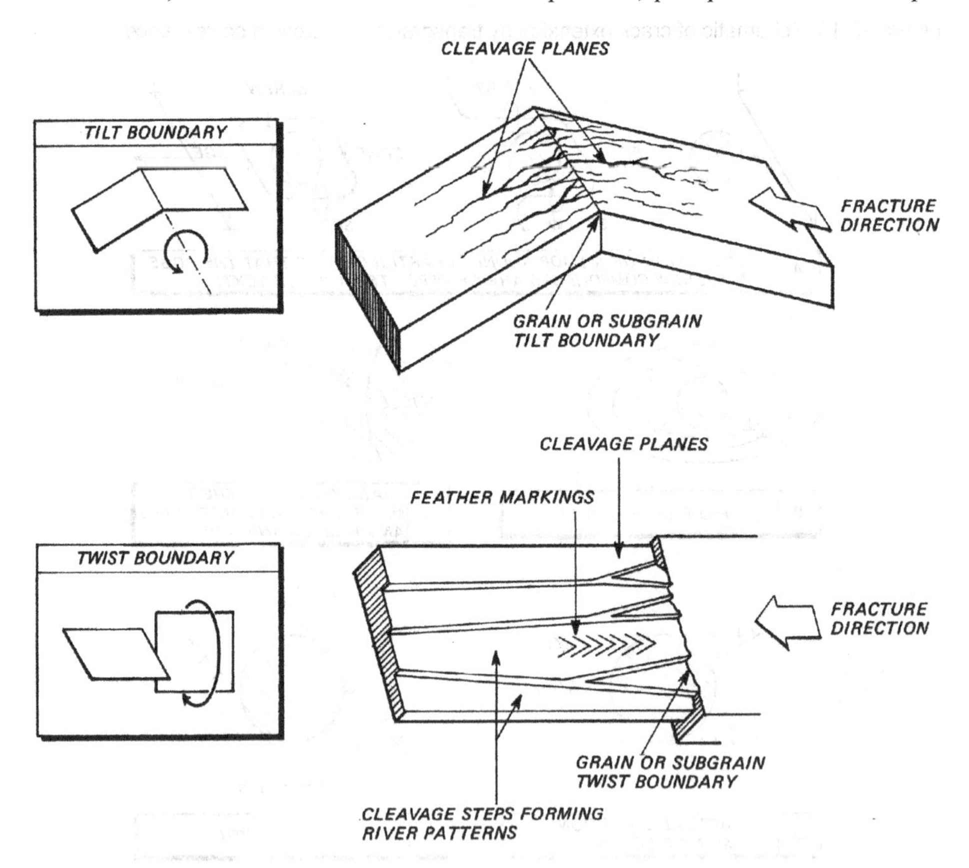

Figure 12.20. Typical features associated with cleavage.

fections that further complicate the fracture path, so that truly featureless cleavage is rare, even within a single grain or subgrain. The changes of orientation between grains and subgrains and the various imperfections produce markings on the fracture surface that are characteristically associated with cleavage.

Figure 12.20 illustrates some typical features associated with cleavage. A principal feature is river patterns, which are steps between cleavage on parallel planes. River patterns always converge in the direction of local crack propagation. If the grains or subgrains are connected by a tilt boundary, which means that they are misoriented about a common axis, the river patterns are continuous across the boundary. But if adjacent grains or subgrains are axially misoriented, *i.e.* they are connected by a twist boundary, the river patterns do not cross the boundary but originate at it. Besides river patterns a distinct feature of cleavage is feather markings. The apex of these fan-like markings points back to the fracture origin, and therefore this feature can also be used to determine the local direction of crack propagation.

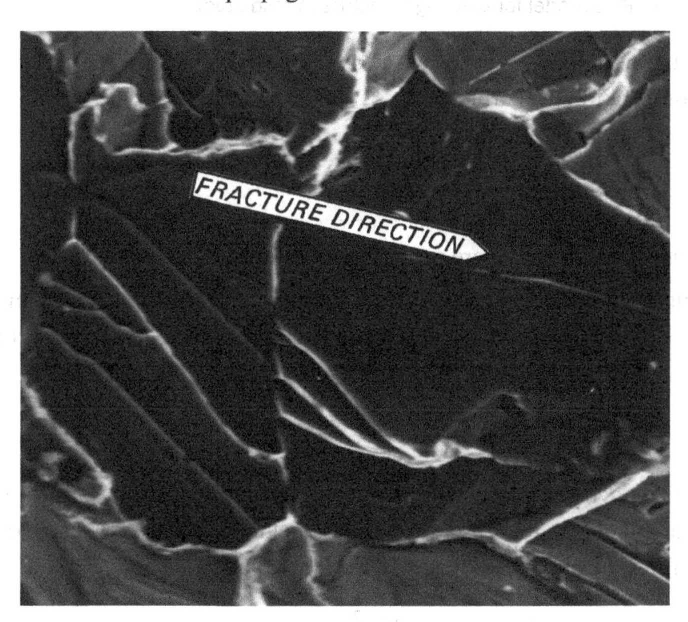

Figure 12.21. Cleavage in BS4360 structural steel. SEM fractograph, × 900.

Cleavage occurs in a number of materials, but it is especially important because it occurs in structural steels. An example is shown in figure 12.21.

Knott's book on fracture mechanics (reference 3) reviews dislocation models of cleavage with particular attention to steels, including a model by Smith (reference 4) that incorporates the important microstructural feature of grain boundary carbides, which are known to greatly influence the fracture toughness of steels.

Smith's model is shown in figure 12.22. As a consequence of a tensile stress a brittle intergranular carbide is subjected to a concentrated shear stress ahead of a dislocation pile-up of length d, the average grain size. The effective shear stress, τ_{eff}, is the maximum stress that can be attained before yielding occurs, *i.e.* before the stress concentra-

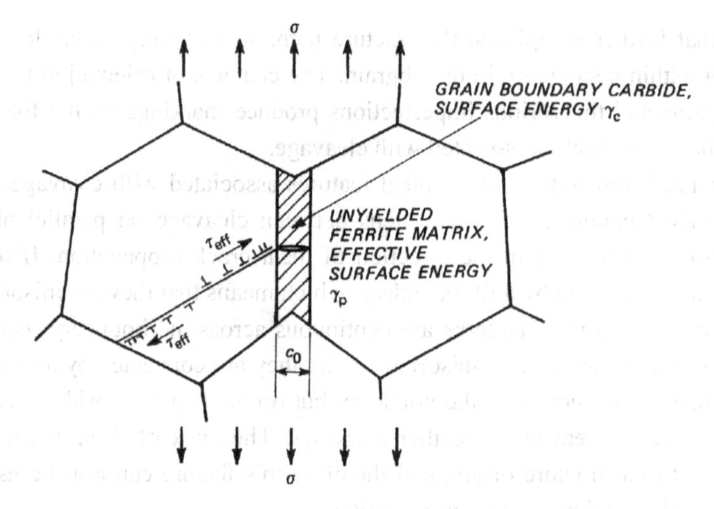

Figure 12.22. Smith's model for cleavage fracture in mild steel.

tion due to the pile-up is relieved by the spreading of slip to a neighbouring grain. The condition for cracking the carbide is

$$\tau_{eff} \geq \sqrt{\frac{4E\gamma_c}{(1-\nu^2)\pi d}}. \tag{12.1}$$

Once the carbide is cracked there are two possibilities:

1) τ_{eff} is high enough to propagate the carbide crack into the ferrite matrix, that is: cleavage of the matrix is nucleation controlled. For this to happen the condition is

$$\tau_{eff} \geq \sqrt{\frac{4E\gamma_p}{(1-\nu^2)\pi d}}. \tag{12.2}$$

2) τ_{eff} at yielding lies between the limits set by equations (12.1) and (12.2). This is much more likely. In this case the carbide thickness, c_o, gives the size of initial crack in a Griffith-type energy balance criterion for crack extension:

$$\sigma_f^2\left(\frac{c_o}{d}\right) + \tau_{eff}^2\left(1 + \frac{4}{\pi}\frac{\tau_i}{\tau_{eff}}\sqrt{\frac{c_o}{d}}\right)^2 \geq \frac{4E\gamma_p}{(1-\nu^2)\pi d}, \tag{12.3}$$

where σ_f is the critical stress for cleavage fracture and τ_i is a so-called friction stress, which includes a number of factors that cause the crystal lattice to resist dislocation movement.

Equation (12.3) shows that larger carbides should result in a lower cleavage fracture stress, as is observed. The model has been tested quantitatively, and gives good predictions of σ_f for steels with large grain sizes. However, it is difficult to check the model for fine grained steels, since fine grains are invariably associated with thin carbides, *i.e.* carbide size variation is not possible.

12.6 Transgranular Fracture by Fatigue

Fatigue results in very distinctive fracture appearances. In general there are three different features that can be observed:

- the location of fatigue initiation
- the fracture surface resulting from fatigue crack growth
- the final fracture due to overload.

Each of these features can give specific information about the fatigue process in a structural component.

Fatigue initiation nearly always occurs at an external surface, though there are important exceptions like jet engine compressor discs where internal fatigue origins have been found. Careful examination of a fatigue initiation site may reveal the cause of fatigue, for example stress concentrations due to a notch, machining grooves or corrosion pitting. This kind of information is of primary importance for analysis of service failures but is less relevant to fracture mechanics studies, although in recent years the problem of short crack growth at notches (see section 9.8) has received widespread attention.

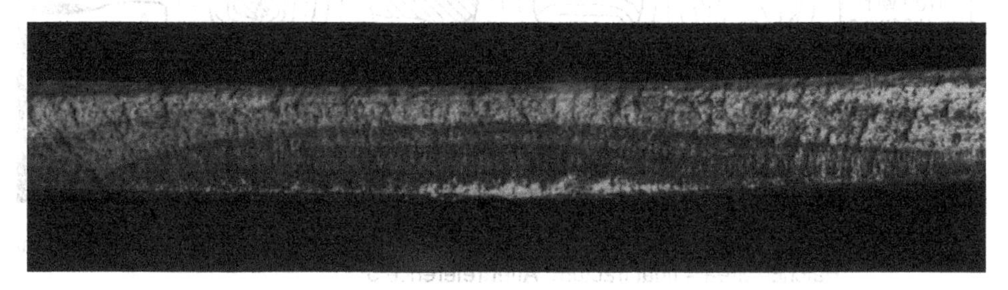

Figure 12.23. Beach markings on a fatigue fracture surface in a thin walled pipe. Optical fractograph, × 5.

Fatigue fracture surfaces tend to be macroscopically flat and smooth, and will often show 'beach markings', which occur owing to variations in the load history. An example is given in figure 12.23.

Beach markings can be very useful, since they give information about the shapes of the fatigue crack front at various stages of growth, and these shapes are diagnostic for the type of fatigue loading. Some typical beach markings for different types of loading imposed on a cylindrical bar are indicated schematically in figure 12.24. Many more examples are presented in reference 5 for both cylindrical and rectangular cross-sections.

The area of final fracture (shown hatched in figure 12.24) gives an indication of the magnitude of the loads. A large final fracture area indicates that K_{Ic} or K_c was exceeded at a relatively short crack length, which means that either the maximum load was high or the fracture toughness low, or both. Knowledge of a material's fracture toughness removes the ambiguity, and providing the crack geometry is not too complex a fair estimation of the maximum stress at failure can be obtained from an approximate expression for the stress intensity factor.

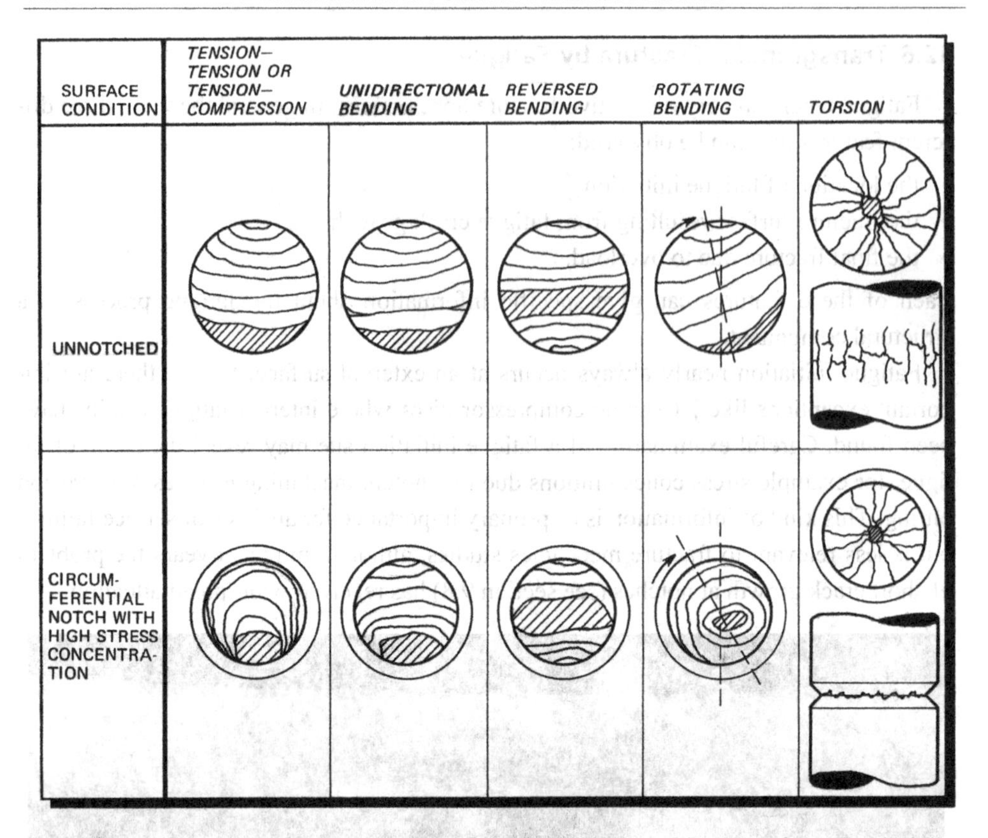

Figure 12.24. Examples of beach markings for different types of loading of a cylindrical bar. Hatched area = final fracture. After reference 5.

On a microscopic scale the most characteristic features of fatigue are the striations that occur during region II (continuum mode) crack growth, see figure 9.4. The striations represent successive positions of the crack front. Each striation is formed during one load cycle but, especially under variable amplitude loading, not every load cycle need result in a striation.

Figure 12.25 shows two different examples of fatigue striations. Aluminium alloys generally give well-defined regular striations, but steels do not. Besides the influence of type of material, the environment also has a strong effect on striation appearance. Note that striations are perpendicular to the local direction of crack growth. This is sometimes helpful in tracing crack growth backwards in order to determine the exact location of fatigue initiation.

Since each striation is formed during one load cycle the spacing between striations is an indication of local crack growth rates, particularly if it is evident from their regularity that they have been formed by constant amplitude loading (typical examples are propellers, helicopter rotor blades and rotating components in power generating equipment). These local crack growth rates can be used to determine the fatigue stresses provided the following procedure is possible:

2024-T3 ALUMINIUM ALLOY

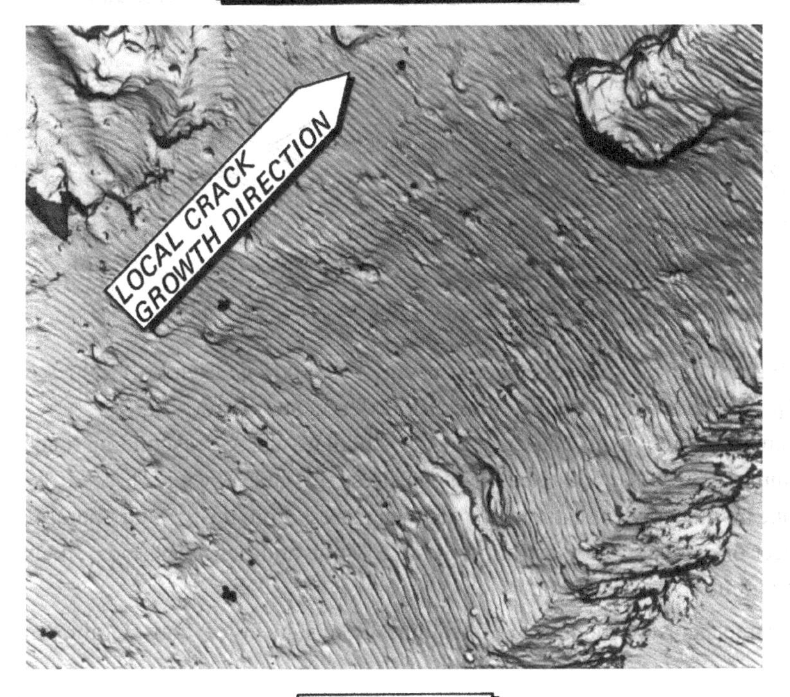

ST E460 STEEL

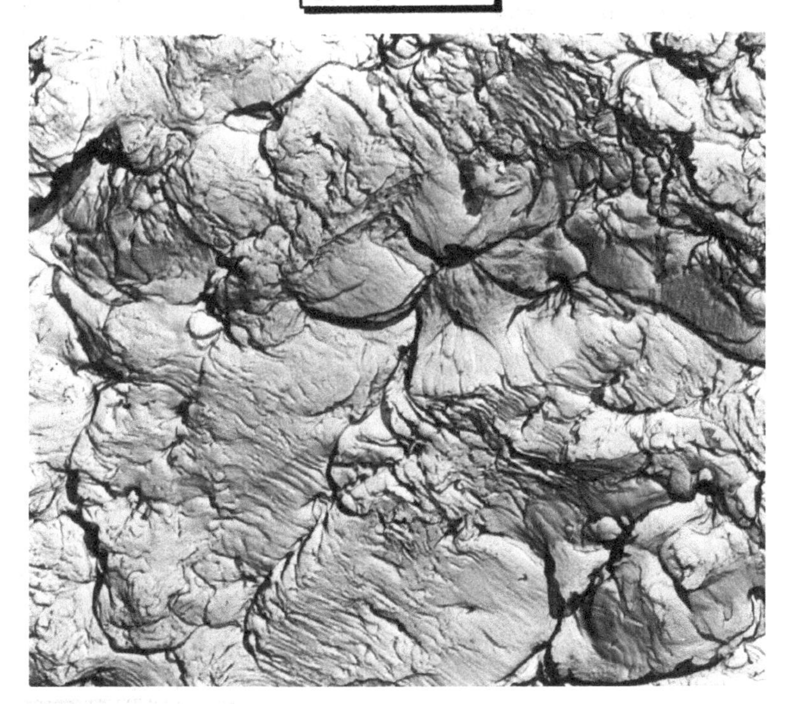

Figure 12.25. Fatigue striations for an aluminium alloy and a structural steel tested in normal air. TEM replica fractographs, $\times$ 4000.

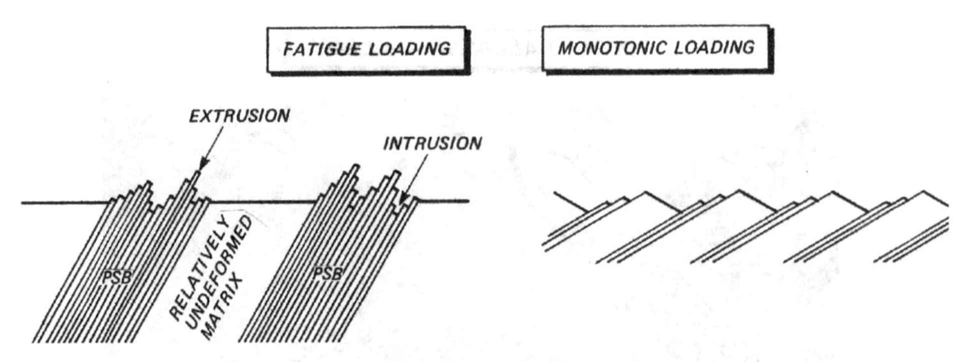

Figure 12.26. Illustration of slip during fatigue and monotonic loading.

1) Assume the local striation spacing is equal to da/dn and read off values of ΔK from da/dn–ΔK crack growth rate curves for different stress ratios, R.
2) From the crack length and geometry calculate the corresponding maximum and minimum stresses, and hence the mean stress, for each R value.
3) Make an independent estimate of the mean stress from design and operating data.
4) Select or estimate the most appropriate R value and hence the most likely values of the fatigue stresses.

Micromechanistic modelling of fatigue is a subject of considerable interest and

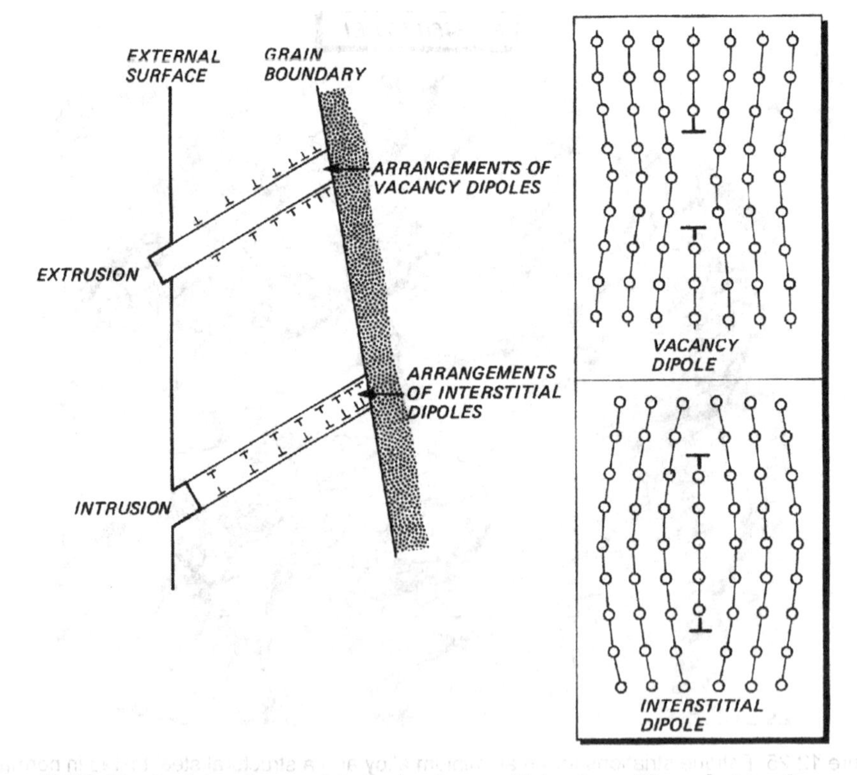

Figure 12.27. Simple dislocation model of extrusions and intrusions. After reference 6.

speculation, especially concerning fatigue striations. There does, however, appear to be a fairly consistent picture of fatigue crack initiation at an external surface.

Under fatigue loading the surface material tends to deform by cyclic slip concentrated in so-called persistent slip bands (PSBs) with irregular notched profiles consisting of extrusions and intrusions. As figure 12.26 shows, this slip distribution is quite different from that produced by monotonic loading.

Continuing cyclic slip leads to deepening of the intrusions and eventually the formation of a crack along the slip plane. This slip plane cracking may extend a few grain diameters into the material, but then changes to a continuum mechanism of crack propagation.

A simple dislocation model for the formation of extrusions and intrusions is shown in figure 12.27. The dislocations pile up at a grain boundary and tend to form pairs (dipoles) owing to interaction of their stress fields. An arrangement of vacancy dipoles means that the material between them has fewer half planes of atoms than the matrix: these half planes have been transported beyond the surface to form an extrusion. On the other hand, an arrangement of interstitial dipoles means that the material between them has more half planes of atoms than the matrix, and this results in an intrusion.

The dislocation arrangements within a PSB are actually much more complicated than those assumed by the model in figure 12.27. Nevertheless, analysis of the model predicted the form of observed relations between fatigue initiation life and cyclic plastic strain and between grain size and fatigue strength in some materials, reference 6.

Various models of fatigue striation formation have been proposed. Most consider only plastic flow at the crack tip and ignore the potential contribution of the environment. One such model, the plastic blunting process, is shown in figure 12.28. During uploading shear deformation concentrates at first at 'ears' on either side of the crack tip. Later the crack tip itself advances and blunts. During unloading the shear deformations

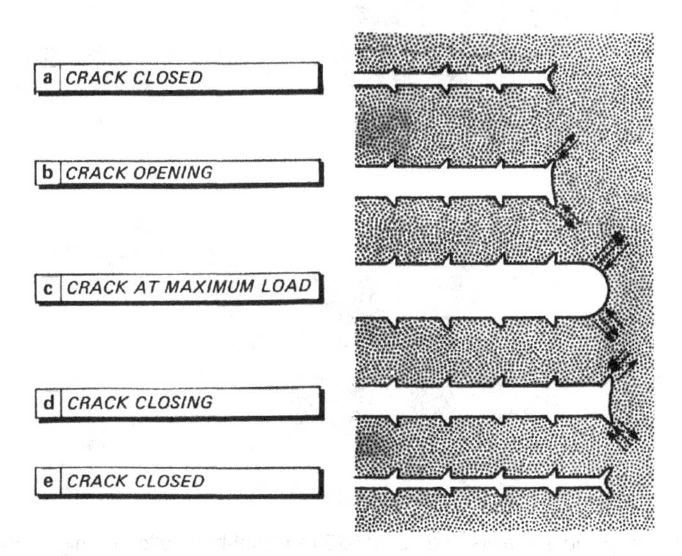

Figure 12.28. The plastic blunting model of fatigue striation formation. After reference 7.

are such that a new pair of ears is formed at the crack tip, thereby producing the characteristic striation marking on the fracture surface. Why this ear formation should happen during unloading has never been explained. However, it does appear to occur in both inert and aggressive environments, reference 8.

12.7 Intergranular Fracture

Intergranular fractures are typically the result of sustained load fracture, discussed in section 12.8, or a lack of ductility in the material owing to segregation of embrittling elements and particles and precipitates to the grain boundaries, for instance in temper-embrittled steels and overaged Al-Zn-Mg-Cu aluminium alloys.

It is not possible to distinguish macroscopically between intergranular fracture and brittle transgranular fracture: both appear faceted. However, metallographic cross-sections through fracture surfaces and cracks will show whether the fracture path is intergranular, see for instance figure 12.29.

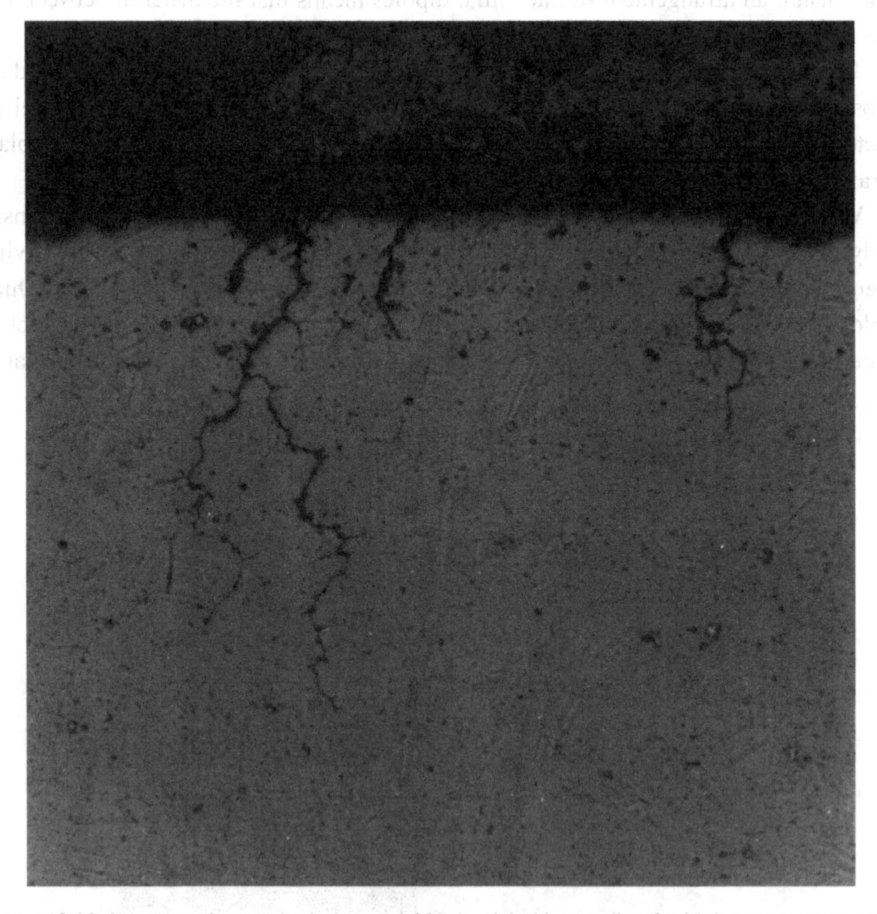

Figure 12.29. Intergranular cracks in Inconel 625 (a nickel-base alloy for high temperature use). Optical metallograph, × 320.

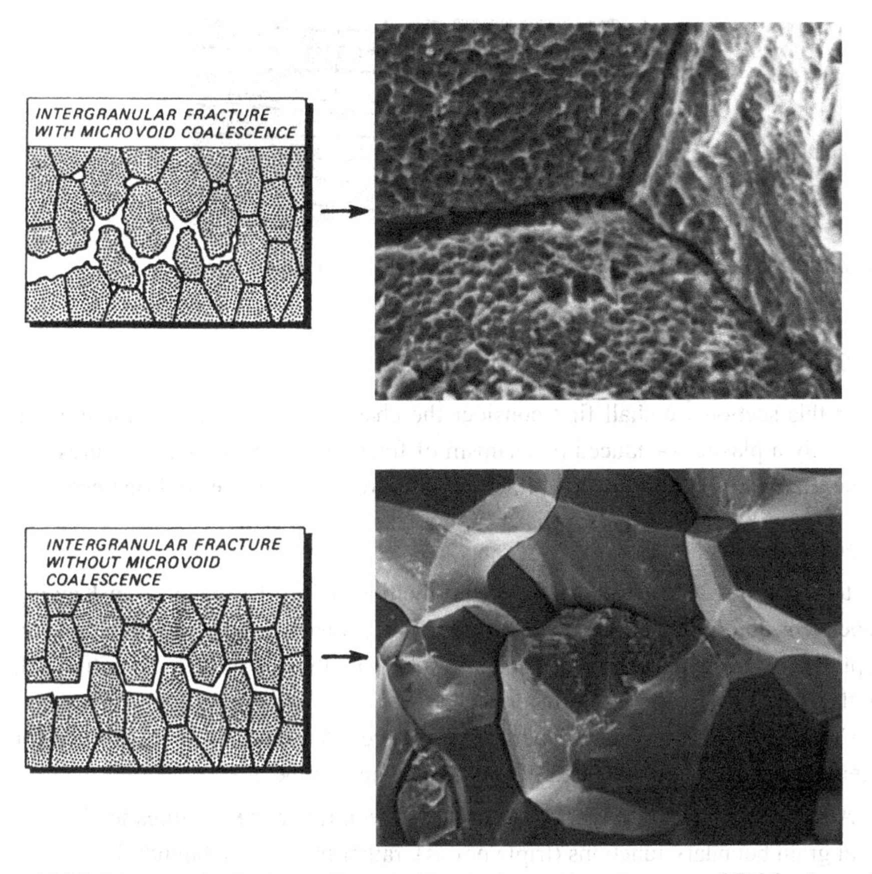

Figure 12.30. Intergranular fracture with and without microvoid coalescence. SEM fractographs, top × 1400, bottom × 500.

There are two main types of intergranular fracture appearance:

1) Grain boundary separation with microvoid coalescence. This type of intergranular fracture occurs during overload failure of some steels and aluminium alloys, and also other materials.
2) Grain boundary separation without microvoid coalescence. This type of intergranular fracture occurs during overload failure of temper-embrittled steels and refractory metals like tungsten, and also during sustained load fracture (creep, stress corrosion cracking, embrittlement by hydrogen and liquid metals).

Examples of both types are given in figure 12.30. The dimples on the grain boundary facets are the main distinguishing feature of intergranular fracture with microvoid coalescence.

Intergranular fractures are not always readily identifiable. Figure 12.31 shows schematically an intergranular fracture along flat elongated grains, which often occur in rolled sheet and plate materials as a consequence of mechanical working. This type of intergranular fracture exhibits few grain boundary junctions and is relatively featureless.

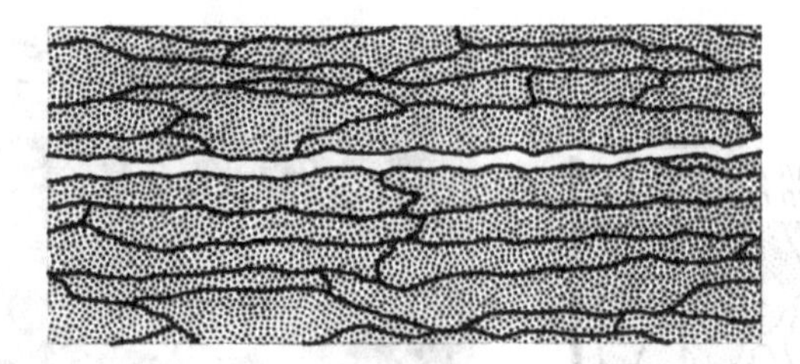

Figure 12.31. Schematic of intergranular fracture along elongated grains.

12.8 Types of Sustained Load Fracture

In this section we shall first consider the characteristics of creep fracture, which is primarily a plasticity-induced mechanism of fracture at elevated temperatures, and then some examples of sustained load fracture induced by aggressive environments.

Creep Fracture

At temperatures in excess of 0.5 T_m (the melting point of a material in Kelvin) time-dependent deformation and rupture (creep) is a primary design consideration. In many applications, such as gas turbines and steam boilers, the operating temperature is limited by the creep characteristics of materials.

Creep fractures in most commercial alloys are intergranular. There are two forms of intergranular separation, depending on the load and temperature:

1) At high loads and low temperatures in the creep range the fractures tend to originate at grain boundary junctions (triple points), rather than on the boundaries.
2) At lower loads and higher temperatures (the typical creep situation) fracture results

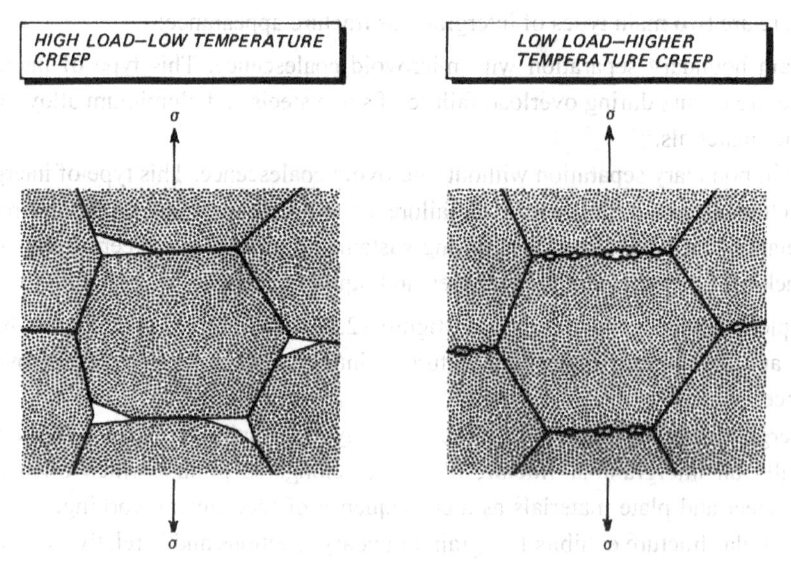

Figure 12.32. Schematic of the two main forms of creep fracture initiation.

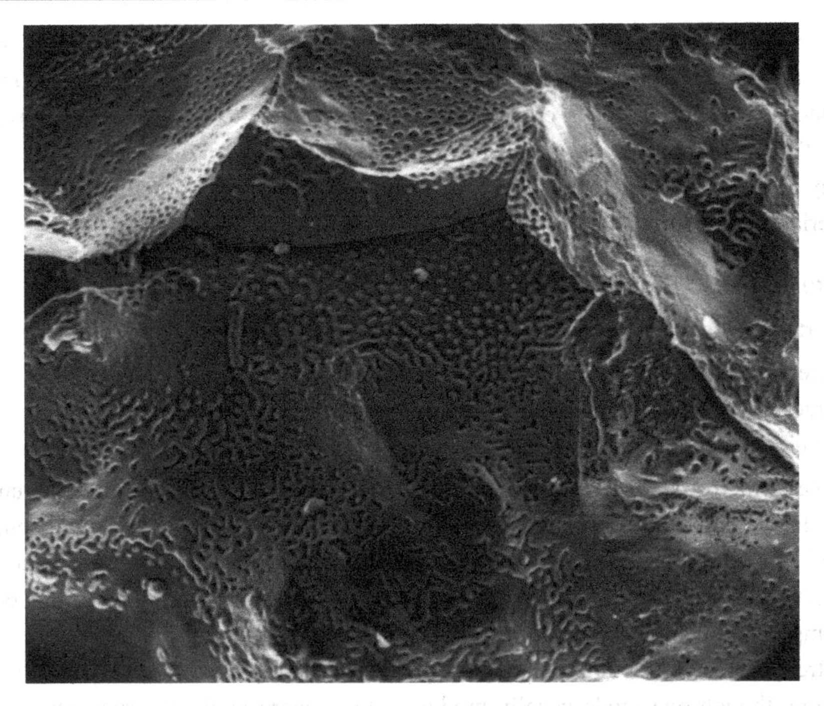

Figure 12.33. Creep fracture in Inconel W (a nickel-base alloy) showing distinctive patterns of void coalescence on grain boundary facets. SEM fractograph, × 400.

more from the formation of voids along grain boundaries, especially those boundaries perpendicular to the loading direction. This process is called cavitation.

The initiation of both types of creep fracture is illustrated schematically in figure 12.32. The fracture surface of a high load–low temperature creep fracture consists of intergranular facets without microvoid coalescence, and is similar in appearance to the lower fractograph in figure 12.30.

On the other hand, the fracture surface of a low load–higher temperature fracture will often exhibit voids on the grain boundary facets. These voids can be observed from a metallographic section, or even by optical microscopy of a replica from a polished surface. (Surface replicas are widely used to assess creep damage in large structures such as steam pipes in power generating plants.) The grain boundary voids can coalesce to resemble dimpled rupture, as in the upper fractograph in figure 12.30, but they can also coalesce in very distinctive patterns. An example of such patterns is shown in figure 12.33.

Nucleation of creep voids most probably occurs by a combination of grain boundary sliding, in which the grain boundary behaves like a slip plane, and stress-assisted diffusion and agglomeration of lattice vacancies. At triple points grain boundary sliding results in geometric incompatibilities (and hence stress concentrations) which can be accommodated by vacancy diffusion to nucleate voids. On the grain boundaries the voids are nucleated by vacancy diffusion, especially to matrix–particle interfaces, and grain boundary sliding can assist this process.

There is less certainty as to the controlling mechanism of void growth and coalescence to fracture, which is a very complicated process. A fairly recent attempt to resolve the inconsistencies in earlier creep fracture models is due to Edward and Ashby, reference 9. They proposed that void growth on cavitated grain boundaries occurs by vacancy diffusion at a rate controlled not by the stress but by deformation of uncavitated material surrounding the voids.

Sustained Load Fracture in Aggressive Environments

Important kinds of sustained load fracture in aggressive environments include:

- stress corrosion cracking
- cracking due to embrittlement by internal or external (gaseous) hydrogen
- liquid metal embrittlement.

The term ***stress corrosion cracking*** covers a very wide range of material–environment interactions, and it is not possible to give more than a brief overview of some proposed fracture mechanisms here. For the interested reader a good background to the subject is provided by references 10, 11, 12 and 13 of the bibliography to this chapter. An important update is provided by reference 14.

Stress corrosion fractures can be transgranular or intergranular. Sometimes they are a mixture, though one mode usually predominates. Macroscopically transgranular stress corrosion cracks are often faceted. On a microscale the fracture surface may show a feather-shaped appearance as in figure 12.34 or can strongly resemble mechanical cleavage (see figure 12.21) as in aqueous stress corrosion of titanium and magnesium alloys.

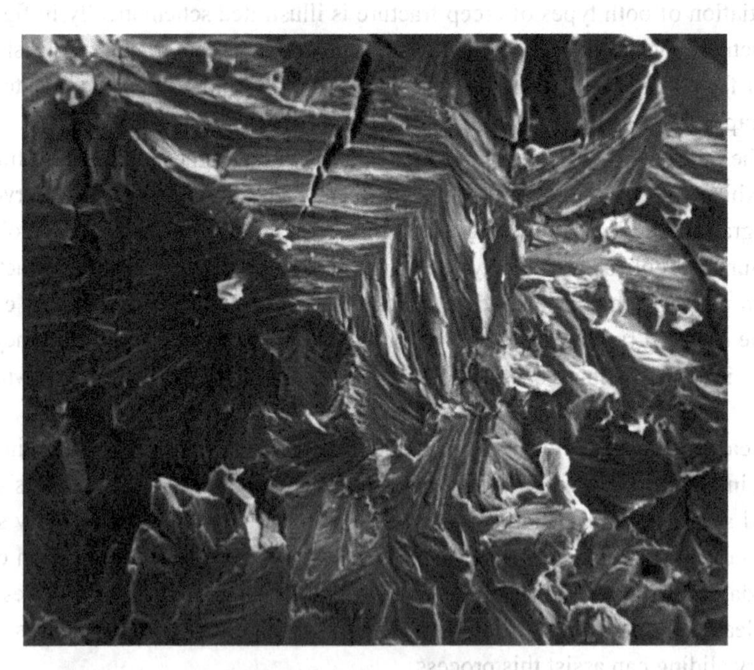

Figure 12.34. Feather-shaped crack surface of a transgranular stress corrosion crack in austenitic stainless steel. SEM fractograph, × 500.

TYPE OF FAILURE	FILM RUPTURE MODEL	STRESS-ASSISTED INTER-GRANULAR CORROSION	TUNNEL MODEL	ADSORPTION (INCLUDING HYDROGEN)	HYDROGEN ABSORPTION + DECOHESION
transgranular cracking of austenitic stainless steels	•		•	•	•
intergranular cracking of low strength ferritic steels	•	•		•	
intergranular cracking of austenitic stainless steels	•	•		•	
intergranular cracking of high strength low alloy steels				•	•
transgranular cracking of low strength ferritic steels	•			•	•
intergranular cracking of titanium alloys	•	•		•	
transgranular cracking of titanium alloys	•			•	•
intergranular cracking of aluminium alloys	•			•	•

Figure 12.35. Classification of some types of stress corrosion and models suggested to explain them. See reference 14 for an update.

In fact, the cleavage-like appearance has led to models of environment-induced cleavage owing to adsorption of specific ions, for instance hydrogen, at the crack tip or to hydrogen absorption followed by internal decohesion which links up with the main crack.

A classification of stress corrosion cracking models proposed for some structural materials is given in figure 12.35. The fact that different models have been suggested to explain each type of failure is an indication of the complexity of stress corrosion cracking. A short description of each model follows.

The *film rupture model* is also sometimes called the slip dissolution model. The model is illustrated in figure 12.36. Emergent slip bands at a surface or crack tip break a passive film and the crack propagates owing to local dissolution of metal.

The *stress-assisted intergranular corrosion* model is sometimes called the brittle film mechanism. It is shown in figure 12.37. The model requires the environment to produce a mechanically weak surface film that grows preferentially along grain bounda-

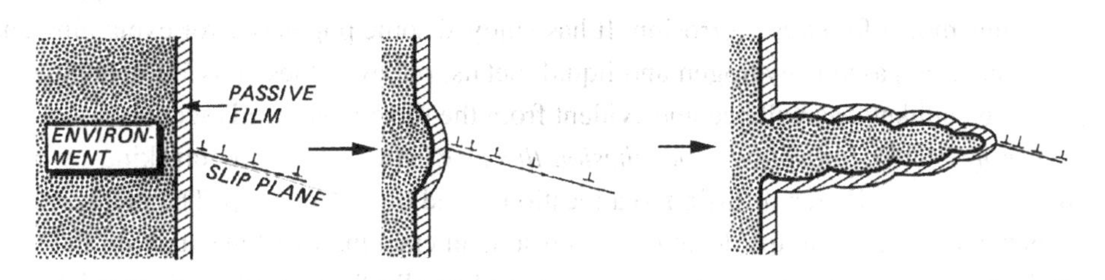

Figure 12.36. The film rupture model of stress corrosion cracking.

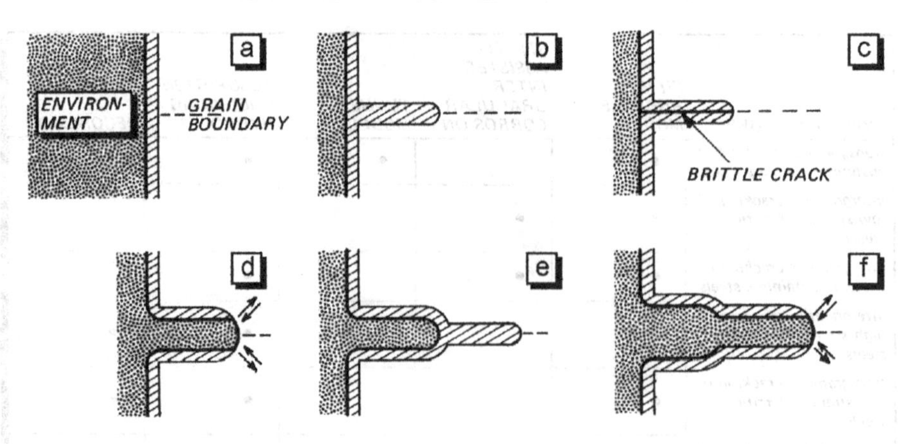

Figure 12.37. The stress-assisted intergranular corrosion model.

ries and eventually cracks under stress. The crack is arrested by plastic deformation of the metal, which then reacts with the environment to reform the film, and so on.

The *tunnel model* was proposed specifically for transgranular cracking of austenitic stainless steels in order to explain the typical feather-shaped appearance shown in figure 12.34. It involves the formation of arrays of corrosion tunnels at slip steps as depicted in figure 12.38. As the tunnels grow the ligaments of metal between them become more highly stressed and eventually fail by ductile rupture. The process is then repeated.

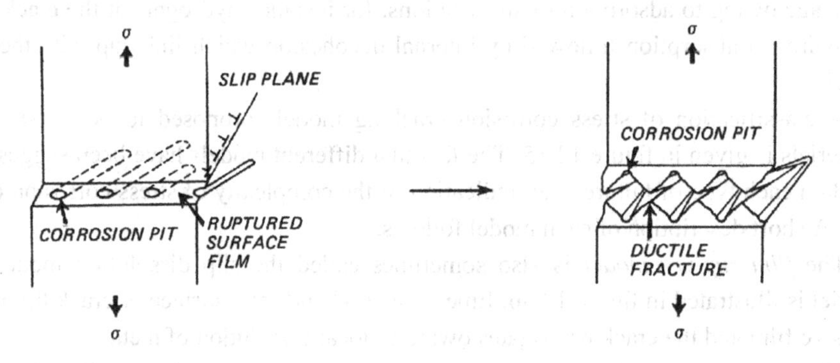

Figure 12.38. The tunnel model of stress corrosion cracking.

The *adsorption model* is a general one proposing that an environmental species can interact with strained crystal lattice bonds at the crack tip to cause a reduction in bond strength and hence brittle crack extension. There are many objections to the applicability of this model for stress corrosion. It has enjoyed some popularity for explaining embrittlement by gaseous hydrogen and liquid metals, but even these possibilities now appear to be unlikely, as will become evident from the remainder of this section.

The *hydrogen absorption + decohesion theory* of stress corrosion cracking involves stress-assisted hydrogen diffusion to a location ahead of the crack tip. The increased hydrogen concentration at this location then results in cracking that links up with the main crack. This model, or a modification of it, may be valid for a number of material environment combinations. In the next chapter, section 13.4, it is shown that the model can be used to quantitatively predict stress corrosion crack growth rates for high strength steels.

Embrittlement by internal or external hydrogen also occurs in a wide range of materials. It is particularly important for steels, and also for high strength nickel-base alloys, titanium alloys and materials used in the nuclear power industry (zirconium, hafnium, niobium, uranium).

Most hydrogen embrittlement fractures are intergranular, but cleavage-like cracking can also occur, especially if transgranular brittle hydrides form and act as crack nucleation sites. Besides hydride formation in the material, hydrogen can remain in the lattice and interact with dislocations and other lattice defects, including segregation to matrix/particle interfaces. Hydrogen may also react to give surface hydrides that lower the fracture stress at crack tips, or else these surface hydrides form brittle films along grain boundaries such that the type of mechanism shown in figure 12.37 applies. In short, hydrogen embrittlement is no one thing. For further reading references 13, 14, 15 and 16 may be consulted.

Because of its importance much attention has been paid to quantitative analyses of hydrogen embrittlement in high strength steels. As in the case of stress corrosion these analyses will be discussed in the next chapter, section 13.4. The results of the analyses indicate that both external and internal hydrogen embrittlement are caused by lattice decohesion ahead of the main crack, in a similar manner to stress corrosion cracking of high strength steels.

The final topic in this section is ***liquid metal embrittlement.*** Again this is a widespread phenomenon. It is usually of much less practical significance than stress corrosion cracking and hydrogen embrittlement, but the most recently proposed mechanism is very interesting since it could have more general applications.

Liquid metal embrittlement results in drastic losses in macroscopic ductility. The fracture path can be intergranular or transgranular, and consists of facets that appear very brittle at low and intermediate magnifications (up to $\approx \times$ 1000). Because of this, it was for a long time thought that liquid metal embrittlement was due to adsorption of liquid metal atoms at the crack tip and a consequent reduction in crystal lattice bond

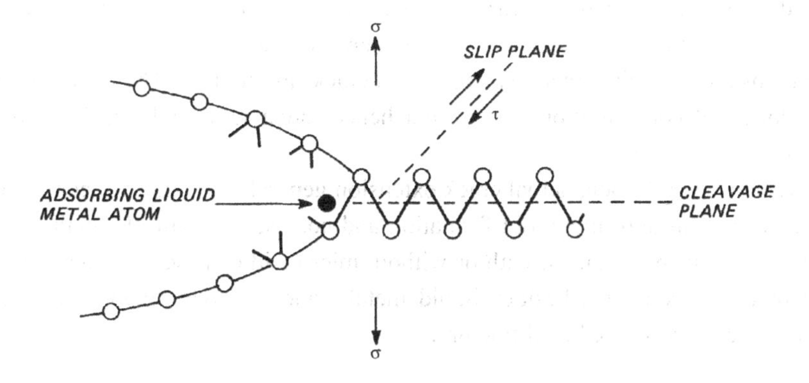

Figure 12.39. Adsorption-induced embrittlement model.

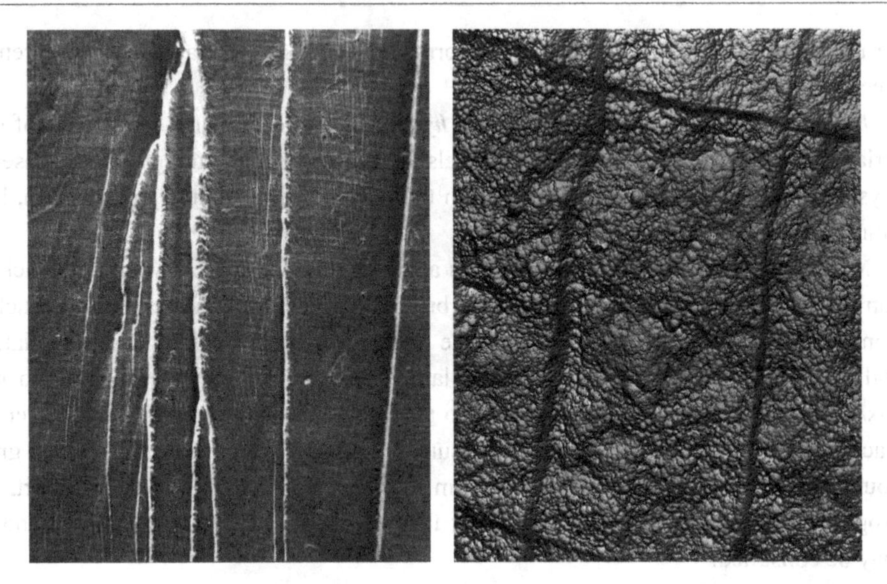

Figure 12.40. Liquid metal embrittlement of an aluminium single crystal. Left: SEM fractograph, × 50. Right: TEM replica fractograph, × 5000.
Courtesy S.P. Lynch.

strength so that brittle crack extension occurs. The model proposed on this basis is shown in figure 12.39. (This model is essentially the same as that suggested, and now largely discounted, to account for stress corrosion cracking and hydrogen embrittlement.) The hypothesis is that adsorbing atoms lower the tensile stress, σ, required to break lattice bonds, but do not influence the shear stress, τ, necessary to move dislocations to cause plastic deformation and crack tip blunting.

Subsequently, however, Lynch (references 17, 18) has shown that the apparently brittle fractures caused by liquid metal embrittlement are in most cases covered by shallow microvoids. An example is given in figure 12.40. This shows macroscopically brittle 'cleavage' in an aluminium single crystal and a TEM replica fractograph of part of the fracture surface. The microscopic ductility is considerable. In view of this microscopic ductility Lynch proposed that adsorbing liquid metal atoms facilitate the nucleation and egress of dislocations near the crack tip. This causes slip at lower stress levels than would be required in an inert environment. These lower stresses activate fewer dislocations ahead of the crack leading to less crack tip blunting. The overall effect is a highly localised concentration of slip and hence much less total plastic deformation during crack extension.

As visualised by Lynch, actual crack extension generally occurs by a combination of intense local shear and microvoid formation and coalescence with the main crack, see figure 12.41. This mechanism, with or without microvoid coalescence, may be applicable to other environments besides liquid metals, and to environmental fatigue crack growth as well as sustained load fracture.

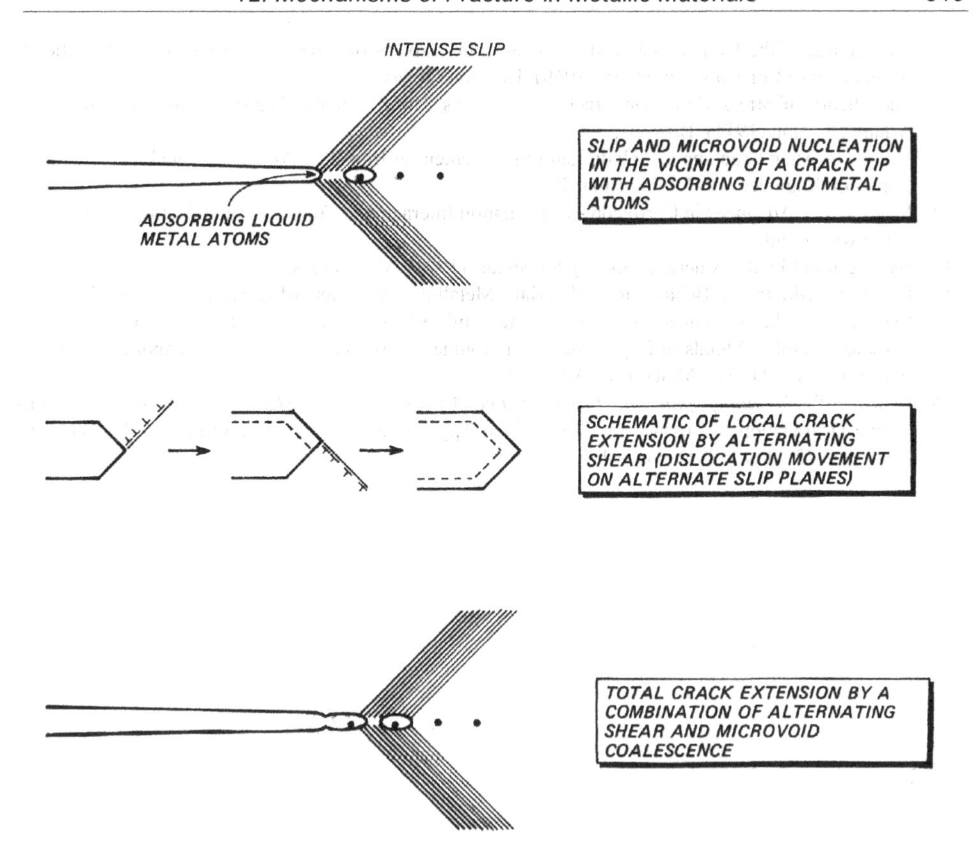

Figure 12.41. Crack extension in liquid metals owing to adsorption-induced intense shear and microvoid coalescence.

12.9 Bibliography

1. Cottrell, A.H., *Dislocations and Plastic Flow in Crystals*, Oxford University Press (1953): London.
2. Broek, D., *A Study on Ductile Fracture*, Ph.D. Thesis, Delft University of Technology, The Netherlands (1971).
3. Knott, J.F., *Fundamentals of Fracture Mechanics*, Butterworths (1973): London.
4. Smith, E., *The Nucleation and Growth of Cleavage Microcracks in Mild Steel*, Proceedings of the Conference on the Physical Basis of Yield and Fracture, Institute of Physics and Physical Society, pp. 36-46 (1966): Oxford.
5. Metals Handbook 8th Edition, Vol. 9, Fractography and Atlas of Fractographs, American Society for Metals (1974): Metals Park, Ohio.
6. Tanaka, K. and Mura, T., *A Dislocation Model for Fatigue Crack Initiation*, Journal of Applied Mechanics, Vol. 48, pp. 97-103 (1981).
7. Laird, C., *The Influence of Metallurgical Structure on the Mechanisms of Fatigue Crack Propagation*, Fatigue Crack Propagation, ASTM STP 415, American Society for Testing and Materials, pp. 131-168 (1967): Philadelphia.
8. Lynch, S.P., *Mechanisms of Fatigue and Environmentally Assisted Fatigue*, Fatigue Mechanisms, ASTM STP 675, American Society for Testing and Materials, pp. 174-213 (1979): Philadelphia.
9. Edward, G.H. and Ashby, M.F., *Intergranular Fracture during Power-Law Creep*, Acta Metallurgica, Vol. 27, pp. 1505-1518 (1979).
10. Logan, H.L., The Stress Corrosion of Metals, John Wiley and Sons, Inc. (1966): New York.

11. Proceedings of the Conference on the Fundamental Aspects of Stress Corrosion Cracking, National Association of Corrosion Engineers (1969): Houston, Texas.
12. The Theory of Stress Corrosion Cracking in Alloys, North Atlantic Treaty Organization Scientific Affairs Division (1971): Brussels.
13. Stress Corrosion Cracking and Hydrogen Embrittlement of Iron Base Alloys, National Association of Corrosion Engineers (1977): Houston, Texas.
14. Magnin, Th., Advances in Corrosion - Deformation Interactions, Trans Tech Publications (1996): Zurich, Switzerland.
15. Hydrogen in Metals, American Society for Metals (1974): Metals Park, Ohio.
16. Effect of Hydrogen on Behaviour of Materials, Metallurgical Society of AIME (1976): New York.
17. Lynch, S.P., The Mechanism of Liquid-Metal Embrittlement-Crack Growth in Aluminium Single Crystals and other Metals in Liquid-Metal Environments, Aeronautical Research Laboratories Materials Report 102 (1977): Melbourne, Australia.
18. Lynch, S.P., *Metallographic and Fractographic Aspects of Liquid-Metal Embrittlement*, Environmental Degradation of Engineering Materials in Aggressive Environments, Virginia Polytechnic Institute. pp. 229-244 (1981): Blacksburg, Virginia.

13 The Influence of Material Behaviour on Fracture Mechanics Properties

13.1 Introduction

In this final chapter we shall provide some insight into the ways in which the actual behaviour of materials influences the fracture mechanics parameters used to describe crack extension. Most of the discussion concerns LEFM parameters, mainly because they are more widely used. Figure 13.1 lists the various types of fracture considered in this course and the relevant fracture mechanics parameters. At present a significant contribution by EPFM is made only in the case of fracture toughness and slow stable tearing.

The influence of material behaviour on fracture mechanics characterization of crack extension will be described in sections 13.2–13.4 as follows:

(13.2) The effects of crack tip geometry:

TYPE OF FRACTURE	FRACTURE MECHANICS	
	• CONDITIONS	• PARAMETERS
• INITIATION OF UNSTABLE CRACK GROWTH (FRACTURE TOUGHNESS)	LEFM	K_{Ic}, K_c
	EPFM	J_{Ic}, COD
• SLOW STABLE TEARING	LEFM	K_R
	EPFM	R-Line, T
• FATIGUE CRACK GROWTH	LEFM	ΔK_{th}, ΔK
	EPFM	ΔK_{ϵ}, ΔJ
• SUSTAINED LOAD FRACTURE (EXCLUDING CREEP)	LEFM	K_{Ith}, K_I, K_{Imax}
• DYNAMIC FRACTURE	LEFM	K_{Id}, K_{Im}

Figure 13.1. Types of fracture and the parameters for describing them.

- blunting
- branching and kinking
- through-thickness irregularity
- change of mode.

(13.3) The effects of fracture path and microstructure:

- transgranular and intergranular fracture; mixed fracture paths
- effects of microstructure
  - second phases
  - particles and precipitates
  - grain size
  - fibering and texturing owing to mechanical working.

(13.4) Fracture mechanics and the mechanisms of fracture:

- fracture by microvoid coalescence
- cleavage in steels
- fatigue crack growth
- sustained load fracture
- superposition or competition of sustained load fracture and fatigue.

This subdivision is convenient in that the subjects are presented more or less in the order of increasing complexity. However, they are often strongly interrelated, as will be seen.

13.2 The Effects of Crack Tip Geometry

LEFM analysis begins with slit-shaped, unbranched cracks of zero tip radius. The latter assumption is obviously unrealistic, owing to the occurrence of plasticity which blunts the crack. Also, most real cracks exhibit at least small amounts of branching, kinking and through-thickness irregularity. These problems have been mentioned in section 10.4.

Changes in mode of loading can occur. The most important are the mode II → mode I and mode I → combined modes I and II transitions during fatigue crack propagation at low stress intensity levels.

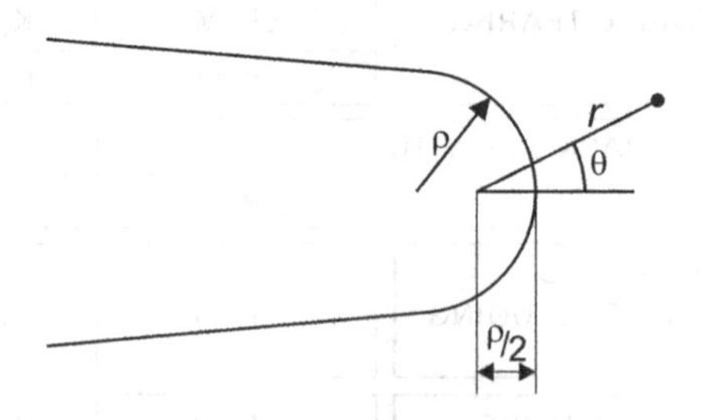

Figure 13.2. Crack with a finite tip radius, ρ.

Crack Tip Blunting

Crack tip blunting always occurs in practice. The blunting may be very limited, *e.g.* during environmental fatigue crack propagation and sustained load fracture in high strength materials. In such cases sharp crack LEFM is usually adequate for characterizing crack extension.

On the other hand, blunting is very important during EPFM fracture toughness testing. This is recognised in the procedures for determining J_{Ic} (section 7.4) and COD (section 7.6).

As stated in section 2.2, the mode I elastic stress field equations for a blunt crack were derived by Creager and Paris, reference 1 of the bibliography to this chapter. The crack tip coordinate system is shown in figure 13.2.

Under mode I loading

$$\sigma_x = \frac{K_I}{\sqrt{2\pi r}}\cos{\theta/2}\left(1 - \sin{\theta/2}\sin{3\theta/2}\right) - \frac{K_I}{\sqrt{2\pi r}}\left(\frac{\rho}{2r}\right)\cos{3\theta/2},$$

$$\sigma_y = \frac{K_I}{\sqrt{2\pi r}}\cos{\theta/2}\left(1 + \sin{\theta/2}\sin{3\theta/2}\right) + \frac{K_I}{\sqrt{2\pi r}}\left(\frac{\rho}{2r}\right)\cos{3\theta/2}, \qquad (13.1)$$

$$\tau_{xy} = \frac{K_I}{\sqrt{2\pi r}}\sin{\theta/2}\cos{\theta/2}\cos{3\theta/2} - \frac{K_I}{\sqrt{2\pi r}}\left(\frac{\rho}{2r}\right)\sin{3\theta/2}.$$

For sharp cracks ρ is small. Unless one is interested in regions very close to the crack tip the terms with (ρ/r) can be neglected and equations (13.1) reduce to the standard form given in equations (2.24).

Blunting has a minor effect on the size and shape of the plastic zone, reference 2. However, for plane strain conditions the distribution of σ_y ahead of the crack tip is greatly altered, as figure 13.3 shows. A blunted crack tip acts as a free surface and locally reduces the triaxiality of the state of stress (at $r = \rho/2$ $\sigma_x = 0$ for $\theta = 0$), see section 3.5. Thus increased blunting decreases the maximum stress and moves its location away from the crack tip towards the elastic-plastic boundary. These effects have been used to assess the effects of blunting on fracture toughness and sustained load fracture, as will be discussed in section 13.4.

Crack Branching and Kinking

Microscopic crack branching and kinking commonly occur. Their significance is often overlooked when the fracture mechanics-related properties of materials are determined. Macrobranching also occurs, but only during region II sustained load fracture and dynamic fracture. The reason for this limitation is that macrobranching depends on there being little or no tendency for one branch to outrun the other(s), and this is only possible when crack growth rates are virtually independent of crack length and stress intensity, as is the case for region II sustained load fracture, see figure 10.4, and for dynamic crack branching, section 11.2.

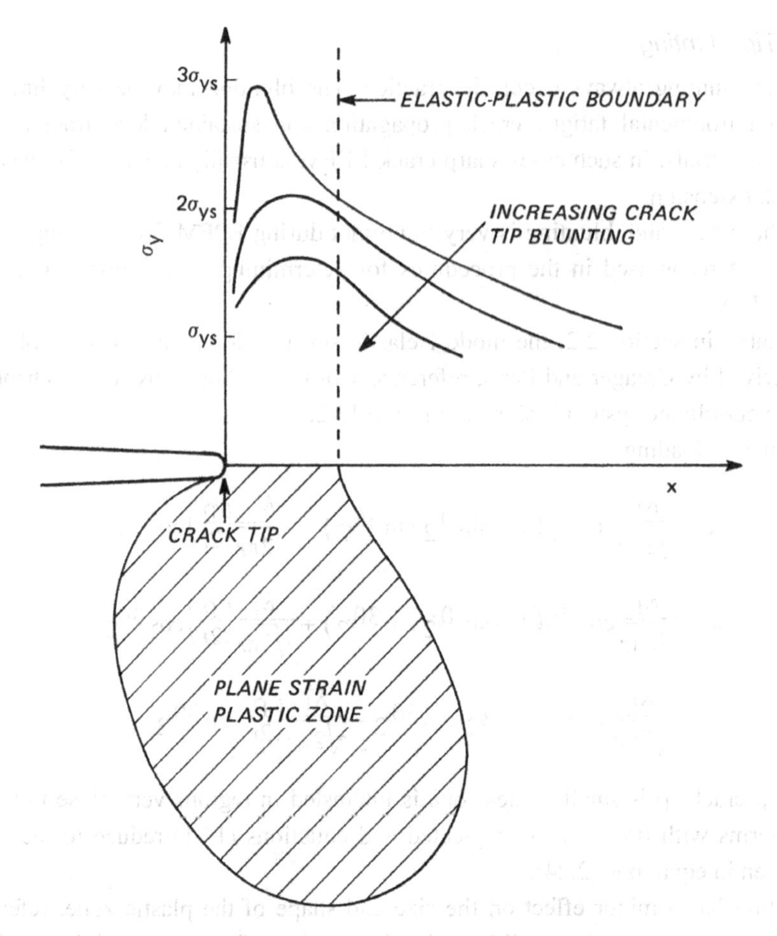

Figure 13.3. Schematic of the effect of crack tip blunting on σ_y for an elastic-perfectly plastic material under plane strain.

Crack branching and kinking lower the mode I stress intensities at the crack tips. This is illustrated in figure 13.4. Consequently, for nominally mode I crack extension it may be expected that microbranching and kinking will result in

- higher fracture toughness
- higher thresholds for fatigue crack growth (ΔK_{th}) and sustained load fracture ($K_{I_{th}}$)
- lower crack growth rates in fatigue and region I sustained load fracture.

Through-Thickness Irregularity

Fracture mechanics analyses of through-thickness crack front irregularity are not available. Nevertheless, for both sustained load fracture (reference 4) and fatigue crack growth (references 5 and 6) it has been found that increased irregularity results in lower crack growth rates.

Change of Mode

Two types of change of mode of crack growth will be considered here:

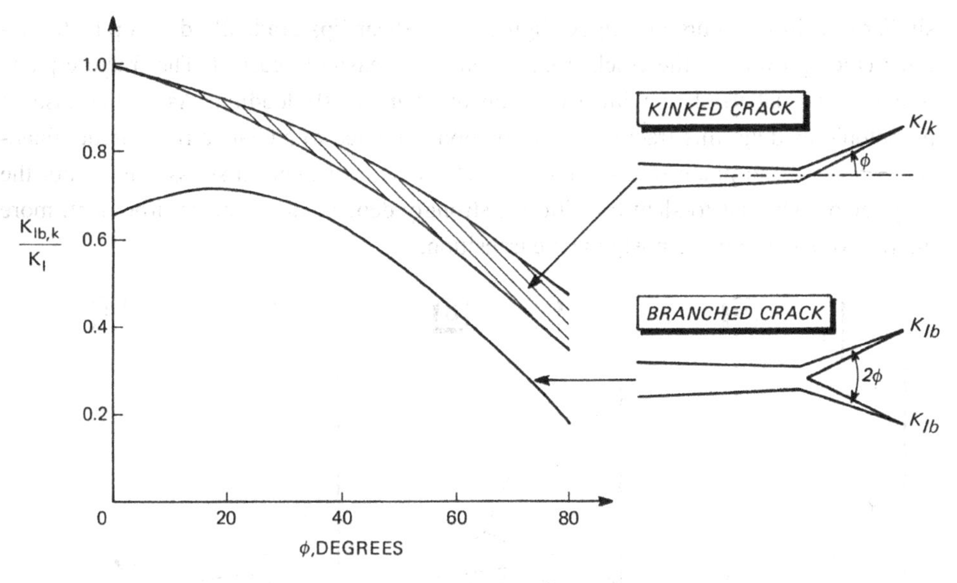

Figure 13.4. Mode I stress intensities for branched and kinked cracks. After reference 3.

1) The flat-to-slant transition at high stress intensities, for both monotonic and fatigue loading.
2) The mode II → mode I and mode I → combined modes I and II transitions during fatigue crack propagation at low stress intensities.

The flat-to-slant transition under monotonic loading was discussed in section 3.6. A

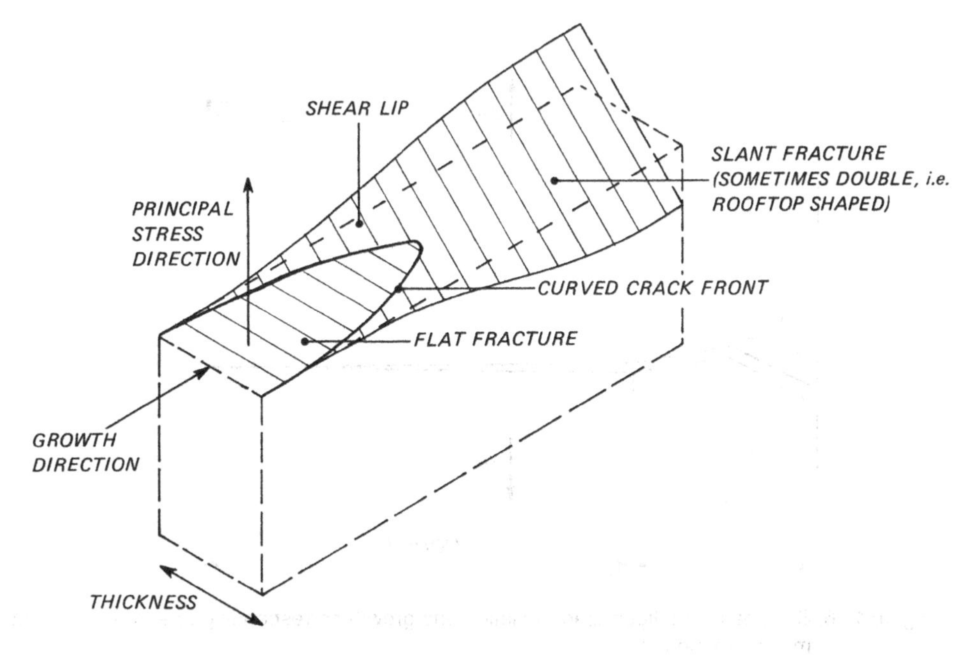

Figure 13.5. Transition from flat to slant fatigue fracture.

similar transition occurs in fatigue, figure 13.5. Shear lips gradually develop as the fatigue crack grows and the crack front becomes increasingly curved. The fracture plane rotates from flat to slant with a component of mode III loading. As in the case of monotonic loading, this fracture plane rotation is related to a change from predominantly plane strain to plane stress conditions. However, a change in stress state is not the only factor. The flat-to-slant transition is strongly dependent on the environment: more aggressive environments postpone the transition.

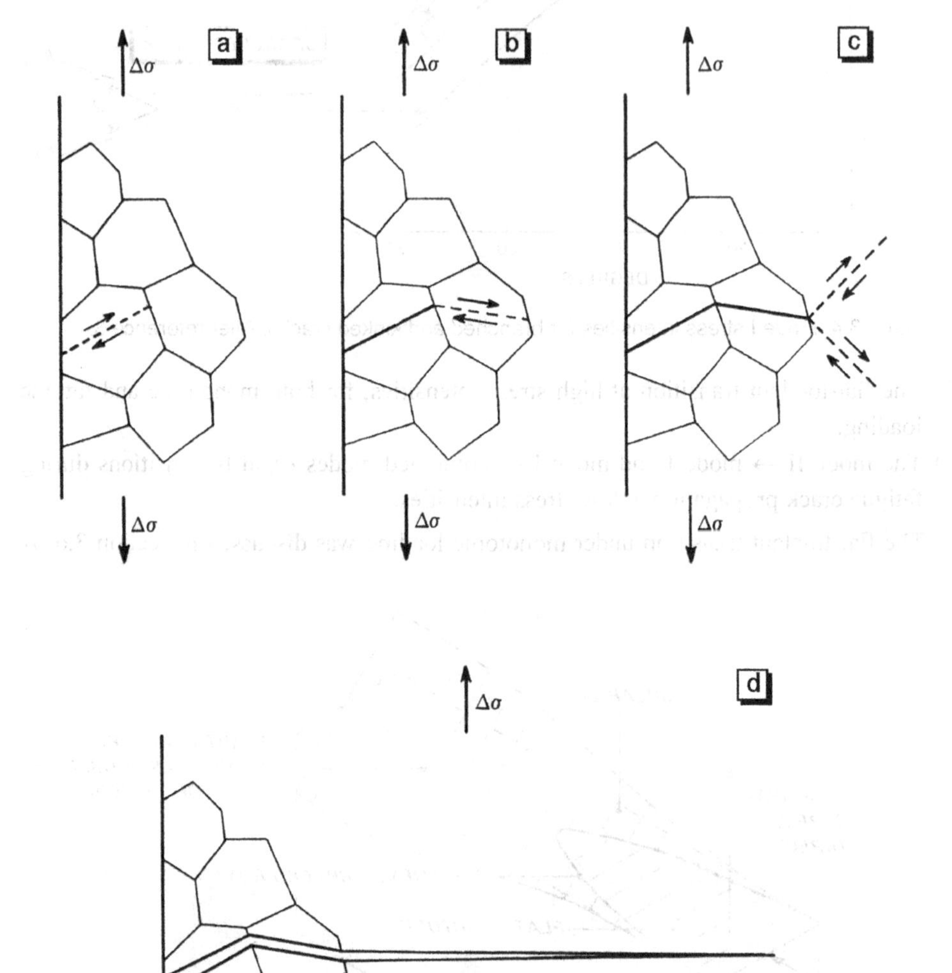

Figure 13.6. Schematic of fatigue crack initiation and growth corresponding to a transition from mode II to mode I.

The flat-to-slant transition affects fatigue crack growth rates. Under constant amplitude loading the slope of the fatigue crack growth rate curve will slightly decrease. Under variable amplitude loading one or more tensile peak loads can induce a change from flat to slant fracture, which then reverts to flat fracture under continued less severe load cycling. Such crack front geometry changes are a potential complication for predicting crack growth, since re-orientation of the crack front back to flat fracture is likely to affect the amount of crack growth retardation due to the peak loads.

The mode II → mode I transition during fatigue crack propagation is of fundamental importance since it often represents the initiation and early growth of fatigue cracks under constant or increasing stress cycling, $\Delta\sigma$. Figure 13.6 gives a schematic of this process as follows:

a) Cyclic slip begins in a surface grain and occurs mainly on one or a few sets of crystal planes.
b) This usually leads to slip plane cracking (mode II), which results in a faceted fracture surface, and spreading of cyclic slip to an adjacent grain. Again the slip is mainly on one or a few sets of crystal planes.
c) The second grain also cracks along the slip plane. Cyclic slip in the interior is now activated on several sets of crystal planes. This enables mode I crack extension.
d) Cyclic slip on several sets of crystal planes results in a continuum mechanism of crack propagation, often characterized by fatigue striations.

(For more details on fatigue crack initiation and propagation the reader is referred to the previous chapter, section 12.6.)

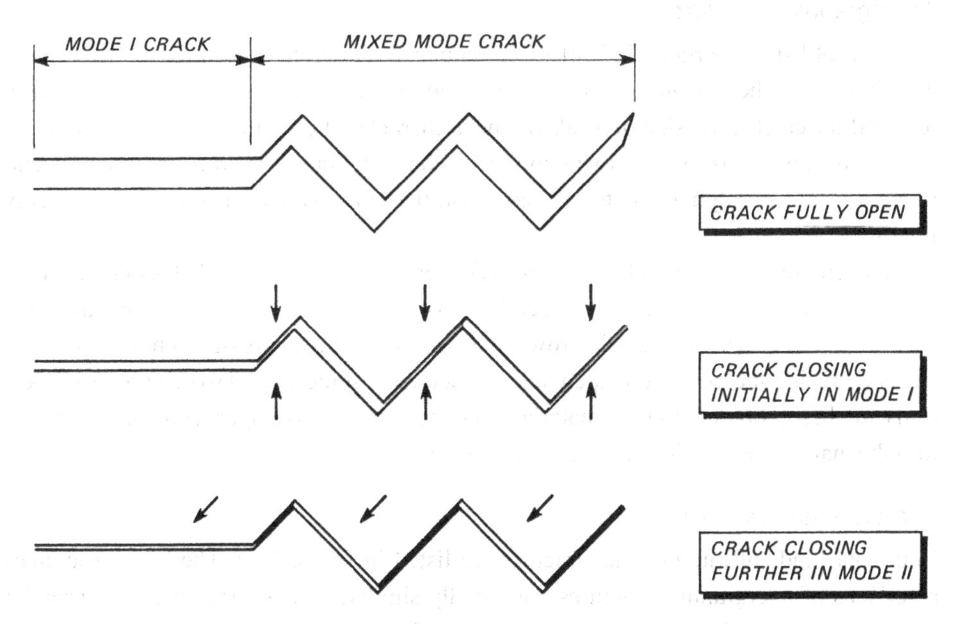

Figure 13.7. Schematic diagram of increased crack closure owing to combined mode I and mode II fatigue crack growth.

The mode I → combined modes I and II transition is also important. It generally occurs when a fatigue crack is growing at low stress intensity levels either when the overall ΔK is gradually decreasing towards the threshold, reference 7, or else after a tensile peak load, reference 8. In both cases faceted fracture (not necessarily slip plane cracking) occurs and the crack path is effectively a series of kinked cracks. This means that apart from a decreasing ΔK or the effects of residual stresses due to a peak load there are two purely geometric effects that contribute to lower crack growth rates:

1) The crack tip mode I stress intensities for kinked cracks are lower than those for pure mode I cracks, *e.g.* figure 13.4.
2) There is an increase in crack closure owing to increased contact of the fracture surfaces. This is illustrated in figure 13.7.

13.3 The Effects of Fracture Path and Microstructure

The effects of fracture path and microstructure on the fracture mechanics properties of materials are based on geometric and inherent characteristics. More specifically, the amounts of plastic deformation and energy required for crack extension depend on both crack tip geometry and the fracture path. In turn, the fracture path is determined by the microstructure and, sometimes, by microstructurally influenced material-environment interactions.

In this section we shall consider transgranular and intergranular fracture and mixtures of different fracture paths before proceeding to the influences of microstructure.

Transgranular Fracture

Figure 13.8 lists important kinds of transgranular fracture, their occurrence and general descriptions of their geometric and mechanical characteristics. The amount of energy required for crack extension depends on these characteristics. At one extreme microvoid coalescence, which is the usual way in which stable and unstable crack extension occur, requires relatively high amounts of energy. On the other hand cleavage is a low energy fracture.

Transgranular fatigue and stress corrosion cracking require much less energy than microvoid coalescence. The highly localised cyclic plasticity characteristic of fatigue is very effective in causing crack growth, especially in an aggressive environment. In stress corrosion the environment enhances fracture in some way. Two common suggestions are dissolution of stressed material, and a reaction to give hydrogen which diffuses into the material and embrittles it, see section 13.4.

Intergranular Fracture

Important kinds of intergranular fracture are listed in figure 13.9. The geometric characteristics of intergranular fractures are broadly similar, namely branched and irregular crack fronts with limited blunting. Intergranular fatigue fractures are usually less branched and less irregular than other kinds of intergranular fracture, particularly stress corrosion cracks.

FRACTURE	TYPICAL OCCURRENCE	CHARACTERISTICS
• MICROVOID COALESCENCE (DUCTILE FRACTURE)	• STABLE AND UNSTABLE FRACTURE • HIGH STRESS INTENSITY FATIGUE	• IRREGULAR BLUNT CRACK FRONTS • CONSIDERABLE LOCAL PLASTICITY
• CLEAVAGE IN STEELS	• UNSTABLE FRACTURE • INTERMEDIATE AND HIGH STRESS INTENSITY FATIGUE • HYDROGEN EMBRITTLEMENT	• SHARP KINKED CRACKS • BRITTLE FRACTURE
• CONTINUUM FATIGUE (STRIATIONS)	• INTERMEDIATE AND HIGH STRESS INTENSITY FATIGUE	• UNIFORM CRACK FRONTS; LIMITED BLUNTING • MODE I FAVOURED • CONSIDERABLE LOCAL (CYCLIC) PLASTICITY
• SLIP PLANE CRACKING (FACETED FRACTURE)	• FATIGUE CRACK INITIATION AND GROWTH AT LOW STRESS INTENSITIES, ESPECIALLY UNDER TORSIONAL LOADING	• BRANCHED, KINKED AND IRREGULAR CRACK FRONTS; LIMITED BLUNTING • MODE II FAVOURED • LIMITED LOCAL (CYCLIC) PLASTICITY
• CLEAVAGE-LIKE FACETED FRACTURE	• LOW STRESS INTENSITY FATIGUE • STRESS CORROSION	• BRANCHED, KINKED AND IRREGULAR CRACK FRONTS; LIMITED BLUNTING • LIMITED LOCAL PLASTICITY

Figure 13.8. Important kinds of transgranular fracture.

Intergranular fractures require only moderate to low amounts of energy, *i.e.* they are generally a sign of weakness.

Mixed Fracture Paths

Combinations of different kinds of transgranular fracture and transgranular and intergranular fractures are common. For example, in high strength steels cleavage often occurs as a secondary kind of fracture during fatigue or in combination with transgranular and intergranular microvoid coalescence during unstable fracture.

Since cleavage is a low energy fracture with sharp and fairly uniform crack fronts its occurrence is always detrimental to the fracture mechanics properties. However, the geometric and mechanical characteristics of some fracture paths can oppose each other,

FRACTURE	TYPICAL OCCURRENCE	MECHANICAL CHARACTERISTICS
• MICROVOID COALESCENCE ALONG GRAIN BOUNDARIES	• STABLE AND UNSTABLE FRACTURE AND HIGH STRESS INTENSITY FATIGUE IN HIGH STRENGTH STEELS AND ALUMINIUM ALLOYS	• LIMITED LOCAL PLASTICITY
• LOW DUCTILITY GRAIN BOUNDARY SEPARATION	• SUSTAINED LOAD FRACTURE (INCLUDING CREEP)	• LITTLE EVIDENCE OF PLASTICITY
• BRITTLE GRAIN BOUNDARY SEPARATION	• UNSTABLE FRACTURE IN TEMPER-EMBRITTLED STEELS AND REFRACTORY METALS (E.G. TUNGSTEN)	• BRITTLE LOW ENERGY FRACTURE

Figure 13.9. Classification of intergranular fracture.

e.g.:

1) Intergranular microvoid coalescence causes a more irregular and branched crack front but the fracture toughness is lower when it occurs in combination with transgranular microvoid coalescence.
2) Transgranular microvoid coalescence causes irregular and blunt crack fronts but accelerates constant amplitude fatigue crack growth when it occurs with fatigue striations.

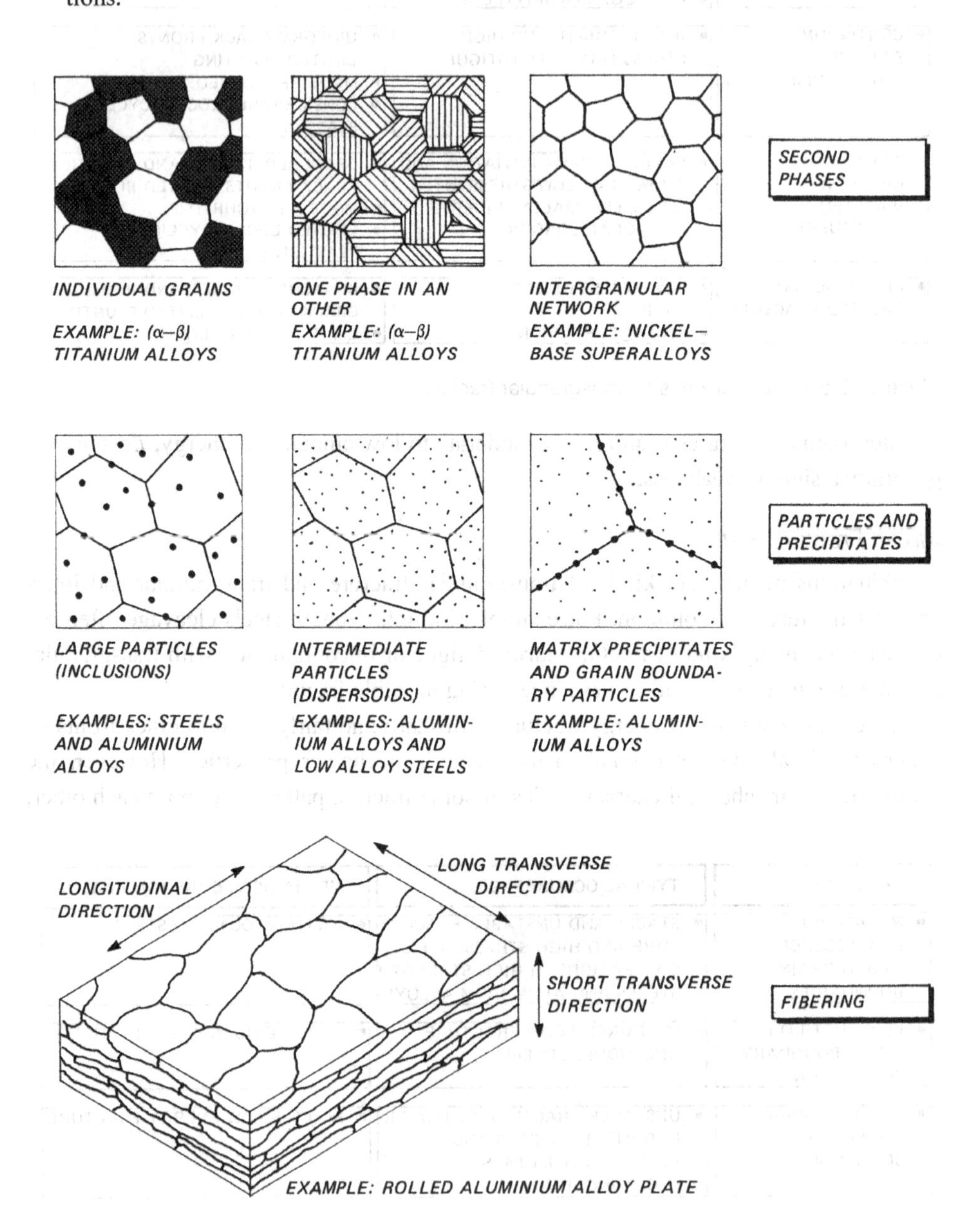

Figure 13.10. Some microstructural and microstructurally-related features in metallic materials.

The first example is apparently straightforward. Fracture toughness is at least partly determined by stable crack extension, see section 5.2. The limited local plasticity characteristic of intergranular microvoid coalescence outweighs potential increases in fracture toughness owing to branching and irregularity. Note, however, that this rationale is incomplete: it does not explain why intergranular or transgranular fracture paths occur or why they differ in energy. The second example is impossible to explain, even partially, without a detailed consideration of the role of the microstructure.

Both examples will be returned to in the discussion on mechanisms of fracture in section 13.4, under the subheading "Fracture by Microvoid Coalescence".

Effects of Microstructure

There are many microstructural and microstructurally-related features that can play a role in determining the fracture path. The most important are:

- second phases
- particles and precipitates
- grain size
- fibering and texturing owing to mechanical working.

Figure 13.10 illustrates some of these features. Not shown are grain size variations and texturing, which is a preferential orientation of crystal planes. Texturing is important for two phase (α–β) titanium alloys (α is hexagonal close packed, β is body centred cubic).

High strength structural materials usually possess several microstructural features that can influence the fracture path and hence the fracture mechanics properties. The fracture process is therefore often complex, and it is impossible to give an overall, unified description.

Nevertheless, in figure 13.11 we have attempted a survey of microstructural influences on the fracture of high strength steels, aluminium alloys and titanium alloys, which represent three of the four major classes of modern structural materials mentioned in chapter 1. The background to this survey is twofold:

1) Relationships between microstructure and fracture properties are better understood for aluminium alloys than for high strength steels and titanium alloys. A review is given in reference 9.
2) There are basic differences between the classes of materials. Titanium alloys contain very few particles and their fracture properties depend mostly on alloy phase morphology, relative amounts of α and β, and the texture.

   Particles are always present in high strength steels and aluminium alloys and have a major effect on fracture toughness, as will be discussed in section 13.4. Also, aluminium alloys are essentially single phase while alloy phases are present only in small amounts in high strength steels, where they are much less important than in titanium alloys.

MICROSTRUCTURAL FEATURE	SPECIFIC ASPECTS	EFFECTS ON FRACTURE PATH	EFFECTS ON FRACTURE MECHANICS PROPERTIES
• PHASE MORPHOLOGY IN α–β TITANIUM ALLOYS	• CHANGE FROM α AND β GRAINS TO α IN β (SEE FIGURE 13.10)	• CRACK BRANCHING AND IRREGULARITY GREATLY INCREASED	• HIGHER K_{Ic} AND K_{Iscc} • INCREASED RESISTANCE TO FATIGUE CRACK GROWTH
• LARGE PARTICLES IN STEELS AND ALUMINIUM ALLOYS	• RANDOMLY DISTRIBUTED	• LARGE MICROVOIDS (MAINLY TRANSGRANULAR)	• LOWER K_{Ic} • ACCELERATION OF HIGH STRESS INTENSITY FATIGUE
	• INTERGRANULAR CARBIDES IN STEELS	• NUCLEATION OF CLEAVAGE	
• DISPERSOIDS IN STEELS AND ALUMINIUM ALLOYS	• RANDOMLY DISTRIBUTED	• SHEETS OF SMALL TRANSGRANULAR MICROVOIDS	• LOWER K_{Ic} • ACCELERATION OF HIGH STRESS INTENSITY FATIGUE
	• INCREASED SPACING BETWEEN DISPERSOIDS IN ALUMINIUM ALLOYS	• FACETED CRACK GROWTH MAINTAINED TO HIGHER ΔK	• TRANSITION FROM REGION I TO REGION II FATIGUE CRACK GROWTH POSTPONED TO HIGHER ΔK
• PRECIPITATES AND PARTICLES IN Al-Zn-Mg-Cu ALLOYS	• LARGER MATRIX PRECIPITATES AND GRAIN BOUNDARY PARTICLES OWING TO OVERAGEING	• INCREASE IN INTERGRANULAR MICROVOID COALESCENCE DURING UNSTABLE FRACTURE	• LOWER K_{Ic} AND K_c
		• NONE FOR STRESS CORROSION	• HIGHER K_{Iscc} AND LOWER REGION II CRACK GROWTH VELOCITY
• LARGER GRAIN SIZE	• OVERAGED Al-Zn-Mg-Cu ALLOYS	• INCREASE IN INTERGRANULAR MICROVOID COALESCENCE	• LOWER K_{Ic} AND K_c
	• STEELS	• FACETED CRACK GROWTH MAINTAINED TO HIGHER ΔK	• TRANSITION FROM REGION I TO REGION II FATIGUE CRACK GROWTH POSTPONED TO HIGHER ΔK; HIGHER ΔK_{th}
	• α–β TITANIUM ALLOYS	• CRACK BRANCHING AND IRREGULARITY MAINTAINED TO HIGHER ΔK	• INCREASED RESISTANCE TO REGION II FATIGUE CRACK GROWTH
• FIBERING IN ALUMINIUM ALLOYS	• ALIGNMENT OF GRAINS AND PARTICLES	• BRANCHING AND IRREGULARITY MUCH LESS WHEN CRACK PLANE NORMAL TO SHORT TRANSVERSE DIRECTION (SEE FIGURE 13.10)	• LOWER K_{Ic} AND K_{Iscc} • DECREASED RESISTANCE TO FATIGUE CRACK GROWTH
• CRYSTALLOGRAPHIC TEXTURE IN α–β TITANIUM ALLOYS	• PREFERRED ORIENTATION OF THE HEXAGONAL α PHASE IN MICROSTRUCTURES WITH INDIVIDUAL α AND β GRAINS (SEE FIGURE 13.10)	• CRACK BLUNTING, BRANCHING AND IRREGULARITY DEPEND ON TEXTURE	• COMPLEX AND STRONG INFLUENCES ON K_{Ic}, K_{Iscc}, ΔK_{th} AND THE RESISTANCE TO FATIGUE

Figure 13.11. Survey of microstructural influences on fracture path and fracture mechanics properties for high strength structural materials.

13.4 Fracture Mechanics and the Mechanisms of Fracture

The mechanisms of fracture in structural materials are incompletely understood. This is particularly true for environmentally influenced crack growth, *i.e.* environmental fatigue crack propagation and sustained load fracture.

There have been many attempts to model fracture processes using fracture mechanics. The models are necessarily based on continuum behaviour. Their successes – and failures – can provide useful insight into the actual mechanisms of crack growth, and it is this aspect of fracture mechanics that is considered in this final section.

The topics that will be discussed are:

- fracture by microvoid coalescence
- cleavage in steels
- fatigue crack growth
- sustained load fracture
- superposition or competition of sustained load fracture and fatigue.

Fracture by Microvoid Coalescence

Transgranular microvoid coalescence is the typical process by which slow stable tearing and unstable ductile fracture occur. As discussed in section 12.4, the microvoids nucleate at various discontinuities. For steels and aluminium alloys the most important nucleation sites are large particles and dispersoids. In titanium alloys the voids nucleate at the boundaries between α and β phases.

In titanium alloys microvoid nucleation, growth and coalescence to cause fracture is a highly complex process that depends greatly on microstructure and anisotropic plastic deformation owing to texturing. This process is not well understood and will not be discussed further. Interested readers should consult references 10, 11 and 12 of the bibliography.

A schematic of transgranular microvoid nucleation at particles and subsequent crack extension was shown in figure 12.18 and is given again as figure 13.12. The voids can initiate both at matrix/particle interfaces and as a result of particle fracture. Large particles provide weak spots for the nucleation of large voids, which link up to the crack and each other via sheets of small voids nucleated at dispersoids. From this qualitative description two factors are evident:

1) Crack tip blunting is important. Blunting will be beneficial to fracture toughness since it lowers the maximum normal stress ahead of the crack, figure 13.3. Therefore void nucleation, growth and coalescence will require higher external stress.
2) The sizes and spacings of particles will greatly affect fracture toughness. Larger particles and less distance between them will be detrimental.

These factors are interrelated. Crack tip blunting depends not only on the matrix ductility but also on the ease with which voids nucleate, grow and coalesce with the crack tip: at that moment blunting ceases.

The influence of particle spacing provides an explanation why intergranular microvoid coalescence lowers fracture toughness and why fibering results in lower fracture toughness when the crack plane is normal to the short transverse direction. In both cases

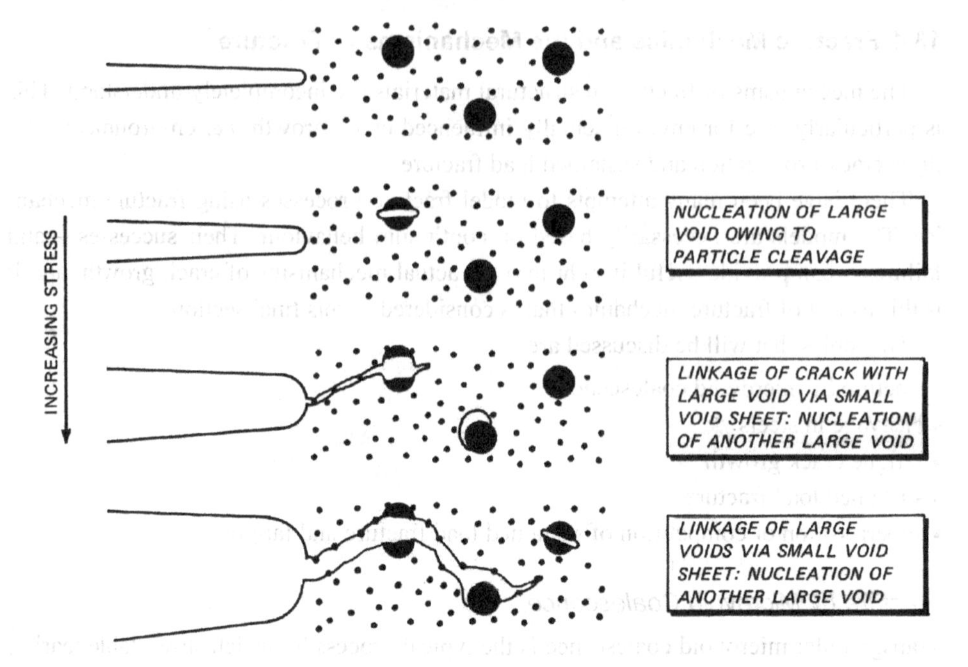

Figure 13.12. Schematic of crack extension by transgranular microvoid coalescence.

the fracture path contains relatively large numbers of particles, *i.e.* the distance between them is small. However, it is the particle size which determines whether intergranular microvoid coalescence will occur instead of transgranular microvoid coalescence. This is demonstrated by the effect of overageing on unstable fracture in Al-Zn-Mg-Cu alloys, mentioned in figure 13.11.

The contribution of particles to transgranular microvoid coalescence makes it possible to explain why its occurrence accelerates high stress intensity constant amplitude fatigue crack growth. During high stress intensity fatigue microvoid coalescence nucleates at large particles ahead of the crack. Localized regions of material between the crack tip and the particles also fracture by microvoid coalescence, as in figure 13.12. This process causes rapid local jumps of the crack front such that the overall crack growth rate increases even though the crack front is locally blunter and more irregular.

Before proceeding to quantitative modelling of fracture toughness a very important practical point that relates to microvoid coalescence will be discussed. It is difficult or impossible to use the highest strength alloys of any material class for fracture resistant structures. This is because there is a general trend for fracture toughness to decrease with increasing yield strength, see figure 13.13. It is most unlikely that any single factor is responsible for this trend. Nevertheless, fractography has shown that the diameters of coalesced microvoids also decrease with increasing yield strength of a class of alloys, and has led to the following simple explanation of the general decrease in fracture toughness. Increasing strength raises the attainable stresses in the crack tip region, resulting in earlier nucleation and growth of microvoids during loading and the nucleation and growth of microvoids at many sites that are inactive in a lower strength alloy.

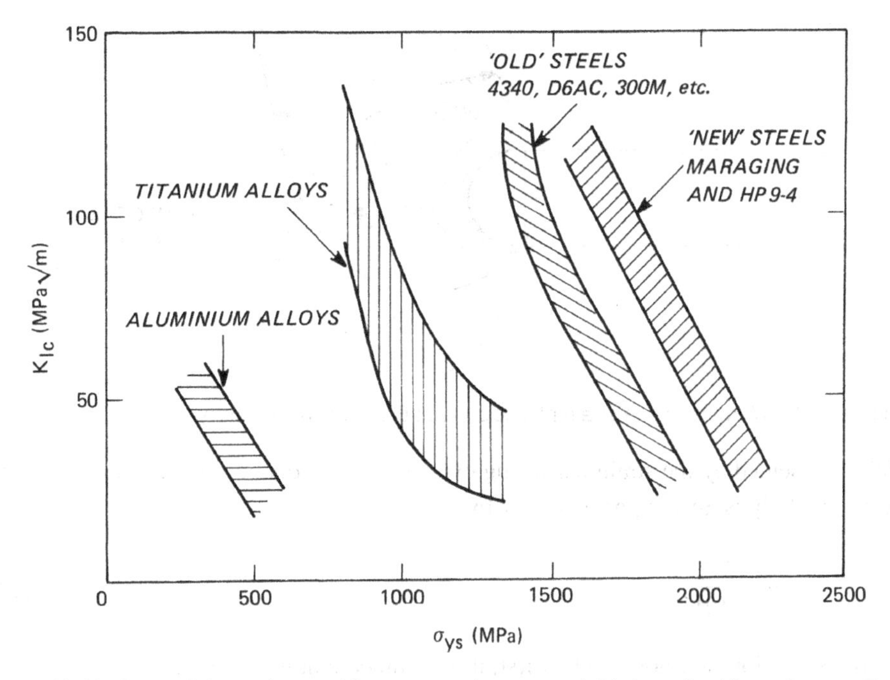

Figure 13.13. General dependency of fracture toughness on yield strength. After reference 9.

Many attempts have been made to calculate the plane strain fracture toughness, K_{Ic}, from other properties. There are several problems that such modelling faces:

1) Stress-strain distributions in the plastic zone ahead of the crack must be known or assumed. In this respect strain hardening is very important.
2) The proper fracture criterion must be chosen. Ductile rupture is strain controlled, *i.e.* the local strains must exceed a critical value. This is sometimes considered to be the uniform elongation strain in a tensile test, and in other cases is considered to be the fracture strain at the crack tip.
3) The critical strain has to be reached or exceeded over a certain distance or volume. A reasonable assumption is that this distance is equal to the particle spacing *d*. However, there is a complication. The critical strain depends strongly on the stress state, which varies significantly near the crack tip.
4) Calculation of K_{Ic} is based on the assumption that unstable fracture occurs when the fracture criterion is satisfied. But actual determination of K_{Ic} involves 2% crack extension which, if stable, can cause a significant increase in stress intensity, as mentioned in section 4.8. Consequently, calculated and measured K_{Ic} values need not be comparable.

In view of these problems it is no surprise that estimates of K_{Ic} are often very inaccurate. One of the more reliable and yet simple models is a semi-empirical one due to Hahn and Rosenfield, reference 13. Figure 13.14 illustrates the basic features of the model. There is a region of intense plastic deformation in the vicinity of the crack tip. The width of this region, λ, depends on the strain hardening characteristics of the mate-

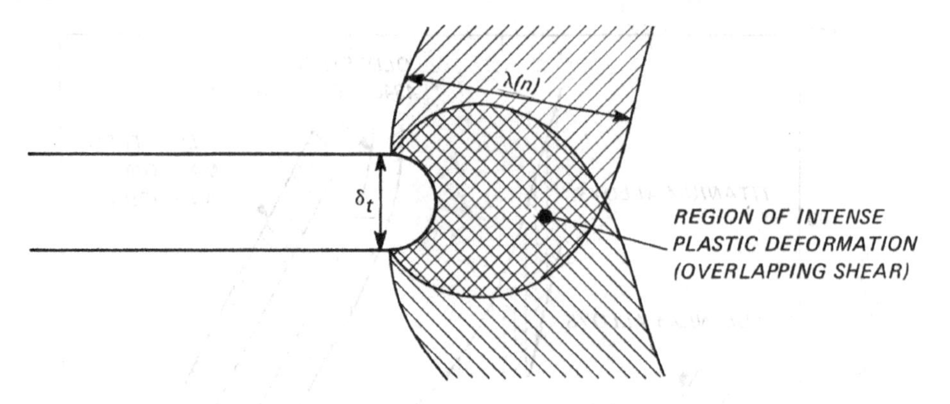

Figure 13.14. Model of plane strain plastic zone at the onset of instability.

rial, represented by the strain hardening exponent n, *cf.* equation (6.32). The shear strain at the crack tip is given approximately by

$$\gamma = \frac{\delta_t/2}{\lambda(n)} . \tag{13.2}$$

Two assumptions are now made. First, it is assumed that the average tensile strain, $\bar{\varepsilon}$, in the region of intense plastic deformation is approximately $\gamma/2$. Second, the strain distribution is assumed to be linear. Then the maximum tensile strain at the crack tip is

$$\varepsilon_{max} = 2\bar{\varepsilon} = \gamma = \frac{\delta_t}{2\lambda(n)} . \tag{13.3}$$

At the onset of fracture $\delta_t = \delta_{t_{crit}}$, $\varepsilon_{max} = \varepsilon_f^*$ and $\lambda(n) = \lambda(n)_{crit}$. Thus

$$\varepsilon_f^* = \frac{\delta_{t_{crit}}}{2\lambda(n)_{crit}} . \tag{13.4}$$

Hahn and Rosenfield measured $\lambda(n)_{crit}$ for a variety of steels and aluminium and titanium alloys. They found that $\lambda(n)_{crit} \approx 0.025\, n^2$ when measured in metres. Also, they argued that the crack tip fracture strain, ε_f^*, can be related to the true strain, ε_f, in a tensile test by $\varepsilon_f^* = \varepsilon_f/3$. Then

$$\delta_{t_{crit}} = \frac{0.05\, \varepsilon_f n^2}{3} . \tag{13.5}$$

Experiments by Robinson and Tetelman (reference 14) have shown that under plane strain conditions the relationship between δ_t and K_I tends to the value predicted by the Dugdale approach (equation 3.20)

$$\delta_t = \frac{K_I^2(1-\nu^2)}{\alpha E \sigma_{ys}} \approx \frac{K_I^2}{E\sigma_{ys}} . \tag{13.6}$$

Substituting for $\delta_{t_{crit}}$ and K_{Ic} in equation (13.5) gives

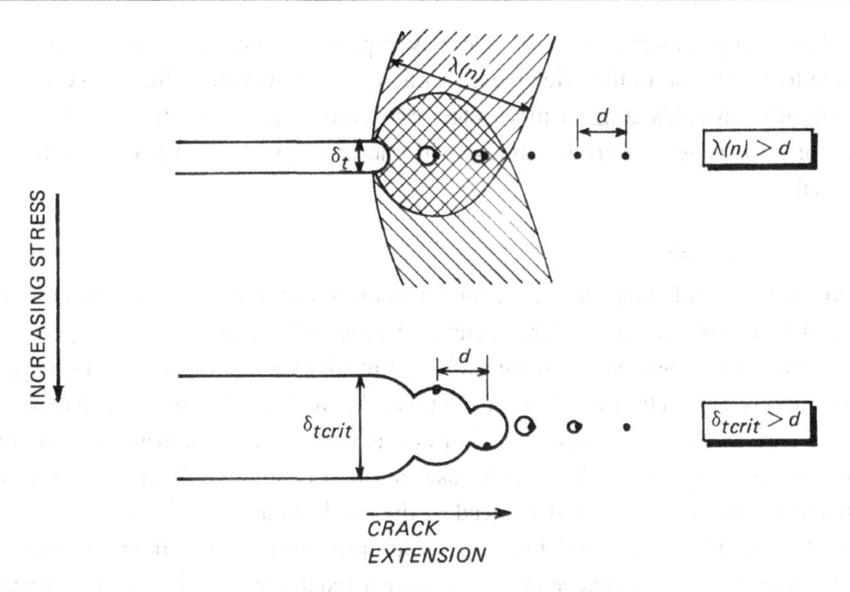

Figure 13.15. Microvoid formation at particles within the region of intense plastic deformation.

$$K_{Ic} = \sqrt{\frac{0.05\, \varepsilon_f n^2 E \sigma_{ys}}{3}} \; (\text{MPa}\sqrt{\text{m}})\,. \tag{13.7}$$

Equation (13.7) was found by Hahn and Rosenfield to be accurate within 30% for eleven different materials.

The derivation of equation (13.7) given here is based on an analysis by Garrett and Knott, reference 15. Alternatively one can arrive at the same result by taking $\varepsilon_f^* = \delta_{t_{crit}}/\lambda_{t_{crit}}$ and $\delta_t \approx K_I^2/2E\sigma_{ys}$, which is the more widely quoted expression for the crack opening displacement under plane strain conditions.

The model of Hahn and Rosenfield contains only macroscopic parameters. The influence of microstructure on fracture toughness is therefore only implicit (*i.e.* by its effect on these parameters) rather than explicit. An obvious extension of the model is to incorporate the observed behaviour of microvoid nucleation and coalescence. This can be done by specifying that the average strain over the distance d between particles must equal ε_f^* for fracture to occur. Then it appears that in order to obtain even very rough agreement with experimentally determined K_{Ic} values the distance d must be considerably less than $\lambda(n)$ and also less than $\delta_{t_{crit}}$. This is illustrated in figure 13.15.

Several relationships between K_{Ic}, particle spacing and other material properties have been derived. Schwalbe (reference 16) lists two of the more successful relationships as:

$$K_{Ic} = \frac{\sigma_{ys}}{1-2\nu}\sqrt{\pi d(1+n)\left(\frac{\varepsilon_f^* E}{\sigma_{ys}}\right)^{1+n}} \tag{13.8}$$

and

$$K_{Ic} = \sqrt{4.55(\varepsilon_f^* + 0.23)\, d\, E\, \sigma_{ys}}\,. \tag{13.9}$$

These equations illustrate that K_{Ic} depends in a complex way on other material properties. For example, as σ_{ys} increases one may expect ε_f^* to decrease, and this decrease is related to a decrease in the effective d (= microvoid diameter) owing to nucleation and growth of microvoids at many more sites. As figure 13.13 shows, the overall result is a trend of decreasing fracture toughness with increasing yield strength for each class of material.

Cleavage in Steels

Cleavage in steels is responsible for the phenomenon of a transition temperature below which brittle, low energy fracture occurs. Because of its great practical importance the occurrence of cleavage has been the subject of much experimental and theoretical work. Knott's book on fracture mechanics, reference 17, reviews the micromechanistic theories of cleavage (see also section 12.5) and a model for the dependence of fracture toughness on temperature. This model uses a semi-quantitative description of cleavage initiated at intergranular carbides ahead of the crack tip and will be discussed with the help of figure 13.16. The model proposes that unstable fracture will occur when a critical fracture stress, σ_f, is exceeded by σ_y over a fixed, characteristic distance ahead of the crack tip. The association of cleavage with cracking of intergranular carbides led Knott and coworkers (reference 18) to choose one or two grain diameters as the characteristic distance.

The dependence of fracture toughness on temperature can be explained as follows.

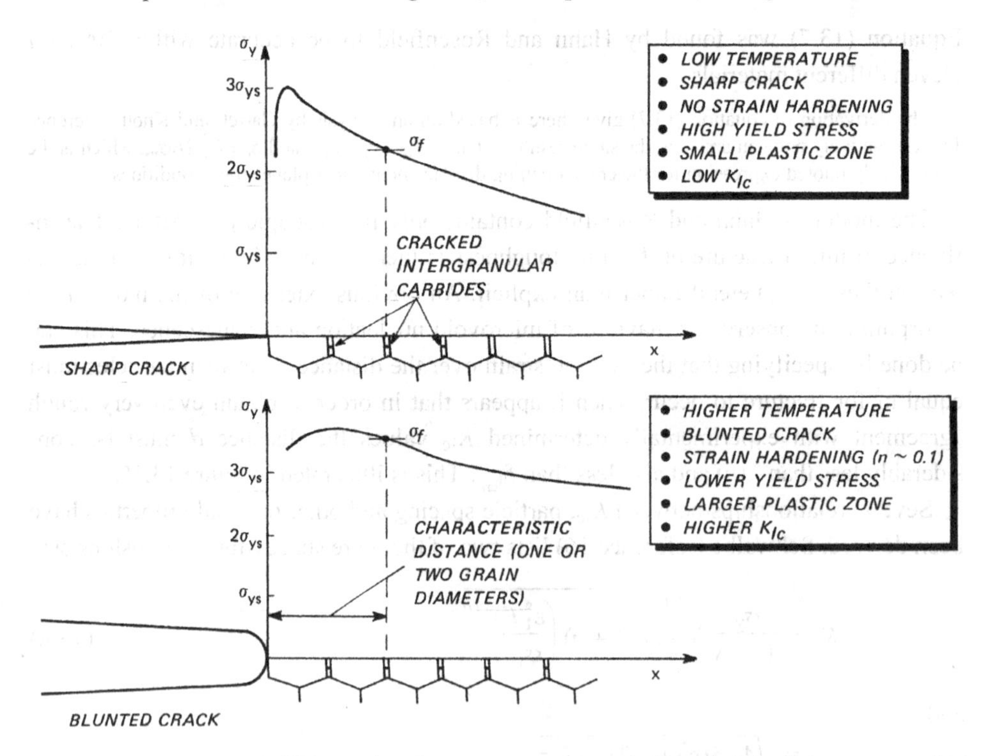

Figure 13.16. Schematic of the critical fracture stress, σ_f, and characteristic distance over which it must be exceeded for unstable cleavage fracture to occur.

At low temperatures the crack is sharp, the material yield stress is high, and little stress intensification is required in order to exceed σ_f over the characteristic distance. Consequently, at failure the plastic zone size is small and K_{Ic} is low. At higher temperatures the crack tip blunts, the material yield stress decreases, and more stress intensification is required for failure, resulting in larger plastic zones and higher K_{Ic} values.

Note that the stress intensification at higher temperatures can exceed $3\sigma_{ys}$. This is because real materials are not elastic – perfectly plastic. Instead crack tip blunting results in strain hardening, which can raise σ_y as high as 4–5 times the nominal yield stress, σ_{ys}. In other words, for real materials the plastic constraint factor, C, (discussed in section 3.5) can be greater than 3.

The model shown in figure 13.16 appears to be broadly correct. There are, however, two additional and important points:

1) Cleavage crack nucleation need not occur directly ahead of the crack, *i.e.* for $\theta = 0°$. This is because the highest values of σ_y occur at $\theta = 50–70°$, see figure 13.17.
   In fact, there is evidence that so-called 'out-of-plane' cleavage does occur ahead of the main crack, reference 19.
2) The characteristic distance over which σ_f must be exceeded by σ_y is not simply related to grain size (*i.e.* one or two grain diameters). Rather, the characteristic distance is a statistically based quantity depending on the volume of material that must be sampled in order to find a cracked carbide greater than the critical size for cleavage nucleation.

Based on this model Curry (reference 20) has given a general expression for the fracture toughness of steels when cleavage occurs:

$$K_{Ic} = \sqrt{\left(\frac{\sigma_f^{n+1}}{\sigma_{ys}^{n-1}}\right)\left(\frac{X}{\beta^{n+1}}\right)}, \qquad (13.10)$$

where X is the characteristic distance, which must be determined empirically, and β is a factor between 3 and 5 that expresses the maximum amount of stress intensification near the crack tip in the actual material, *i.e.* β includes the effects of strain hardening

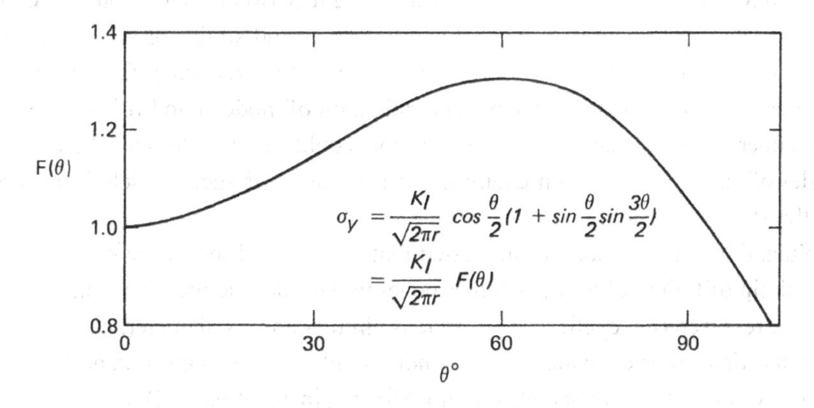

Figure 13.17. Variation of σ_y with θ according to equations (2.24).

and crack blunting.

Equation (13.10) can be used to predict the cleavage-controlled fracture toughness of a steel at different temperatures and with different microstructures provided that

- the dependencies of σ_y and X on microstructure (principally grain size) are known;
- the dependencies of σ_{ys} and n on temperature and microstructure are known;
- the value of β can be estimated from elastic-plastic analysis of crack tip stresses;
- some K_{Ic} data are available as empirical checks on σ_f and X.

These requirements restrict the predictive usefulness of equation (13.10) to steels for which a substantial data bank already exists, for example mild steel.

Fatigue Crack Growth

In the previous chapter the mechanisms of fatigue crack initiation and propagation were discussed in some detail and the concept of dislocation movement to cause slip was introduced. This micromechanistic background is necessary to understanding some of the attempts to model fatigue crack growth in fracture mechanics terms.

There are three aspects of fatigue crack growth that will be considered here:

1) The fatigue crack propagation threshold, ΔK_{th};
2) Relations between cyclic plastic zone size, microstructure and regions I and II fatigue crack growth;
3) Prediction of continuum mode region II fatigue crack growth.

1) The ***fatigue crack propagation threshold, ΔK_{th}***, depends markedly on

- the elastic modulus;
- the microstructure, especially grain size in steels, and phase morphology and texture in titanium alloys;
- the environment;
- the stress ratio, R.

This list is discouraging. Nevertheless, some simplified models of the threshold condition have achieved remarkably good agreement between prediction and experiment. The models consider either critical fracture stresses and strains or critical shear stresses for slip as the criteria for crack extension. In view of the tendency for near-threshold fatigue crack growth to take place by a combination of modes I and II it seems reasonable that either type of crack extension criterion could apply. We shall here discuss one model of each type and then examine the usefulness of such models in understanding the threshold condition.

Yu and Yan (reference 21) suggested that at threshold the cyclic plastic strain, ε_p^c, at a crack tip of finite radius, ρ, becomes equal to the true fracture strain, ε_f. They assumed that the reversed (*i.e.* cyclic) plastic zone is circular and its diameter, r_y^c, can be obtained from the first order estimate of the monotonic plane stress plastic zone by substituting ΔK for K_I and $2\sigma_{ys}$ (reversed plastic flow) for σ_{ys} in equation (3.2), *i.e.*

monotonic plastic zone $$r_y = \frac{1}{2\pi}\left(\frac{K_I}{\sigma_{ys}}\right)^2 , \tag{13.11}$$

cyclic plastic zone $$r_y^c = \frac{1}{2\pi}\left(\frac{\Delta K}{2\sigma_{ys}}\right)^2 . \tag{13.12}$$

The plastic strain distribution within the cyclic plastic zone was taken to be

$$\varepsilon_p^r = 2\varepsilon_{ys}\left(\frac{r_y^c}{r+\rho}\right)^{\frac{1}{1+n}} , \tag{13.13}$$

where r is the distance from the crack tip, n is the strain hardening exponent and $2\varepsilon_{ys}$ represents reversed plastic flow. Substitution of equation (13.12) into equation (13.13) gives

$$\varepsilon_p^c = 2\varepsilon_{ys}\left[\frac{\Delta K^2}{8\pi\sigma_{ys}^2(r+\rho)}\right]^{\frac{1}{1+n}} . \tag{13.14}$$

At threshold $\Delta K = \Delta K_{th}$ and $\varepsilon_p^c = \varepsilon_f$ at the crack tip ($r = 0$). Thus

$$\varepsilon_f = 2\varepsilon_{ys}\left[\frac{\Delta K_{th}^2}{8\pi\sigma_{ys}^2\rho}\right]^{\frac{1}{1+n}} \tag{13.15}$$

or

$$\Delta K_{th} = 2\sigma_{ys}\left(\frac{\varepsilon_f}{2\varepsilon_{ys}}\right)^{\frac{1+n}{2}}\sqrt{2\pi\rho} . \tag{13.16}$$

Since $\varepsilon_{ys} = \sigma_{ys}/E$,

$$\Delta K_{th} = 2\sigma_{ys}\left(\frac{E\varepsilon_f}{2\sigma_{ys}}\right)^{\frac{1+n}{2}}\sqrt{2\pi\rho} . \tag{13.17}$$

Yu and Yan then assumed $n = 1$, which is true only for elastic straining. Then equation (13.17) becomes

$$\Delta K_{th} = E\varepsilon_f\sqrt{2\pi\rho} . \tag{13.18}$$

Equation (13.18) is supposed to be valid for $R = 0$ and a minimum crack tip radius, ρ_{min}, which cannot be less than the Burgers vectors of the dislocations causing plastic deformation. ρ_{min} turns out to be 0.25–0.29 nm (nanometres) depending on the material.

Figure 13.18 compares the predictions of equation (13.18) with experimental values of ΔK_{th} for a wide variety of materials. The agreement is very good. This is remarkable in view of the assumption that the cyclic plastic zone is circular. Actually it is nothing of the kind, as figure 13.19 shows.

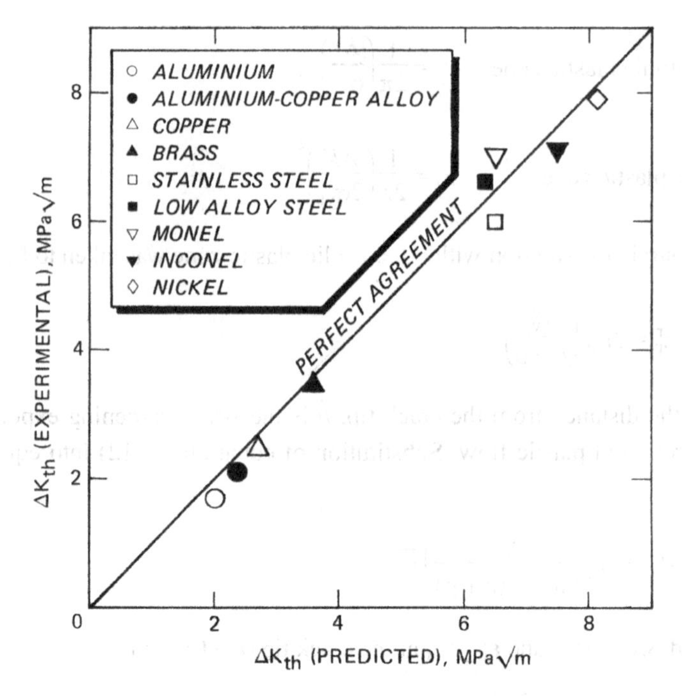

Figure 13.18. Comparison of predicted and experimental values of ΔK_{th} from reference 21.

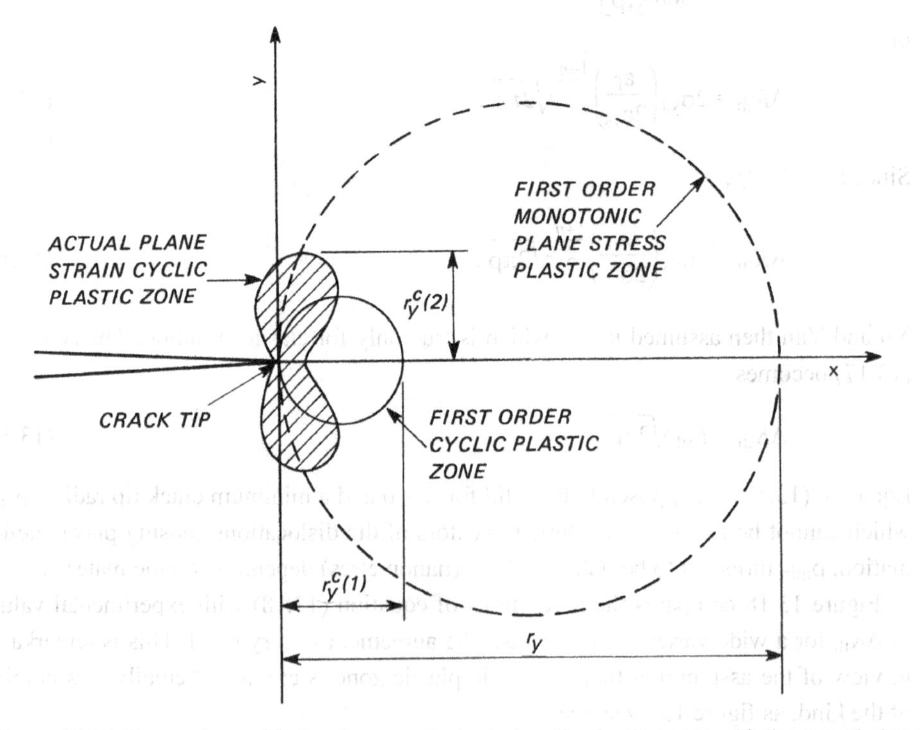

Figure 13.19. Comparison of first order monotonic and cyclic plastic zones with the actual plane strain cyclic plastic zone.

However, the major dimension of the actual cyclic plastic zone is close to the diameter of the first order cyclic plastic zone. Specifically, from reference 22:

$$r_y^c(2) \approx 0.033\left(\frac{\Delta K}{\sigma_{ys}}\right)^2 = 0.83\, r_y^c(1) \; . \tag{13.19}$$

Substitution of $r_y^c(2)$ for $r_y^c(1)$ in the analysis of Yu and Yan increases the predicted ΔK_{th} values by $1/\sqrt{0.83} = 1.1$. The agreement is still good, and so it appears that the major dimension of the cyclic plastic zone is more important than its shape.

Note also that Yu and Yan suggested that the crack tip fracture strain is equal to ε_f, the true fracture strain in a tensile test, whereas Hahn and Rosenfield argued that for unstable crack extension, equation (13.7), the crack tip fracture strain is $\varepsilon_f/3$. This illustrates the uncertainties in the understanding of deformation and fracture at crack tips.

A very different threshold model is the dislocation model of Sadananda and Shahinian, reference 23. A simplified version of the model is shown in figure 13.20. Since fatigue crack growth is usually a consequence of plastic deformation by slip due to the nucleation and movement of dislocations, it is assumed that the threshold condition corresponds to the minimum shear stress required to nucleate and move a dislocation from the crack tip. Sadananda and Shahinian derived a general expression for this shear stress. For a dislocation moving directly away on a slip plane with $\theta = 45°$ the general expression reduces to

$$\tau = \frac{\mu}{4\pi(1-\nu)}\left(\ln\frac{4r}{b} + \frac{b}{r}\right) + \frac{\gamma_e}{b\sqrt{2}} + \frac{\sigma_{ys}}{2} \; , \tag{13.20}$$

where μ is the shear modulus, γ_e is the surface energy and b is the Burgers vector. Now for a sharp crack we can write from equations (2.24):

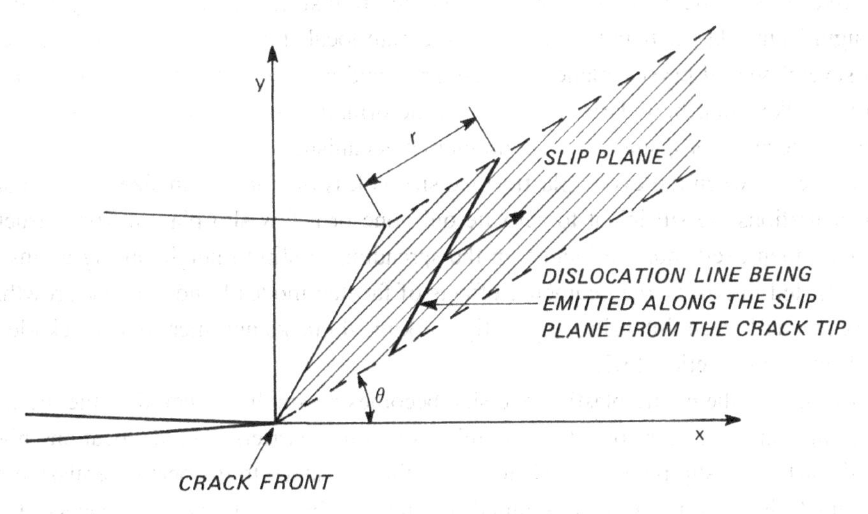

Figure 13.20. Schematic of dislocation emission from a crack tip.

$$\tau_{xy} = \frac{K_I}{\sqrt{2\pi r}} \sin\theta/2 \cos\theta/2 \cos 3\theta/2 . \qquad (13.21)$$

For a dislocation to be nucleated at the crack tip τ_{xy} should be at least equal to τ and r should be at least b. Hence the threshold condition is given by

$$K_{I_{th}} = \frac{\tau\sqrt{2\pi b}}{\sin\theta/2 \cos\theta/2 \cos 3\theta/2} . \qquad (13.22)$$

For $R = 0$ the quantity $K_{I_{th}}$ can be replaced by ΔK_{th}. Also, $\theta = 45°$. Then equation (13.22) reduces to

$$\Delta K_{th} = 18.5\tau\sqrt{b} \qquad (13.23)$$

and for different R values Sadananda and Shahinian suggest

$$\Delta K_{th} = K_{I_{th}}(1 - R) . \qquad (13.24)$$

The model gives good agreement between predicted and experimental values of ΔK_{th}, although there is some uncertainty as to the actual values of surface energy, γ_e, for many metals and alloys, and hence the value of τ.

Both of the models just described show that ΔK_{th} depends strongly on the elastic moduli (E or μ), in agreement with many experimental results and with other models. Also, the cyclic plastic zone size is an important parameter. This is emphasized by a model due to Taylor (reference 24) in which the cyclic plastic zone size is set equal to the grain size and thereby allows reasonable prediction of the dependence of ΔK_{th} on grain size in steels.

With insight provided by the models it is possible to explain the existence of fatigue thresholds. Figure 13.21 is a schematic of how fatigue crack growth depends on the relative sizes of the cyclic plastic zone and the grain size. When the cyclic plastic zone is significantly larger than the grain size the high local stress concentrations induce slip on several sets of crystal planes in each grain and also ensure that slip spreads across barriers like grain boundaries. The plastic deformation is therefore homogeneous and the fatigue crack propagates by a continuum mechanism.

However, when the cyclic plastic zone size is less than the grain size the local stress concentrations are sufficient to activate only one or a few slip planes. Microstructural barriers then exert more influence on the spreading of slip to neighbouring grains, and are assisted in this by the geometric effects of faceted modes I and II crack growth, *i.e.* lower mode I stress intensities at the tips of kinked cracks and increased crack closure, as discussed in section 13.2.

Eventually the cyclic plastic zone size becomes so small compared to the grain size that slip can no longer spread across microstructural barriers. The dislocations pile up on the activated slip plane(s). This increases the resistance to dislocation emission from the crack tip until no more are emitted: the fatigue threshold has been reached. Cyclic slip can still occur, but is confined to to-and-fro movement of dislocations along the slip

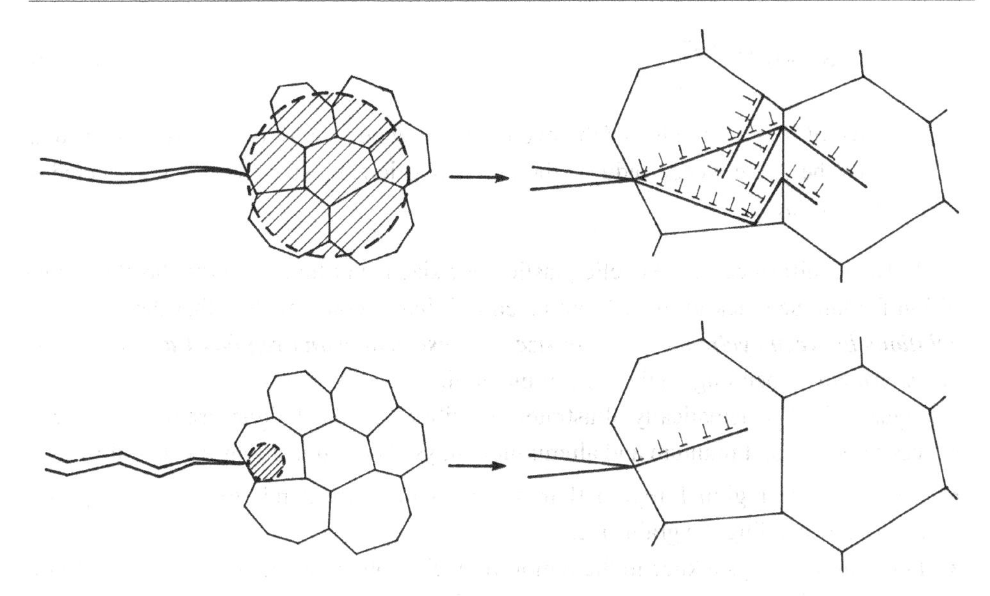

Figure 13.21. Dependence of fatigue crack growth on cyclic plastic zone size and grain size.

plane as far as a microstructural barrier (typically a grain boundary, but in aluminium alloys dispersoid particles may also act as barriers, reference 25).

The foregoing explanation is mechanical and takes no account of the environment. An aggressive environment usually lowers ΔK_{th}, but sometimes there is little or no change or even an increase in ΔK_{th}. As mentioned in section 9.4, in some instances (mainly for steels) an increase in ΔK_{th} is due to corrosion contributing to crack closure, *i.e.* the environmental effect is largely mechanical. The way this occurs is illustrated in figure 13.22. In the near threshold region of crack growth there is an enhanced build-up of corrosion product (oxide) on the crack surfaces close behind the crack tip. This oxide build-up acts as a wedge and reduces the cyclic crack tip opening displacement δ_t^c, which under plane strain conditions is approximately given by

$$\delta_t^c = 0.5\frac{\Delta K^2}{E2\sigma_{ys}^c}, \tag{13.25}$$

where σ_{ys}^c is the cyclic yield stress. For many materials $\sigma_{ys}^c \approx \sigma_{ys}$, *i.e.*

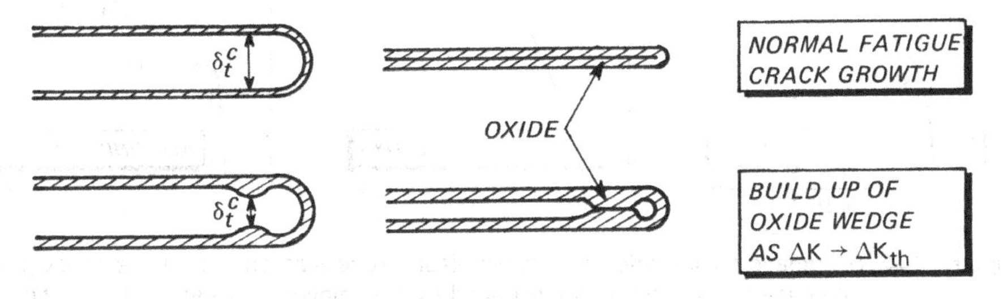

Figure 13.22. Oxide wedge mechanism of raising ΔK_{th}. After reference 26.

$$\delta_t^c \approx 0.25 \frac{\Delta K^2}{E\sigma_{ys}}. \tag{13.26}$$

Irrespective of the exact value of the cyclic plastic zone size, r_y^c is directly proportional to δ_t^c. Thus the oxide wedge reduces the cyclic plastic zone size and this is reflected in an increase in ΔK_{th}.

2) The significance of the cyclic plastic zone size is not limited to the threshold condition for fatigue crack growth. More recently it has become evident that there are also ***relations between cyclic plastic zone size, microstructure and regions I and II fatigue crack growth***, depending on the type of material.

Figure 13.23 schematically illustrates transitions in the fatigue crack growth rate curves for steels and titanium and aluminium alloys. These transitions are as follows:

- For steels the region I–region II transition point corresponds to the cyclic plastic zone size equalling the grain size.
- For titanium alloys a knee in the region II crack growth rate curve corresponds to the cyclic plastic zone size equalling the grain size.
- For aluminium alloys there appear to be at least three transition points. At the region I–region II transition point (T_1) the cyclic plastic zone size is approximately equal to the spacing between dispersoid particles. For the region II transition points T_2 and T_3 the cyclic plastic zone size correlates with subgrain size and grain size respectively. (Subgrains are regions within a grain that differ slightly in crystal orientation with respect to each other. The boundaries between subgrains are dislocation arrays that are barriers to slip, though less effective than grain boundaries.)

The transition points in steels and conventional titanium alloys represent a change from faceted fracture at lower ΔK to a continuum mechanism of crack propagation. This is as expected from the correlation between the cyclic plastic zone size and the grain

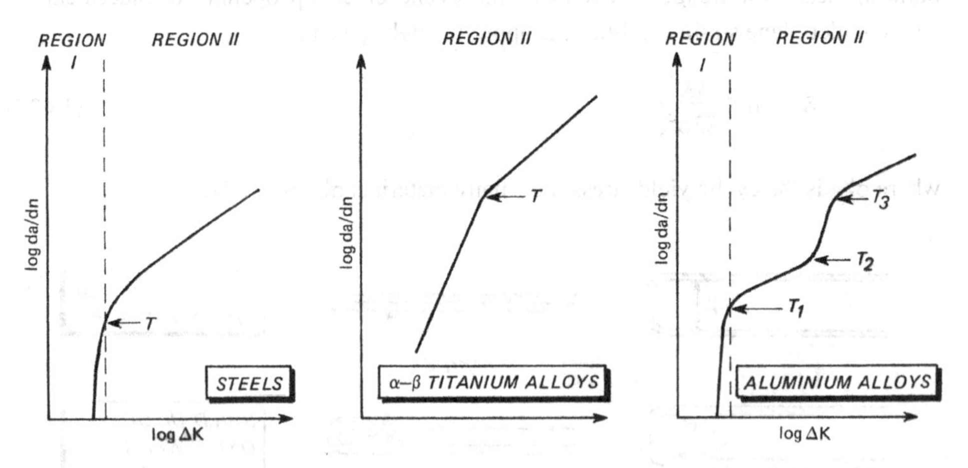

Figure 13.23. Schematic for the effects of cyclic plastic zone size on the fatigue crack growth rate curve: T = transitions influenced by the relation between cyclic plastic zone size and microstructure (see text). After references 25, 27 and 28.

size at these transitions. However, for aluminium alloys the transition point T_3 does not correspond to a change from faceted to continuum-type crack growth. Instead, this change occurs at T_2, reference 29. A possible explanation is that even though the cyclic plastic zone is smaller than the grain size below T_3, the misorientations of the subgrains cause slip to be fairly homogeneous. Only when the cyclic plastic zone is smaller than the subgrain size, below T_2, is slip confined to one or a few sets of crystal planes.

3) Finally, we shall consider briefly the ***prediction of continuum mode region II fatigue crack growth***. Many empirical models have been formulated (see section 9.2) but they do not contribute to understanding the crack growth process. Only a few attempts have been made at absolute predictions of fatigue crack growth rates. The problems to be faced are similar to those for modelling fracture toughness: stress-strain distributions in the plastic zone ahead of the crack must be known or assumed, and the proper criteria for crack advance must be chosen. In this respect all models are severely limited, since they consider purely mechanical crack extension, *i.e.* environmental effects cannot be accounted for. This point is often overlooked when model predictions are compared with experimental data. Nearly always the data are for fatigue tests in air, which has a strong environmental influence compared to vacuum or dry gases.

The models fall into two main categories based on the criteria for crack advance, which is assumed to occur during each load cycle by either

- Alternating shear at the crack tip, *i.e.* incompletely or totally irreversible slip on alternate sets of crystal planes.
- Exceedance of the true fracture strain over a certain distance.

Alternating shear models predict relations of the form

$$\frac{da}{dn} = C\frac{\Delta K^2}{E\sigma_{ys}}, \tag{13.27}$$

i.e. linear log-log behaviour with a slope $m = 2$. In fact m is rarely 2, nor is m always constant during region II fatigue crack growing. Note that equation (13.27) predicts that da/dn is inversely proportional to E. This relation is supported by experimental data. In figure 13.24 the crack growth rate data of figure 9.6 are plotted as functions of $\Delta K/E$ instead of ΔK. The widely varying crack growth rate curves for various materials are brought together into a single (wide) scatterband.

The simplest model is that of Pelloux, reference 30. In this model all the slip takes place at the crack tip and contributes to crack advance, as shown in figure 13.25.

It is evident that the crack advance, δ_a, is one-half the cyclic crack tip opening displacement, δ_t^c. Thus from equation (13.25) crack advance under plane strain conditions is given by

$$\delta_a = \frac{\delta_t^c}{2} = \frac{\Delta K^2}{8E\sigma_{ys}^c}, \tag{13.28}$$

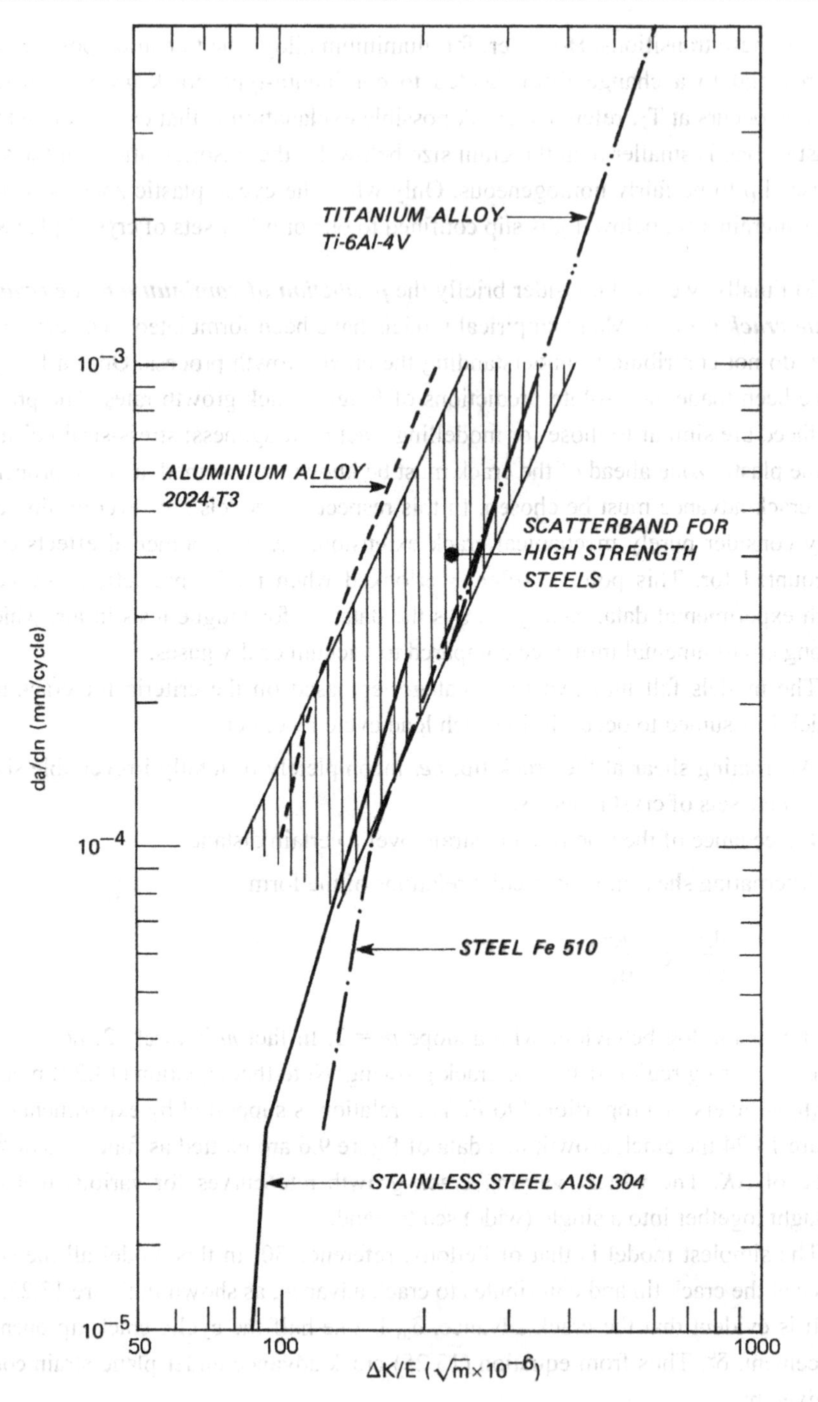

Figure 13.24. Fatigue crack growth rates as functions of $\Delta K/E$ for the data of figure 9.6.

where δ_a is the crack advance per cycle, *i.e.* $\delta_a = da/dn$.

Other alternating shear models allow only part of the slip to contribute to crack advance: the remainder causes crack blunting. Kuo and Liu (reference 31) proposed a

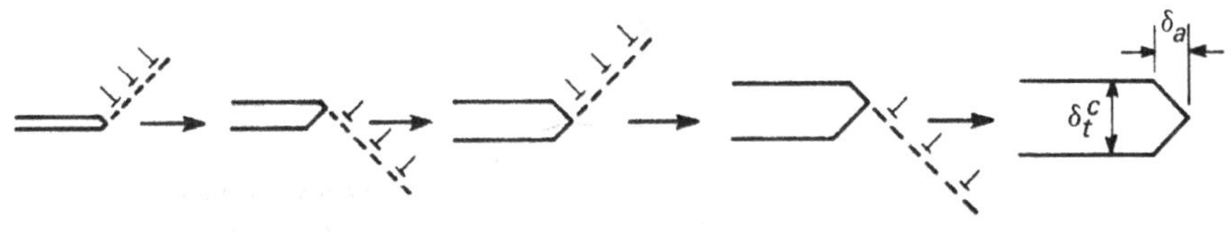

Figure 13.25. Alternating shear model of fatigue crack growth.

model based on both theoretical considerations and experimental determinations of COD and strains near the crack tip. They obtained the following semi-empirical formula:

$$\frac{da}{dn} = \frac{0.019(1-\nu^2)\Delta K^2}{E\sigma_{ys}^{c}}. \tag{13.29}$$

Figure 13.26 compares the models of Pelloux and Kuo and Liu with data for two titanium alloys tested *in vacuo, i.e.* purely mechanical crack extension. The model of Kuo and Liu is clearly in better agreement with the test data. This shows that slip in the crack tip region mainly causes crack blunting. Only a small amount of slip contributes to actual crack advance.

Of the models using a crack advance criterion based on true fracture strain, that of Antolovich *et al.* (reference 32) provides the most insight into the mechanisms of region II fatigue crack growth. The model will be discussed using figure 13.27, which shows that the fatigue crack is assumed to be blunt and that within the cyclic plastic zone there is a 'process zone'. Note also that r_y^c is defined as the plane strain cyclic plastic zone

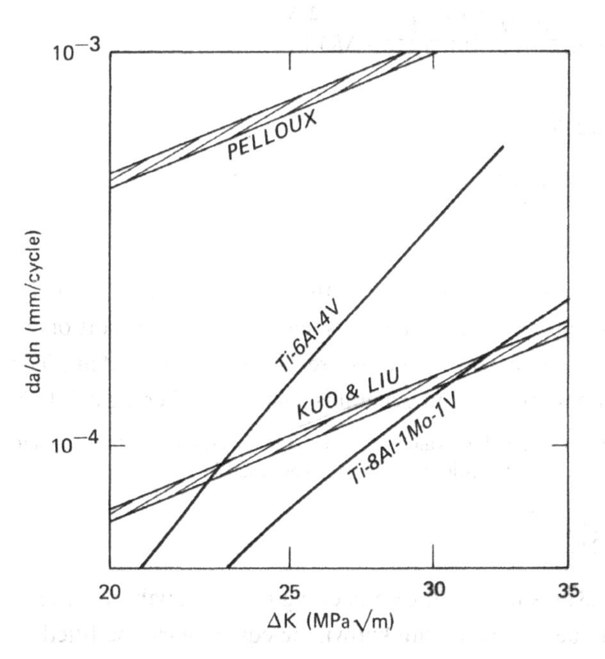

Figure 13.26. Comparison of alternating shear models with test data.

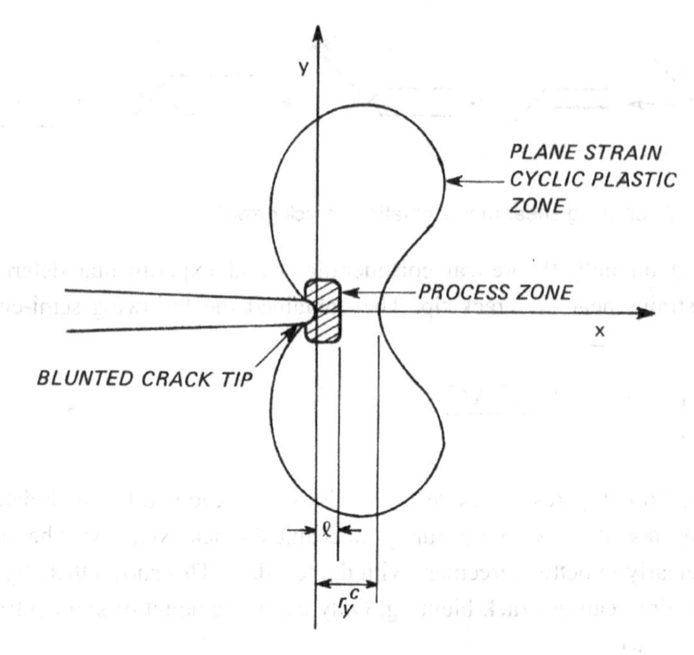

Figure 13.27. Schematic of cyclic plastic and process zones ahead of a propagating fatigue crack.

size along the x axis.

Antolovich *et al.* derived a rather complicated expression for the fatigue crack growth rate:

$$\frac{da}{dn} = 4\left(\frac{0.7\alpha}{E\sigma_{ys}^{1+S}\varepsilon_f}\right)^{\frac{1}{\beta}}\left(\frac{1}{\ell^{1/\beta-1}}\right)(\Delta K)^{\frac{2+S}{\beta}}, \tag{13.31}$$

where S and α are defined by

$$r_y^c = \alpha\left(\frac{\Delta K}{\sigma_{ys}}\right)^{2+S} \tag{13.32}$$

and β is the Coffin-Manson low cycle fatigue exponent, which usually lies between 0.4 and 0.6. Equation (13.32) shows that r_y^c is not taken to be proportional to $(\Delta K/\sigma_{ys})^2$. This is because crack tips in real materials are blunt to some extent. The parameter S accounts for crack blunting, and in general is a small number $\approx \pm 0.1$–0.2.

For a sharp crack $S = 0$, and the value of r_y^c can be obtained directly from equation (3.27) in chapter 3 by substituting $2\sigma_{ys}$ (reversed plastic flow) for σ_{ys}. The result is

$$r_y^c = \frac{1}{24\pi}\left(\frac{\Delta K}{\sigma_{ys}}\right)^2. \tag{13.33}$$

Equation (13.31) is the result of modelling crack growth in the process zone in terms of low cycle fatigue. If S and β are known the equation can be fitted to actual data in order to determine the process zone size, ℓ. This acts as a check on the model: there should be a reasonable physical interpretation of ℓ to support the assumption that continuum mode region II fatigue crack growth is the result of a low cycle fatigue process occurring near the crack tip. In fact, ℓ is about the size of dislocation cells, which are polygonal arrangements of dense tangles of dislocations that form during both low cycle fatigue and region II fatigue crack growth. The assumptions of the model therefore appear to be justified (for a more detailed treatment see reference 33).

Sustained Load Fracture

A comprehensive treatment of mechanisms of sustained load fracture is a forbidding task well beyond the scope of this course. The discussion will be restricted to a few examples where fracture mechanics concepts have been used to study such mechanisms. These examples are:

1) Time-to-failure (TTF) tests in modes I and III.
2) Cleavage during stress corrosion of titanium alloys.
3) Susceptibility of steels to hydrogen embrittlement.
4) Analysis and modelling of crack growth rates.

(1) The background to ***TTF testing in modes I and III*** is that sustained load fracture for some material-environment combinations may occur by hydrogen embrittlement as a result of diffusion of hydrogen to a location ahead of the crack. The increased hydrogen concentration at this location then results in cracking that links up with the main crack.

In such cases any factor that promotes hydrogen diffusion to a location ahead of the main crack should increase the susceptibility to sustained load fracture. One possibility is dilatation (three-dimensional expansion) of the crystal lattice owing to a state of hydrostatic stress, *i.e.* a state in which, strictly speaking, all three principal stresses are equal. In the elastic stress fields of cracks the three principal stresses are not all equal to each other, even in plane strain. However, it is convenient to split the triaxial stress state into a hydrostatic component

$$\sigma_h = \frac{1}{3}(\sigma_1 + \sigma_2 + \sigma_3) \qquad (13.34.a)$$

and a stress deviator s

$$s_1 = \sigma_1 - \sigma_h, \quad s_2 = \sigma_2 - \sigma_h, \quad s_3 = \sigma_3 - \sigma_h. \qquad (13.34.b)$$

From equations (2.33) and (2.43) in chapter 2 it follows that

$$\text{mode I plane strain} \quad \sigma_h = \frac{2(1+\nu)K_I}{3\sqrt{2\pi r}} \cos \theta/2, \qquad (13.35)$$

$$\text{mode III} \quad \sigma_h = 0. \qquad (13.36)$$

In other words, for mode I loading σ_h reaches a maximum directly ahead of the crack, but for mode III loading there is no hydrostatic stress component.

Thus if stress-assisted hydrogen diffusion and embrittlement play a role in sustained

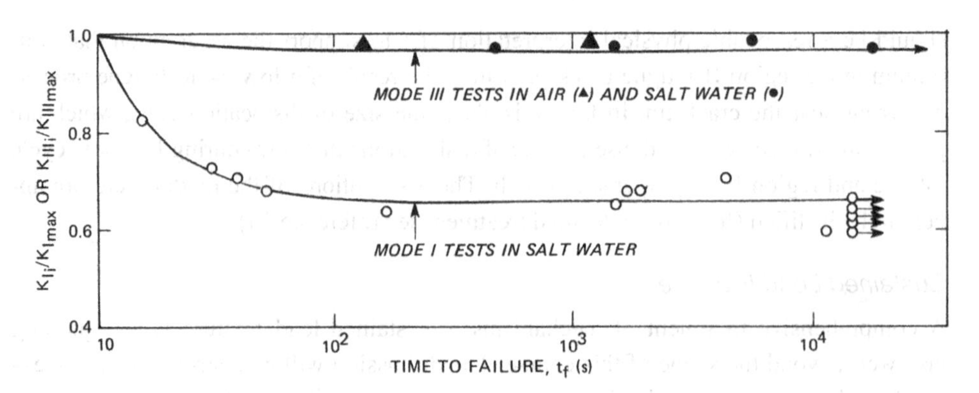

Figure 13.28. Stress corrosion susceptibility of precracked titanium alloy Ti-8Al-1Mo-1V specimens under mode I and mode III loading. After reference 34.

load fracture, then comparison tests in mode I and mode III should show differences in susceptibility. This does, in fact, appear to be the case for stress corrosion cracking of steels and aluminium and titanium alloys. An example is given in figure 13.28, which shows that the susceptibility of a titanium alloy to stress corrosion cracking was considerable in mode I but negligible in mode III.

It should be noted that there are also hydrogen embrittlement models for which the concentration of hydrogen in the crystal lattice does not depend on a hydrostatic stress component, *e.g.* transport of hydrogen by dislocations or the hindering of dislocation movement by hydrogen. In such cases there need not be any difference in susceptibility under mode I and mode III loading.

(2) The concept of stress-assisted hydrogen diffusion and embrittlement has also been examined with respect to α-phase ***cleavage during stress corrosion of titanium alloys***, reference 35. In highly textured alloys a mode I crack can be approximately parallel to the cleavage planes in most grains. Under these conditions the stress corrosion fracture path is very flat, *i.e.* cleavage occurs directly ahead of the crack. Figure 13.29 shows that this is evidence for hydrogen embrittlement control of the cleavage location rather than mechanical control, since σ_h is a maximum directly ahead of the crack ($\theta = 0°$) but σ_y is a maximum at $\theta = 65°$ Mechanical control would be expected to cause out-of-plane cleavage, discussed earlier in this section with respect to steels (see the discussion to figure 13.17).

(3) Apart from the more generally applicable TTF comparison tests in modes I and III, fracture mechanics analyses of the susceptibility to sustained load fracture are concerned mostly with ***hydrogen embrittlement of steels***, an important practical problem. For example, steel pipelines and pressure vessels are used for transport and storage of hydrogen-containing environments, and power generating equipment may have to operate in a hydrogen atmosphere. These are examples where *external* hydrogen may be a problem. *Internal* hydrogen, *i.e.* residual hydrogen within the metal, is also important, especially for welds in high strength structural steel, and ultra high strength steels as used in most aircraft landing gear and many other components where high strength is essential.

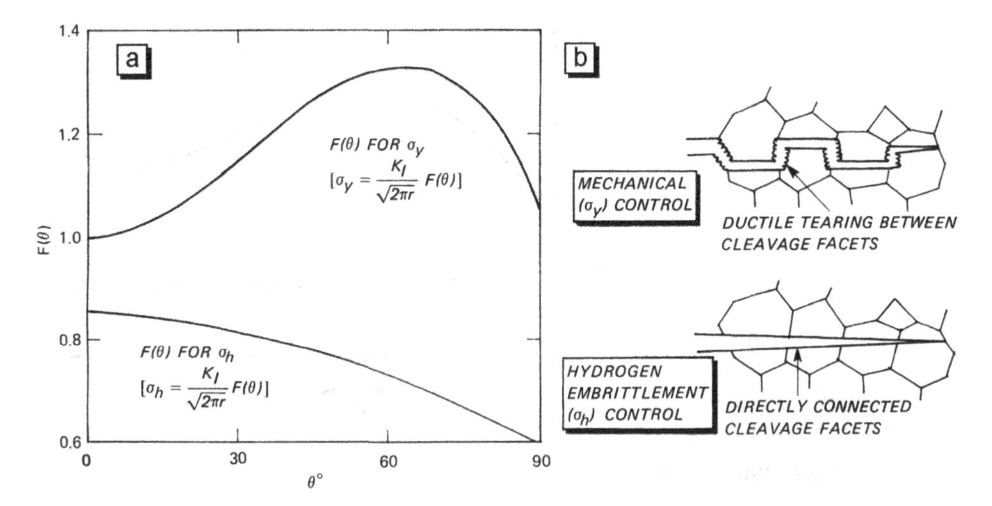

Figure 13.29. Variation of σ_y and σ_h with θ and predicted stress corrosion fracture paths for a mode I crack parallel to the cleavage planes in a highly textured titanium alloy.

With respect to external hydrogen embrittlement, fracture mechanics stress field solutions have been used to distinguish between two types of theory:

- The adsorption theory, which assumes that hydrogen is adsorbed at the crack tip and changes the surface energy of the crystal lattice, thereby lowering the crack resistance. The surface energy change, and hence the amount of embrittlement, depend on the external gas pressure.
- The absorption + decohesion theory. This is another name for stress-assisted hydrogen diffusion ahead of the crack tip and the reaching of a sufficient concentration of hydrogen to cause cracking that links up with the main crack. The theory is applicable not only to hydrogen embrittlement per se, but also to stress corrosion of steels (discussed later in this section). Again, the external gas pressure determines the amount of embrittlement: higher pressure decreases $K_{I_{th}}$.

Details of the very complex analyses of external hydrogen embrittlement will not be given here: reference 36 of the bibliography contains a comprehensive, though dated, review. However, the results are important and appropriate:

- the adsorption theory fails to predict realistic $K_{I_{th}}$ values;
- absorption + decohesion theories which require lattice decohesion to occur very close to the crack tip provide the best predictions of $K_{I_{th}}$ as a function of external gas pressure.

The physical significance of the second result is that the experimental data with which the models were compared must have been obtained from materials with sharp cracks, such that the maximum value of σ_h (and hence σ_x, σ_y and σ_z) was well within the plastic zone, compare figure 13.3. There is thus a clear correlation between crack tip sharpness and the susceptibility to hydrogen embrittlement. This agrees with the fact that steels with higher yield strengths and less capacity for plastic deformation are more susceptible.

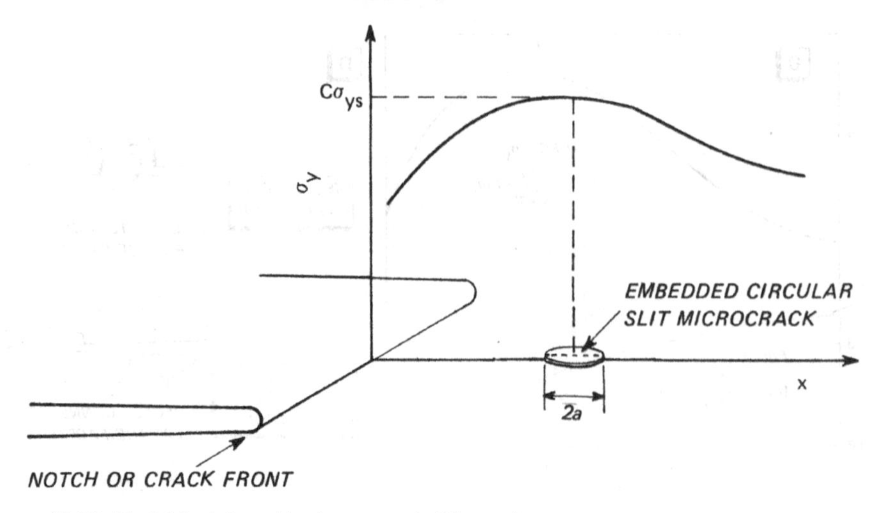

Figure 13.30. Model for internal hydrogen embrittlement.

Internal hydrogen embrittlement seems at first to be a different phenomenon from external hydrogen embrittlement: after all, there is no apparent source of gaseous hydrogen. However, van Leeuwen (reference 37) has provided a model to show that the absorption + decohesion theory can be applied to internal as well as external hydrogen embrittlement. The model is illustrated in figure 13.30. Ahead of a notch or crack a microcrack forms by decohesion at the location of maximum hydrogen concentration. This is also the location of maximum hydrostatic stress and the maximum value of σ_y, $C\sigma_{ys}$ (C is the plastic constraint factor, previously mentioned in this section with respect to cleavage in steels).

Formation of the microcrack results in a local decrease in hydrostatic stress. Hydrogen can no longer be held in the lattice at the same concentration, and therefore enters the microcrack as gas, building up to an equilibrium pressure, P_e. The stress intensity factor for the microcrack can be obtained from the well-known solution in section 2.8, *i.e.*

$$K_I = \frac{2}{\pi}\sigma\sqrt{\pi a} = \frac{2}{\pi}(C\sigma_{ys} + P_e)\sqrt{\pi a}\,. \tag{13.37}$$

Equation (13.37) shows that an increase in P_e will increase K_I. At the same time, by analogy with external hydrogen embrittlement, an increase in P_e should decrease $K_{I_{th}}$ for the microcrack. Thus critical combinations of K_I and P_e will result in exceedance of $K_{I_{th}}$ for the microcrack, which then will propagate back to the main crack or notch.

When the microcrack links up with the main crack or notch the gas pressure drops to zero. Thus crack growth stops until another microcrack is nucleated ahead of the new crack front. This means that the total crack propagation process must take place in discrete steps, and this is exactly what happens.

(4) The last topic to be discussed is the ***analysis and modelling of sustained load crack growth rates***. As in the case of fatigue crack growth, of the many models proposed for sustained load crack growth only a few are quantitative in fracture mechanics terms.

The difficulties of analysis are formidable, and the phenomenological behaviour is often complex. Models predicting a dependence of da/dt on K_I include:

- stress-assisted dissolution of the crack tip material (region I crack growth);
- stress-assisted diffusion of hydrogen (absorption + decohesion) for regions I and II crack growth in steels owing to stress corrosion and external and internal hydrogen embrittlement;
- the capillary model of stress corrosion.

The stress-assisted dissolution model predicts a linear dependence of $\log(da/dt)$ on K_I, *i.e.* region I crack growth. The type of equation obtained is

$$\log\frac{da}{dt} = \log A + \frac{2.3}{RT}\left(\frac{2V^* K_I}{\sqrt{\pi\rho}} - E^*\right), \qquad (13.38)$$

where A is a constant, R is the gas constant, T is temperature, ρ is the crack tip radius, E^* is the activation energy of dissolution at zero load, and V^* is an 'activation volume' whose precise physical meaning is unknown. A more serious objection to the model is that it requires extremely high stresses $\approx E/20$ to exist in the crack tip region, reference 38. This means that the model cannot be applied to metallic materials, but it may be appropriate to ceramics.

A fairly successful model for sustained load crack growth due to stress-assisted diffusion of hydrogen has been derived by van Leeuwen (reference 39) for stress corrosion of high strength steels. Van Leeuwen proposed that the concentration of hydrogen reaches a maximum at the elastic-plastic boundary directly ahead of the crack. When the maximum hydrogen concentration reaches a critical value, H_{cr}, decohesion occurs at the elastic-plastic boundary to cause cracking that links up with the main crack, as shown schematically in figure 13.31. Overall crack propagation is treated as a series of such steps, as in the case of internal hydrogen embrittlement.

Note that it is an assumption of the model that the crack tip is sufficiently blunt for the hydrostatic stress to be a maximum at the elastic-plastic boundary directly ahead of the crack. This is a significant difference from what were found to be the requirements for predicting $K_{I_{th}}$ in gaseous hydrogen, *i.e.* sharp crack tips and lattice decohesion commencing well within the plastic zone. However, there is ample evidence that a stress corrosion crack can be blunted not only by plasticity but also by corrosion.

To account for crack tip blunting van Leeuwen used the mode I elastic stress field equations derived by Creager and Paris, *i.e.* equations (13.1) discussed at the beginning of this chapter. The following general expression for the stress intensity factor K_I was obtained, assuming $\theta = 0$ in equation (13.1):

$$K_I = \sigma_y\sqrt{2\pi r}\left(\frac{2r}{2r+\rho}\right). \qquad (13.39)$$

Substituting $C\sigma_{ys}$ for σ_y and $r = r_y + \rho/2$ gives

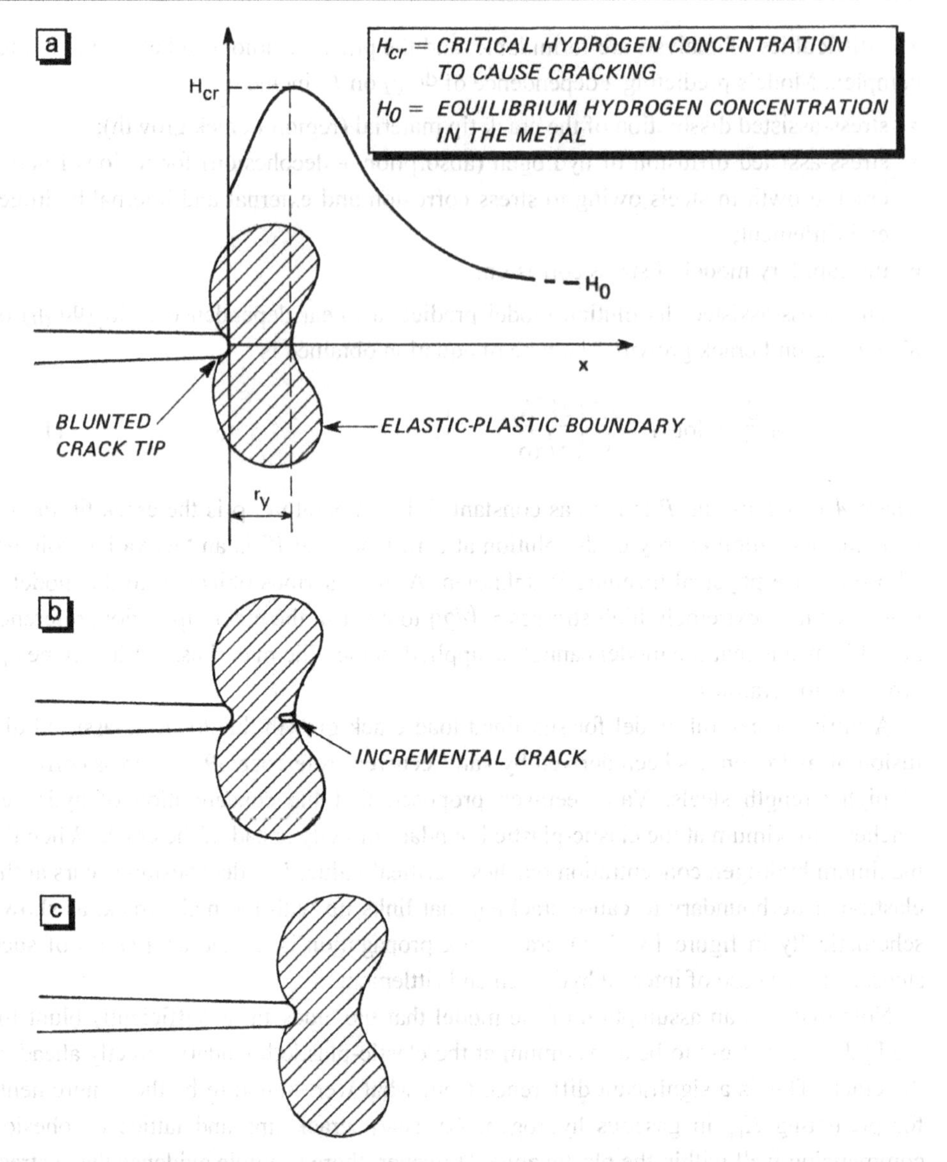

Figure 13.31. Hydrogen absorption + decohesion model of stress corrosion cracking in high strength steels.

$$\text{for a blunt crack} \quad K_I = \frac{C\sigma_{ys}\sqrt{2\pi}}{r_y + \rho}\left(r_y + \frac{\rho}{2}\right)^{\frac{3}{2}},$$

$$\text{for a sharp crack} \quad K_I = C\sigma_{ys}\sqrt{2\pi r_y}, \tag{13.40}$$

where r_y is the distance along the x axis to the elastic plastic boundary and is also the increment of crack growth. A problem is that equation (13.40) will have to be solved by iteration. The actual value of r_y is not known since it will depend on the plastic constraint factor C (compare section 3.5). But C is not known and can only be estimated by substituting $r = r_y + \rho/2$ in equations (13.1) and calculating the principal stresses σ_1 and σ_2 for $\theta = 0°$. Substituting these stresses and $\sigma_3 = \nu(\sigma_1 + \sigma_2)$ in the Von Mises yield criterion results in

$$C = \frac{2(r_y + \rho)}{\sqrt{3\rho^2 + (2r_y + \rho)^2(1 - 2\nu)^2}} . \tag{13.41}$$

Thus C in its turn depends on what value of r_y is substituted.

Van Leeuwen determined a very complicated expression for the relation between da/dt and the hydrogen concentration. Since the critical hydrogen concentration, H_{cr}, depends on r_y and ρ, da/dt will depend on r_y and ρ as well. Choosing various values of r_y and ρ, solving equations (13.40) and (13.41) iteratively, and calculating da/dt from the values of r_y and ρ resulted in a series of curves relating da/dt and K_I. An example is given in figure 13.32. Good fits to experimental data are possible, especially for region II crack growth.

Note that the choice of ρ is always somewhat arbitrary. However, the value of r_y which is used in calculating K_I also determines da/dt, *i.e.* r_y is not a second arbitrary parameter for data fitting.

Van Leeuwen also showed that the da/dt–K_I curve can be estimated remarkably well using equations

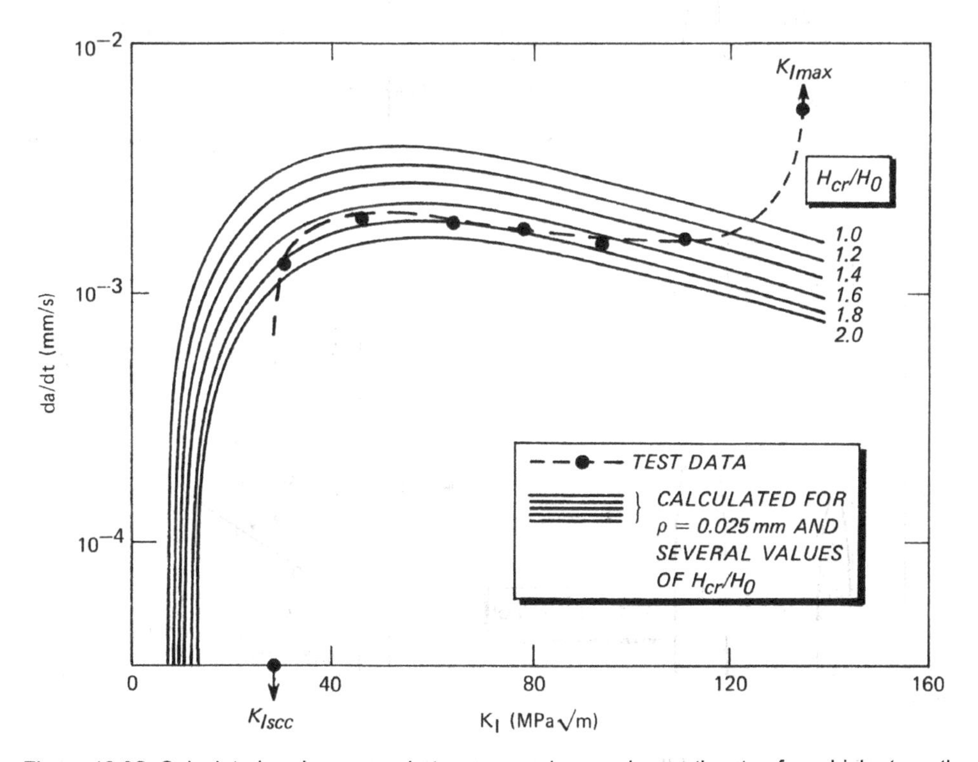

Figure 13.32. Calculated and measured stress corrosion crack growth rates for a high strength steel in salt water. After reference 39.

(13.40) and the simple expression

$$\frac{da}{dt} = \frac{4Dr_y}{(r_y + \rho/2)^2}, \tag{13.42}$$

where D is the diffusion constant for hydrogen in steel ($\approx 2 \times 10^{-7}$ cm^2/s at room temperature).

Since the foregoing discussion was written, in the 1980s, there have been considerable advances in the understanding of hydrogen embrittlement, though not necessarily in the context of fracture mechanics. The interested reader is referred to a special issue of the well-known journal Engineering Fracture Mechanics (reference 40).

The capillary model of stress corrosion (reference 41) proposes that under certain conditions a crack may be incompletely penetrated by liquid, so that if crack propagation occurs the liquid must flow down the crack to maintain the propagation. If this dy-

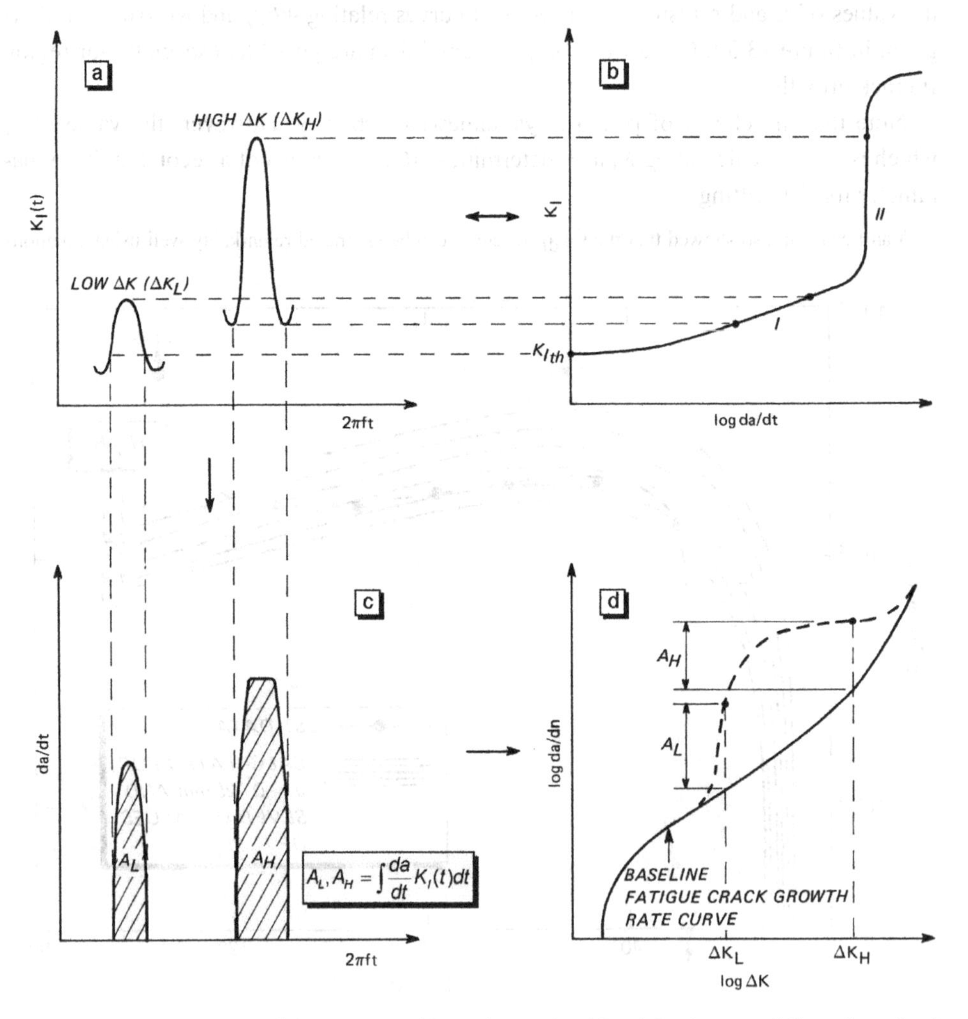

Figure 13.33. Schematic of the superposition of sustained load fracture on fatigue in order to construct the overall crack growth rate curve.

namic capillary fluid flow is rate controlling the model predicts the existence of regions I and III crack growth. The model also predicts dependencies of the crack growth rate curve on fluid viscosity and external gas pressure. Limited testing of the model has been done. The most that can be said is that capillary flow is likely to be a factor in stress corrosion cracking, hut it is not known whether it can be rate controlling. It is worth noting, however, that capillary flow could be rate controlling for crack growth due to liquid metal embrittlement, since the velocities of cracking are often extremely high, ≈1–10 $^{cm}/_{s}$.

Superposition or Competition of Sustained Load Fracture and Fatigue

As mentioned in section 9.4, fatigue crack growth in aggressive environments can be enhanced by sustained load fracture during each load cycle. There are two fairly simple models which try to account for this effect:

- the superposition model, reference 42;
- the process competition model, reference 43.

The superposition model proposes that the overall crack growth rate is the sum of a baseline fatigue component and a component due to sustained load fracture. On the other hand, the process competition model assumes that fatigue and sustained load fracture are mutually competitive and that the crack will grow at the fastest available rate, whether that is the baseline fatigue crack growth rate or the crack growth per cycle owing to sustained load fracture.

The models can be expressed formally in the following way:

superposition model
$$\left(\frac{\mathrm{d}a}{\mathrm{d}n}\right)_{\mathrm{tot}} = \left(\frac{\mathrm{d}a}{\mathrm{d}n}\right)_{\mathrm{B}} + \int \frac{\mathrm{d}a}{\mathrm{d}t} \cdot K_{\mathrm{I}}(t)\,\mathrm{d}t\,, \tag{13.43}$$

process competition model
$$\left(\frac{\mathrm{d}a}{\mathrm{d}n}\right)_{\mathrm{tot}} = \left(\frac{\mathrm{d}a}{\mathrm{d}n}\right)_{\mathrm{B}} \quad \text{for} \left(\frac{\mathrm{d}a}{\mathrm{d}n}\right)_{\mathrm{B}} > \int \frac{\mathrm{d}a}{\mathrm{d}t} \cdot K_{\mathrm{I}}(t)\,\mathrm{d}t$$
$$= \int \frac{\mathrm{d}a}{\mathrm{d}t} \cdot K_{\mathrm{I}}(t)\,\mathrm{d}t \quad \text{for} \left(\frac{\mathrm{d}a}{\mathrm{d}n}\right)_{\mathrm{B}} > \int \frac{\mathrm{d}a}{\mathrm{d}t} \cdot K_{\mathrm{I}}(t)\,\mathrm{d}t\,, \tag{13.44}$$

where the integral in equations (13.43) and (13.44) is taken over one cycle of the fatigue loading and incorporates the effects of frequency, f, and stress ratio, R, via $K_I(t)$.

Figure 13.33 is a schematic of the way in which the superposition model can be used to predict the overall crack growth rate curve for constant amplitude sinusoidal loading. For each ΔK value of interest the integral in equation (13.43) can be obtained as the area A under the curve relating $^{da}/_{dt}$ and the load cycle wave form and frequency characteristics (= $2\pi ft$ for sinusoidal loading). Addition of each A value to the baseline fatigue crack growth rate at the same ΔK results in the predicted overall crack growth rate curve.

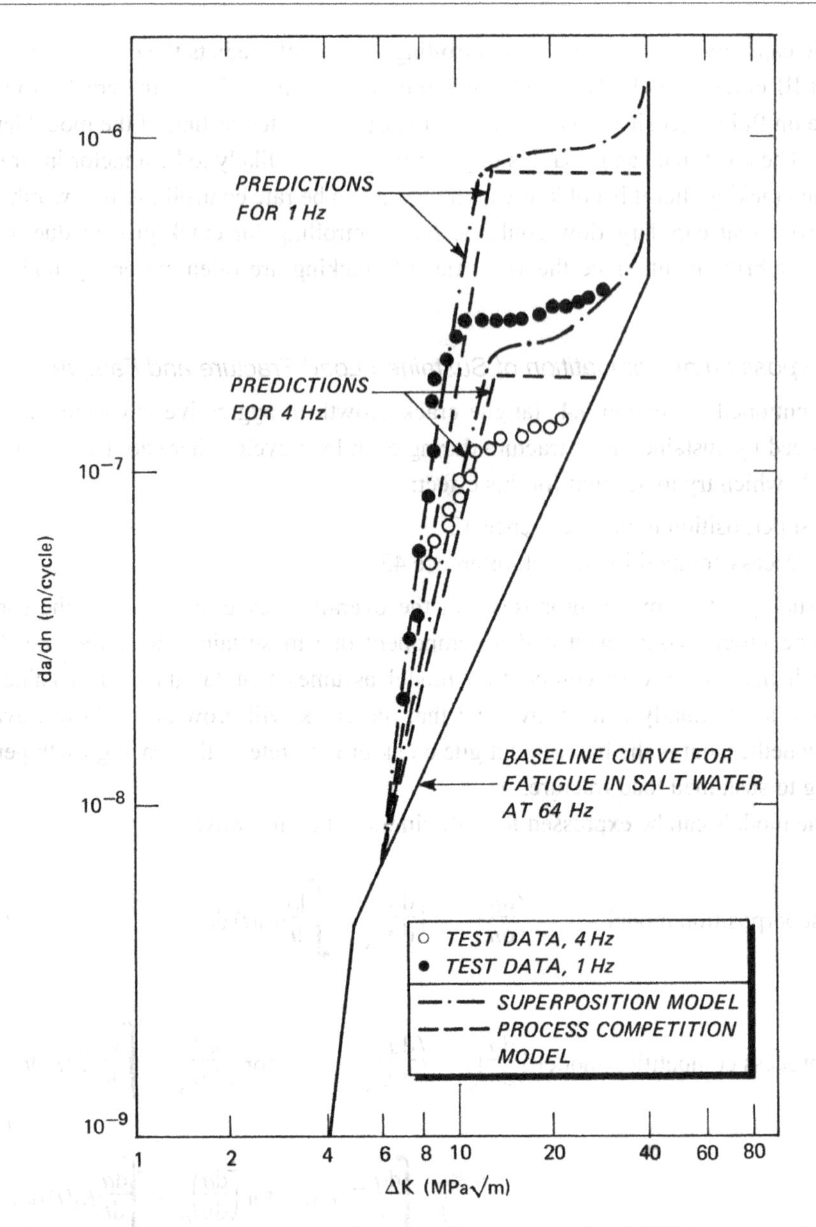

Figure 13.34. Comparison of model predictions with test data for 835M30 steel undergoing stress corrosion during fatigue in salt water at $R = 0.5$.

A similar procedure can be used for the process competition model, except that the predicted overall crack growth rate curve is given either by the baseline fatigue crack growth rate or by the integral in equations (13.44).

A comparison of the predictions of both models with experimental data is given in figure 13.34. The process competition model gives slightly better predictions, but both models significantly overestimate the overall crack growth rates at each frequency. This is a general trend when the environment is liquid rather than gaseous, *i.e.* when stress corrosion cracking occurs during each load cycle. The reason is that cyclic closing and opening of the crack provide a pumping action that mixes the liquid near the crack tip, and when the crack is open it takes a finite time for the liquid to re-establish the conditions necessary for stress corrosion.

There is no general trend of overestimation or underestimation when the models are used to predict overall crack growth rates in gaseous environments, and agreement with test data can be good or poor. Thus, in summary it seems fair to state that both models give an indication of the effect of cycle frequency on the overall crack growth rates, but neither is sufficiently accurate for determining whether superposition or competition of sustained load fracture and fatigue actually occur.

13.5 Bibliography

1. Creager, M. and Paris, P.C., *Elastic Field Equations for Blunt Cracks with Reference to Stress Corrosion Cracking*, International Journal of Fracture Mechanics, Vol. 3, pp. 247-252 (1967).
2. McGowan, J.J. and Smith, C.W., *A Plane Strait, Analysis of the Blunted Crack Tip Using Small Strain Deformation Plasticity Theory*, Advances in Engineering Science, NASA, Vol. 2. pp. 585-594 (1976): Hampton, Virginia.
3. Gerberich, W.V. and Jatavallabhula, K., *Quantitative Fractography and Dislocation Interpretations of the Cyclic Cleavage Crack Growth Process*, Acta Metallurgica, Vol. 31, pp. 241-255 (1983).
4. Dorward, R.C. and Hasse, K.R., *Flaw Growth of 7075. 7475. 7050 and 7049 Aluminium Plate in Stress Corrosion Environments*, Final Report Contract NAS8-30890, Kaiser Aluminium and Chemical Corporation Centre for Technology (1976): Pleasanton, California.
5. Schijve, J., *The Effect of an Irregular Crack Front on Fatigue Crack Growth*, Engineering Fracture Mechanics, Vol. 14, pp. 467-475 (1981).
6. Forsyth, P.J.E. and Bowen, A.W., *The Relationship between Fatigue Crack Behaviour and Microstructure in 7178 Aluminium Alloy*, International Journal of Fatigue, Vol. 3, pp. 17-25 (1981).
7. Minakawa, K. and McEvily. A.J., *On Crack Closure in the Near-Threshold Region*, Scripta Metallurgica, Vol. 15, pp. 633-636 (1981).
8. Suresh, S., *Crack Growth Retardation due to Micro-Roughness: a Mechanism for Overload Effects in Fatigue*, Scripta Metallurgica. Vol. 16, pp. 995-999 (1982).
9. Wanhill, R.J.H., *Microstructural Influences on Fatigue and Fracture Resistance in High Strength Structural Materials*, Engineering Fracture Mechanics, Vol. 10, pp. 337-357 (1978).
10. Greenfield, M.A. and Margolin, H., *The Mechanism of Void Formation, Void Growth and Tensile Fracture in an Alloy Consisting of Two Ductile Phases*, Metallurgical Transactions, Vol. 3, pp. 2649-2659 (1972).
11. Bowen, A.W., *The Influence of Crystaliographic Orientation on the Fracture Toughness of Strongly Textured Ti-6Al-4V*, Acta Metallurgica, Vol. 26, pp. 1423-1433 (1978).
12. Tchorzewski, R.M. and Hutchinson, W.B., *Anisotropy of Fracture Toughness in Textured Titanium-6Al-4V alloy*, Metallurgical Transactions A, Vol. 9A, pp. 1113-1124 (1978).
13. Hahn, G.T. and Rosenfield, A.R., *Sources of Fracture Toughness: the Relation between K_{Ic} and the Ordinary Tensile Properties of Metals*, Applications Related Phenomena in Titanium Alloys, ASTM STP 432, American Society for Testing and Materials, pp. 5-32 (1968): Philadelphia.
14. Robinson, J.N. and Tetelman, A.S., *The Determination of K_{I_c} Values from Measurements of the Critical Crack Tip Opening Displacement at Fracture Initiation*, Paper II-421, Third International Conference on Fracture, Munich (1973).
15. Garrett, G.G. and Knott, J.F., *The Influence of Compositional and Microstructural Variations on the Mechanism of Static Fracture in Aluminium Alloys*, Metallurgical Transactions A, Vol. 9A, pp. 1187-1201(1978).
16. Schwalbe, K.-H., *On the Influence of Microstructure on Crack Propagation Mechanisms and Fracture Toughness of Metallic Materials*, Engineering Fracture Mechanics, Vol. 9, pp. 795-832 (1977).
17. Knott, J.F., Fundamentals of Fracture Mechanics, Butterworths (1973); London.

18. Ritchie, R.O., Knott, J.F. and Rice, J.R., *On the Relationship between Critical Tensile Stress and Fracture Toughness in Mild Steel*, Journal of Mechanics and Physics of Solids, Vol. 21, pp. 395-410 (1973).
19. Stonesifer, F.R. and Cullen, W.H., *Fractographs Furnish Experimental Support for Cleavage Fracture Model*, Metallurgical Transactions A, Vol. 7A, pp. 1803-1806 (1976).
20. Curry, D.A., *Cleavage Micromechanisms of Crack Extensions in Steels*, Metal Science, Vol. 14, pp. 319-326 (1980).
21. Yu, C. and Yan, M., *A Calculation of the Threshold Stress Intensity Range for Fatigue Crack Propagation in Metals*, Fatigue of Engineering Materials and Structures, Vol. 3, pp. 189-192 (1980).
22. Yoder, G.R., Cooley, L.A. and Crooker, T.W., *A Micromechanistic Interpretation of Cyclic Crack-Growth Behaviour in a Beta-Annealed Ti-6Al-4V Alloy*, US Naval Research Laboratory Report 8048, November 1976.
23. Sadananda, K. and Shahinian, P., *Prediction of Threshold Stress Intensity for Fatigue Crack Growth Using a Dislocation Model*, International Journal of Fracture, Vol. 13, pp. 585-594 (1977).
24. Taylor, D., *A Model for the Estimation of Fatigue Threshold Stress intensities in Materials with Various Different Microstructures*, Fatigue Thresholds Fundamental and Engineering Applications, Engineering Materials Advisory Services, pp. 455-470 (1982): Warley, West Midlands, UK
25. Yoder, G.R., Cooley, L.A. and Crooker, T.W., *On Microstructural Control of Near-Threshold Fatigue Crack Growth in 7000-Series Aluminium Alloys*, US Naval Research Laboratory Memorandum Report 4787, April 1982.
26. Suresh, S., Parks, D.M., and Ritchie, R.O., *Crack Tip Oxide Formation and its Influence on Fatigue Thresholds*, Fatigue Thresholds Fundamental and Engineering Applications, Engineering Materials Advisory Services, pp. 391-408 (1982): Warley, West Midlands, U.K.
27. Yoder, G.R., Cooley, L.A. and Crooker, T.W., *Observations on the Generality of the Grain Size Effect on Fatigue Crack Growth in Alpha Plus Beta Titanium Alloys*, US Naval Research Laboratory Memorandum Report 4232, May 1980.
28. Yoder, G.R., Cooley, L.A. and Crooker, T.W., *A Critical Analysis of Grain Size and Yield Strength Dependence of Near-Threshold Fatigue-Crack Growth in Steels*, US Naval Research Laboratory Memorandum Report 4576, July 1981.
29. Stofanak, R.J., Hertzberg, R.W., Leupp, J. and Jaccard, R., *On the Cyclic Behaviour of Cast and Extruded Aluminium Alloys. Part B: Fractography*. Engineering Fracture Mechanics, Vol. 17, pp. 541-554 (1983).
30. Pelloux, R.M.N., *Crack Extension by Alternating Shear*, Engineering Fracture Mechanics, Vol. 1, pp. 697-704 (1970).
31. Kuo, A.S. and Liu, H.W., *An Analysis of Unzipping Model for Fatigue Crack Growth*, Scripta Metallurgica, Vol. 10, pp. 723-728 (1976).
32. Antolovich, S.D., Saxena, A. and Chanani, G.R., *A Model for Fatigue Crack Propagation*, Engineering Fracture Mechanics, Vol. 7, pp. 649-652 (1975).
33. Saxena, A. and Antolovich, S.D., *Low Cycle Fatigue, Fatigue Crack Propagation and Substructures in a Series of Polycrystalline Cu-Al Alloys*, Metallurgical Transactions A, Vol. 6A, pp. 1809-1828 (1975).
34. Green, J.A.S., Hayden, H.W. and Montague, W.G., *The Influence of Looding Mode on the Stress Corrosion Susceptibility of Various Alloy-Environment Systems*, Effect of Hydrogen on Behaviour of Materials, Metallurgical Society of AIME, pp 200-217 (1976): New York.
35. Wanhill, R.J.H., *Application of Fracture Mechanics to Cleavage in Highly Textured Titanium Alloys*, Fracture Mechanics and Technology, Sijthoff and Noordhoff, Vol. 1, pp. 563-572 (1977): Alphen aan den Rijn, The Netherlands.
36. Van Leeuwen, H.P., *Quantitative Models of Hydrogen-Induced Cracking in High -Strength Steel*, Reviews on Coatings and Corrosion, Vol. IV, pp. 5-93 (1979).
37. Van Leeuwen, H .P., *A Failure Criterion for Internal Hydrogen Embrittlement*, Engineering Fracture Mechanics, Vol. 9, pp. 291-296 (1977).
38. Beck, T.R., Blackburn, M.J. and Speidel, M.O., *Stress Corrosion Cracking of Titanium Alloys: SCC of Aluminium Alloys, Polarization of Titanium Alloys in HCl and Correlation of Titanium and Aluminium SCC Behaviour*, Quarterly Progress Report No. 11 Contract NAS 7-489, Boeing Scientific Research Laboratories (1969): Seattle, Washington.
39. Van Leeuwen, H.P., *Plateau Velocity of SCC in High Strength Steel – A Quantitative Treatment*, Corrosion-NACE, Vol. 31, pp. 42-50 (1975).

40. Sofronis, P. (Editor), *Special Issue on Recent Advances in the Engineering Aspects of Hydrogen Embrittlement*, Engineering Fracture Mechanics, Vol. 68, No. 6 (2001).
41. Smith, T., *A Capillary Model for Stress-Corrosion Cracking of Metals in Fluid Media*, Corrosion Science, Vol. 12, pp. 45-56 (1972).
42. Wei, R.P. and Landes, J.D., *Correlation between Sustained-load and Fatigue Crack Growth in High-Strength Steels*, Materials Research and Standards, Vol. 9, pp. 25-27, 44, 46 (1969).
43. Austen, I.M. and Walker, E.F., *Quantitative Understanding of the Effects of Mechanical and Environmental Variables on Corrosion Fatigue Crock Growth Behaviour*, The Influence of Environment on Fatigue, Institution of Mechanical Engineers, pp. 1-10 (1977): London.

Index

9780415346221